Tangled Magnetic Fields in Astro- and Plasma Physics (1) Quantised Flux in Tightly Knotted and Linked Systems (2)

A selection of papers arising from two workshops organised within the six-month programme Topological Dynamics in the Physical and Biological Sciences

Journal of Physics: Conference Series Volume 544

Edinburgh, United Kingdom
15-19 December 2012

Cambridge, United Kingdom
3-7 December 2012

ISBN: 978-1-63439-476-5
ISSN: 1742-6588

Printed from e-media with permission by:

Curran Associates, Inc.
57 Morehouse Lane
Red Hook, NY 12571

Some format issues inherent in the e-media version may also appear in this print version.

Printed by Curran Associates, Inc. (2014)

For permission requests, please contact the Institute of Physics
at the address below.

Institute of Physics
Dirac House, Temple Back
Bristol BS1 6BE UK

Phone: 44 1 17 929 7481
Fax: 44 1 17 920 0979

techtracking@iop.org

Additional copies of this publication are available from:

Curran Associates, Inc.
57 Morehouse Lane
Red Hook, NY 12571 USA
Phone: 845-758-0400
Fax: 845-758-2634
Email: curran@proceedings.com
Web: www.proceedings.com

Table of contents

Volume 544

Tangled Magnetic Fields in Astro- and Plasma Physics (1)
Quantised Flux in Tightly Knotted and Linked Systems (2)
15–19 October 2012, Cambridge (1) and 3–7 December 2012, Edinburgh (2)

Accepted papers received: 1 October 2014
Published online: 20 October 2014

Preface

Papers

Journal of Physics: Conference Series **544** (2014) 011001 doi:10.1088/1742-6596/544/1/011001

Photo courtesy of Victor L'vov, May 2011

Konrad Bajer

(13 October 1956–29 August 2014)

These Proceedings are dedicated to our co-editor Konrad Bajer, who carried the lion's share of the work involved until his sudden and untimely death from cancer on 29th August 2014. Konrad was one of the key Organisers of the program "Topological Dynamics in the Physical and Biological Sciences" held at the Isaac Newton Institute for Mathematical Sciences in Cambridge from July to December 2012. During that time, he was on leave from the University of Warsaw, and held a Visiting Fellow Commonership at Trinity College, Cambridge. He had particular responsibility for the satellite workshop Tangled Magnetic Fields in Astro- and Plasma Physics, held at ICMS, Edinburgh, in October 2012.

Konrad graduated in Physics from the University of Warsaw in 1980, and took his PhD in Fluid Dynamics at DAMTP, Cambridge, 1984-89. He was active in EUROMECH, having been a principal organiser of the European Turbulence Conference (ETC13) in Warsaw 2011, and the meeting Turbulence: the Historical Perspective that followed it. He has been equally active in IUTAM; his distinction was recognised by his very recent election to the Congress Committee of IUTAM.

Konrad's research interests were in magnetohydrodynamics and vortex dynamics in classical and quantum systems; also in problems of gasifcation, in which he played a coordinating role at the University of Warsaw. He impressed all with his exceptionally warm and friendly personality, and will be greatly missed in the whole scientifc community.

Preface

This online volume contains a selection of papers arising from two workshops organised within the six-month programme Topological Dynamics in the Physical and Biological Sciences held at the Isaac Newton Institute for Mathematical Sciences, Cambridge, from July to December 2012.

The first of these was a 'satellite workshop' held at the International Centre for Mathematical Sciences (ICMS), Edinburgh, 15-19 October 2012, under the title Tangled Magnetic Fields in Astro- and Plasma Physics, and with Scientific Organising Committee: Konrad Bajer (Warsaw), Mitchell Berger (Exeter), Steve Cowley (Culham Centre for Fusion Energy), Andrew Gilbert (Exeter), Gunnar Hornig (Dundee), and Clare Parnell (St Andrews). The second was the workshop Quantised Flux in Tightly Knotted and Linked Systems held at the Newton Institute, 3- 7 December 2012, with Scientific Organising Committee: Natalia Berloff (DAMTP, Cambridge), Anne-Christine Davis (DAMTP, Cambridge), Jason Cantarella (University of Georgia), Thomas Kephart (Vanderbilt University), Paul Sutcliffe (Durham University), and Tanmay Vachaspati (Arizona State University). Videos of the lectures given at this second workshop can be viewed at http://www.newton.ac.uk/webseminars.

The papers published here follow a natural progression through the following topics: helicity and related invariants of magnetic fields in ideal MHD; relaxation under topological constraints; lower bounds on magnetic energy; current and vortex filaments; applications in the solar corona, tokamak plasmas, and cyclone dynamics; higher-order invariants; topology of curves and surfaces, and energy measures; tight knots; applications to Bose-Einstein condensates, QCD, and cosmic superstring theory. Some of the papers span more than one of these areas.

We owe a great debt of gratitude to Konrad Bajer, who was one of the guiding spirits behind the whole Newton Institute program, and who took particular responsibility for the Satellite Workshop at ICMS, Edinburgh. Konrad fell seriously ill in June 2014, having completed much of the editorial work for these Proceedings. Following an operation at the end of July, he remained optimistic of a complete recovery, but sadly this was not to be, and he died from a particularly malignant form of cancer on 29th August. The funeral was held in Warsaw on 5th September. Konrad's warmth of personality and generosity of spirit will be remembered by all who participated in these workshops and in the wider Newton Institute program.

The Isaac Newton Institute provided an ideal environment for informal interactions before and after both these workshops. We wish to express our thanks to the Director and staff of the Institute for their tireless efforts to ensure the success of the whole program; also to the staff of ICMS, Edinburgh, for their expert hosting of the satellite workshop.

Keith Moffatt
Thomas Kephart
22 September 2014

Journal of Physics: Conference Series **544** (2014) 011003 doi:10.1088/1742-6596/544/1/011003

Peer review statement

All papers published in this volume of *Journal of Physics: Conference Series* have been peer reviewed through processes administered by the proceedings Editors. Reviews were conducted by expert referees to the professional and scientific standards expected of a proceedings journal published by IOP Publishing.

Helicity, linking, and writhe in a spherical geometry

J. Campbell[1], **M.A. Berger**[1]
[1] CEMPS, University of Exeter
Harrison Building, North Park Road, Exeter EX4 4QF, U.K.

Abstract. Linking numbers, helicity integrals, twist, and writhe all describe the topology and geometry of curves and vector fields. The topology of the space the curves and fields live in, however, can affect the behaviour of these quantities. Here we examine curves and fields living in regions exterior to a sphere or in spherical shells. The winding of two curves need not be conserved because of the topology of a spherical shell. Avoiding the presence of magnetic monopoles inside the sphere is very important if magnetic helicity is to be a conserved quantity. Considerations of parallel transport are important in determining the transfer of helicity through the foot of a magnetic flux tube when it is in motion.

1. Introduction

In 1833 the German mathematician Carl Friedrich Gauss mentioned the idea of linking number as a brief note in his diary (Ricca & Nipoti, 2011). Gauss was considering the path of asteroids across the sky when he realised that an asteroid's path would be restricted to a range of latitudes unless its orbit linked the Earth's orbit. It is not clear how Gauss formalised this idea into the Gauss linking integral, but Ricca and Nipoti suggest that he might have used his knowledge of terrestrial magnetism for this purpose.

As Gauss noted (Epple, 1998), linking number is – most importantly – invariant to smooth deformation of the curves as long as they are not permitted to pass through each other: it tells you about the topology of a curve rather than its geometry. Linking number is an important and widely used tool in many areas of study including molecular biology (Fuller, 1978; White & Bauer, 1986), fluid dynamics (Ricca, 1998) and super-fluid dynamics (Barenghi *et al.*, 2001).

Many applications of topological and geometrical measures of linking, twisting, and writhing involve geometries that have a curved nature to them (Priest, 1982). For example, the magnetic coronal-loops in the sun's atmosphere share a spherical boundary (the photosphere). Also, biologists are interested in long molecules such as DNA, RNA, and proteins which can be connected to roughly spherical objects (e.g. cell nuclei Bates & Maxwell (2005)). The buckling and twisting of elastic rods also depends on the geometry of external restraints (Hannay, 1998; Neukirch & Starostin, 2008). The purpose of this work is to extend the existing concepts of linking number, twist and writhe, so that they work in a consistent manner in a spherical geometry.

Spherical geometry differs in several ways to flat space. First, the usual spherical coordinate system (ρ, θ, ϕ) contains points that cannot be uniquely described; if $\rho = 0$ or $\theta = 0$ or $\theta = \pi$, then one or more of the other components can be changed without altering a point's position. Another difference is that on a sphere, Euclid's fifth postulate breaks down: two parallel lines need not stay a fixed distance apart. As a consequence the idea of parallel vectors must be carefully defined; one must first specify how to move a vector around – using a connection – before this is possible. A third difference is topological: the space outside of a sphere has non-trivial topology; it has a hole. The hole can play havoc with helicity integrals because magnetic monopoles can be hidden inside. Helicity in triply periodic geometries (i.e. a three-torus) can also be affected by the non-trivial topology (Berger, 1997).

We first review the expressions for self and mutual helicity of flux tubes. We then examine the helicity of two tubes in a spherical shell, taking care to avoid hidden monopoles. In section 4 we calculate the flux of helicity into the tubes due to boundary motions.

Figure 1. A loop extending out of a sphere. If we fix the intersection points with the inner and outer sphere, then the helicity in the shell between is conserved in the absence of resistivity.

2. The self and mutual helicity of two flux tubes

Magnetic helicity measures the net linking of pairs of fieldlines in a magnetic field (Moffatt, 1969). The total helicity of two tubes in a volume $\mathcal{V}$ can be decomposed into a contribution from the self helicities of the individual tubes, plus a mutual helicity due to their winding number. Let $\mathcal{T}_1$ and $\mathcal{T}_2$ be the self helicities of tubes 1 and 2, and let w_{12} measure the winding of the tubes about each other. The self helicity of tube 1 arises from twisting T_{w1} of field lines inside the tube about its central axis, as well as the writhe W_{r1} of the axis itself:

$$\mathcal{T}_1 = T_{w1} + W_{r1} \tag{1}$$

(Berger & Prior, 2006). Let the magnetic fluxes of the two tubes be Φ_1 and Φ_2. Then in total we have (Berger, 1986, 2009)

$$H = \mathcal{T}_1\Phi_1^2 + \mathcal{T}_2\Phi_2^2 + 2w_{12}\Phi_1\Phi_2. \tag{2}$$

Next consider a volume with planar boundaries (see figure 2). Suppose two flux tubes stretch from bottom to top. The winding number between the two tubes w_{12} is given by $w_{12} = \alpha_{12}/(2\pi)$, where α_{12} is the net angle through which the tubes wind about each other.

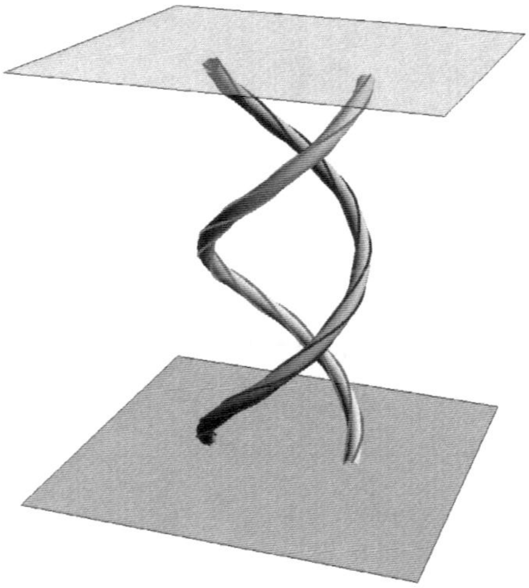

Figure 2. Two flux tubes extending between two parallel planes. The twist plus writhe of each individual tube is conserved, as well as the winding number between the two tubes.

If the boundaries are held fixed, then the helicity will be conserved if the tubes are distorted between the boundaries. The fluxes Φ_1 and Φ_2 will also be conserved. As the two fluxes can be chosen independently, we may conclude that $\mathcal{T}_1$, $\mathcal{T}_2$ and w_{12} are separately conserved.

Next suppose that the footpoints of the tubes on the lower boundary move. Then the fluxes do not change, but the topological measures evolve. Suppose the angular rotation rate of footpoint 1 is $\dot{\alpha}_1$ and similarly for footpoint 2. The relative position vector between the centres of the two footpoints rotates with a rate $\dot{\alpha}_{12}$. Then

$$\frac{\mathrm{d}H}{\mathrm{d}t} = \frac{-1}{2\pi}\left(\dot{\alpha}_1\Phi_1^2 + \dot{\alpha}_2\Phi_2^2 + 2\dot{\alpha}_{12}\Phi_1\Phi_2\right). \quad (3)$$

3. The helicity of two tubes in a spherical region

Suppose a loop of magnetic field rises out of a spherical boundary, for example, the surface of a star (see figure 1). We can build an external magnetic field to the sphere by summing over many loops, but here we will consider just one. We may wish to consider just the part of the loop inside a shell between two spherical boundaries, as in figure 1. The magnetic helicity of the entire loop will equal the helicity inside the shell between the two boundary spheres, plus the helicity above. Motions which vanish at the two spheres will separately conserve the helicities above and below.

The nature of spherical geometries leads to a strange effect. Can we deform the blue curve in figure 1 so as to reverse the sign of the crossing? It turns out you can – see figure 3. Here the end points of the curves have remained fixed, but the winding angle between the two curves has changed by minus one: $w_{12} \rightarrow w_{12} - 1$. In terms of the sign of the crossing between the tubes, going from the diagram on the left to the diagram on the right, a negative crossing has been replaced by a positive crossing.

It might seem that the helicity has changed; however, this is not the case: the helicity is still equal to Φ^2. This will be easiest to see if we consider a cross-section extending along the tube in the shape of a ribbon (see figure 4). In the process of moving the middle part of the curve twist has been injected. As the curve is deformed winding is converted into twist. Figure 4 illustrates how a ribbon would change its twist by -2 under such a transformation. (This effect can be readily verified by experiment with a length of ribbon and a ball!) Thus from equation 2

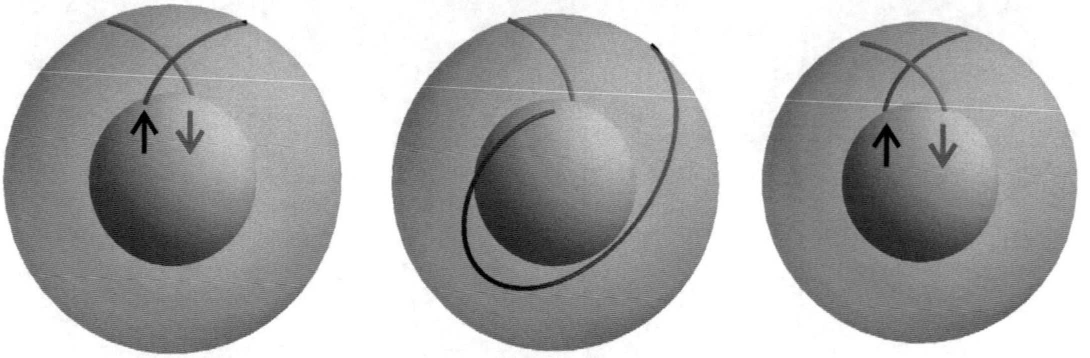

Figure 3. The foreground curve on the left curve smoothly deforms to sweep around the sphere, becoming the background curve in the right panel. In the left panel the curves wind about each other by $\alpha_{12} = \pi$ as they rise from the inner to the outer sphere; thus $w_{12} = 1/2$. On the other hand, as one tube has flux pointing inwards and one outwards, the crossing is negative giving a negative helicity contribution. In the right panel the winding is negative (alternatively, the crossing has become positive). Thus, despite the fixed endpoints, the winding of the two curves has changed from approximately 1/2 to -1/2.

the helicity will be conserved, *provided the two fluxes are equal and opposite in magnitude.* Let the fluxes be $\Phi_1 = \Phi$, $\Phi_2 = -\Phi$. Then

$$H_{before} = \Phi^2 (\mathcal{T}_1 + \mathcal{T}_2 - 2w_{12}); \tag{4}$$

$$H_{after} = \Phi^2 ((\mathcal{T}_1 - 2) + \mathcal{T}_2 - 2(w_{12} - 1)). \tag{5}$$

Thus $H_{before} = H_{after}$.

Here an important distinction is apparent between the planar slice and spherical shell topologies. In the plane geometry of figure 2, the self and mutual terms are separately conserved even if the two fluxes differ. In the spherical case, however, the terms are not separately conserved; but total helicity conservation requires equal and opposite fluxes. Even though we do not directly enquire about what is happening inside the inner spherical boundary, we run into trouble if there is net magnetic flux coming out of it. Thus we must require that there are no monopoles inside the sphere (or at least equal numbers of positive and negative monopoles). For the planar case, net flux is allowed. We can always assume return fluxes at an infinite distance away from our tubes, so that there is no mutual helicity between our tubes and the return flux.

4. Change in helicity due to surface motions

Let the radius of the inner boundary be ρ_0 and consider two flux tubes crossing this boundary. We seek expressions for the change in helicity due to their two footpoints on the boundary moving about each other. A field with more flux tubes, or even a continuous field, can be analyzed by summing over many pairs of tubes.

For flux moving on a boundary due to motions parallel to the boundary with velocity $\boldsymbol{V}$, the helicity change can be written as a surface integral (Berger, 1986)

$$\frac{\mathrm{d}H}{\mathrm{d}t} = \int (\boldsymbol{A}_P \cdot \boldsymbol{V})\boldsymbol{B} \cdot \hat{\boldsymbol{n}}\ \mathrm{d}^2S, \tag{6}$$

where $\hat{\boldsymbol{n}}$ is the unit normal and $\boldsymbol{A}_P$ is the unique Coulomb gauge (divergence-free) vector potential of a potential (vacuum) magnetic field with the same boundary flux.

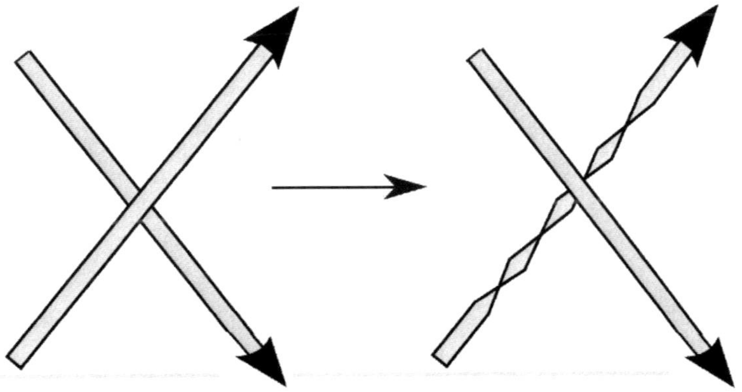

Figure 4. Before deforming: $\mathcal{T} = 0$ and $\mathcal{W} = -1$. Post deformation: $\mathcal{T} = -2$ and $\mathcal{W} = 1$. Hence in both cases $H = -\Phi^2$ as expected.

In some situations, Stoke's theorem provides an easy method of determining $\boldsymbol{A}_P$. For a planar boundary, a circular footpoint centred at the origin with normal flux Φ has

$$\boldsymbol{A}_P = \frac{\Phi}{2\pi r}\hat{\phi}, \tag{7}$$

where polar coordinates are (r, ϕ) and we assume r is greater than the radius of the footpoint. We can then build up $\boldsymbol{A}_P$ from linearly summing over footpoints. But things are not so simple with spherical geometries!

In defining the problem we want to avoid the possibility of magnetic monopoles. If there is net flux coming through the surface then $\nabla \cdot \boldsymbol{B} \neq 0$ inside the sphere. The two flux tubes cutting our layer must have the same magnitude but oppositely directed radial fluxes to avoid this problem. The net radial flux would then be zero, so $\nabla \cdot \boldsymbol{B} = 0$ as required. We need to find the vector potential on the surface corresponding to these two tubes. Suppose footpoint 1 is at the North pole and footpoint 2 is at the location (θ_2, ϕ_2). If we naively follow the procedure of equation 7, then we would say $\boldsymbol{A}_P = \boldsymbol{A}_1 + \boldsymbol{A}_2$ where

$$\boldsymbol{A}_1 = \frac{\Phi_1}{2\pi\rho_0 \sin\theta}\hat{\phi}, \tag{8}$$

and $\boldsymbol{A}_2$ is found by an appropriate rotation of this vector field (so that the $\boldsymbol{A}_2$ lines encircle spot 2). This approach fails however: the expression in equation 8 has a singularity at the South pole, which is not removed by adding A_2. It is better to only consider fields which explicitly have zero net flux before attempting to calculate the vector potential.

The problem with this approach is that in a sphere, it is no longer possible to study each tube individually. Instead, we will introduce a net return flux as in figure 5. We assume the areas $\mathcal{A}_1$ and $\mathcal{A}_2$ of the footpoints are small compared to the area of the sphere. Then for each flux tube i

$$B_{i_\rho} = \begin{cases} \frac{\Phi_i}{\mathcal{A}_1} - \frac{\Phi_i}{4\pi\rho_0^2} & \text{inside spot} \\ -\frac{\Phi_i}{4\pi\rho_0^2} & \text{outside spot} \end{cases} \tag{9}$$

If the two tubes have equal and opposite flux, then the return fluxes exactly cancel.

If footpoint 1 is at the North pole, then we have (e.g. employing Stoke's theorem) the vector potential due to its flux Φ_1,

$$\boldsymbol{A}_{P1}(\theta, \phi) = \left(\frac{1 + \cos\theta}{2}\right)\frac{\Phi_1}{2\pi\rho_0 \sin\theta}\hat{\phi}. \tag{10}$$

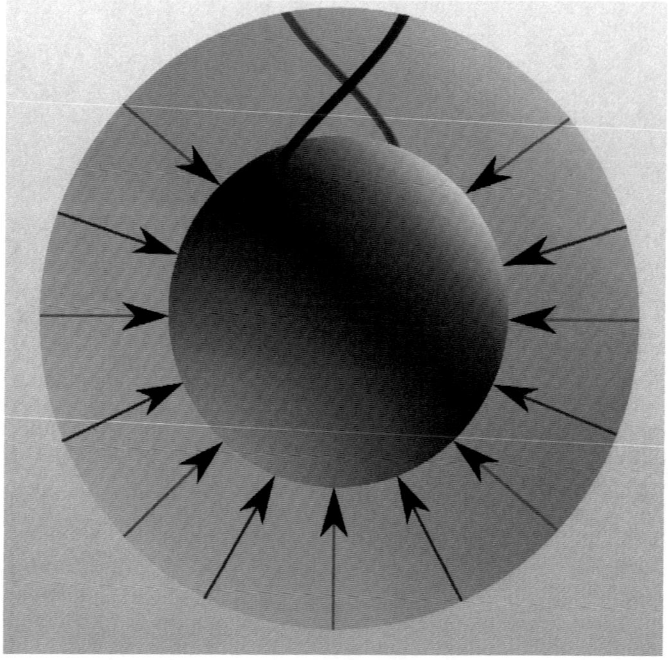

Figure 5. We introduce a net return flux to ensure that there are no magnetic monopoles. Each tube's B-field is shared out evenly and directed back into the surface.

Note that no singularity remains. The vector potential for $\boldsymbol{A}_{P2}$ will have a similar form, but with Φ replaced by $-\Phi$ and with a rotation of the vector field to centre on footpoint 2.

Suppose that footpoint 1 on the lower sphere is located at the North pole and that the second footpoint is at co-latitude θ_0. We wish to examine the change in helicity as footpoint 2 moves around on the bottom sphere. Let footpoint 2 move though a angle of $\delta\phi$ on the circle of co-latitude defined by θ_0.

The change of helicity is:

$$\delta H = \Phi^2 \left(\delta \mathcal{T}_1 + \delta \mathcal{T}_2 + 2\delta w_{12}\right). \tag{11}$$

We will examine the linking term first. As tube 2 moves it will wind with tube 1 and also with the return flux. The amount of return flux it winds with will be proportional to the surface area of the polar cap that it circumscribes:

$$\begin{aligned} \delta w_{12} &= \frac{\delta\phi}{2\pi}\left(1 - \frac{\text{area of polar cap at } \theta_0}{4\pi\rho_0^2}\right) & (12) \\ &= \frac{\delta\phi}{2\pi}\left(\frac{1}{2} + \frac{\cos(\theta_0)}{2}\right). & (13) \end{aligned}$$

Next we will need to work out the self-linking terms; these refer to the internal twist inside each flux tube. To measure the twist, we choose a field line on the axis of the tube and a field line on the boundary. The boundary field line will twist about the centre field line by T_w turns. The twist number T_w grows as the footpoint of the tube rotates - but what causes it to rotate? We will consider a vector which points from the centre of the footpoint to the chosen boundary line. We now ask how much this vector rotates.

Imagine walking around a circular path on a flat surface. If you always face the direction of travel, you will turn through an angle of 2π. Facing in the direction of travel corresponds to facing in the direction of the tangent vector $\hat{\boldsymbol{t}}$ to the path, which rotates by 2π. Similarly, a footpoint proceeding in this fashion will rotate by 2π. Suppose instead that you always face North, sometimes walking sideways, sometimes walking backwards. The vector $\hat{\boldsymbol{p}}$ representing the direction you are facing is the constant North vector, and there is no turning.

On a curved surface, the analogy of the non-turning constant vector $\hat{\boldsymbol{p}}$ is called the *parallel transport* vector. We will compare the forward-facing vector $\hat{\boldsymbol{t}}$ to the parallel transport vector $\hat{\boldsymbol{p}}$. After following a path, calculating the angle between these two vectors will give the net rotation of the footpoint. We will need to introduce some additional tools beforehand.

4.1. Parallel transport

Suppose we would like to transport a vector $\boldsymbol{v}$ along a parametrised curve $\gamma(t)$, whose tangent vector is $\boldsymbol{u}$. Basis vectors are labelled $\hat{\mathbf{e}}_\beta$. Then standard differential geometry gives the covariant derivative of $\boldsymbol{v}$ along the curve:

$$\frac{D\boldsymbol{v}}{Dt} = \left(\frac{\mathrm{d}v^\beta}{\mathrm{d}t} + u^\alpha v^\gamma \Gamma^\beta_{\gamma\alpha}\right)\hat{\mathbf{e}}_\beta. \tag{14}$$

Here the last term in this equation includes the Christoffel symbols $\Gamma^\gamma_{\beta\alpha}$, which describe how the basis vectors change.

This expression is known as the covariant derivative $\nabla_{\boldsymbol{u}}\boldsymbol{v}$. This derivative describes how a vector varies as it is transported along a curve. Suppose that we wish to transport $\boldsymbol{v}$ without adding any extra rotation. More precisely, we require that for each infinitesimal amount $\boldsymbol{v}$ is transported, it remains locally parallel to its previous direction. Then we say that $\boldsymbol{v}$ is parallel transported along $\boldsymbol{u}$ if

$$\nabla_{\boldsymbol{u}}\boldsymbol{v} = 0. \tag{15}$$

A geodesic is a curve whose tangent vector remains parallel when transported along the curve. For a sphere, these are curves that follow a great circle. Along other curves, there will be some acceleration or rotation of the tangent vector.

Suppose $\boldsymbol{v}$ is a vector pointing from the centre of a footpoint to some point on its boundary. Keeping track of how this vector rotates tells us how much the footpoint rotates. So, when transporting the footpoint (and hence $\boldsymbol{v}$) along a curve which is not a great circle, the acceleration of the tangent vector will induce a corresponding rotation of the footpoint.

In particular, if the footpoint moves along a latitude line other than the equator, the rotation will inject helicity into the tube above. Of course, additional rotation can occur in addition to the transport contribution, for example if vorticity at the surface rotates the footpoint.

4.2. Twist calculation

We can now calculate how much $\boldsymbol{v}$ rotates. The method here generalises from spherical geometries to other curved geometries. Expanding equation 15 gives:

$$0 = \left(\frac{\mathrm{d}v^\beta}{\mathrm{d}t} + u^\alpha v^\gamma \Gamma^\beta_{\gamma\alpha}\right), \tag{16}$$

where the metric connection gives

$$\Gamma^\beta_{\gamma\alpha} = g^{\beta\delta}\frac{1}{2}(g_{\delta\gamma,\alpha} + g_{\delta\alpha,\gamma} - g_{\gamma\alpha,\delta}). \tag{17}$$

Journal of Physics: Conference Series **544** (2014) 012001 doi:10.1088/1742-6596/544/1/012001

There are only three non-zero Christoffel symbols:

$$\Gamma^{\phi}_{\phi\theta} = \Gamma^{\phi}_{\theta\phi} = \cot(\theta_0); \tag{18}$$

$$\Gamma^{\theta}_{\phi\phi} = -\sin(\theta_0)\cos(\theta_0). \tag{19}$$

For motion along a latitude line $u^{\phi} = 0$, we obtain

$$\partial_{\phi} v^{\theta} + v^{\phi}\Gamma^{\theta}_{\phi\phi} = 0; \tag{20}$$

$$\partial_{\phi} v^{\phi} + v^{\theta}\Gamma^{\phi}_{\theta\phi} = 0. \tag{21}$$

Differentiating the first equation, putting it into the second and doing the same the other way around we obtain

$$\frac{\partial^2 v^{\theta}}{\partial\phi^2} = -v^{\theta}\cos(\theta_0)^2; \tag{22}$$

$$\frac{\partial^2 v^{\phi}}{\partial\phi^2} = -v^{\phi}\cos(\theta_0)^2. \tag{23}$$

Letting $\Omega = \cos(\theta_0)$, these equations have solutions of the form:

$$v^{\theta}(\phi) = A\sin(\Omega\phi) + B\cos(\Omega\phi); \tag{24}$$

$$v^{\phi}(\phi) = C\sin(\Omega\phi) + D\cos(\Omega\phi). \tag{25}$$

We now consider a vector $\hat{\boldsymbol{p}}$ which initially points tangent to the latitude line at θ_0. Thus it has the initial condition $\hat{\boldsymbol{p}}_0 = (p^{\theta}_0, p^{\phi}_0) = (0, 1/\sin\theta_0)$ (the factor of $1/\sin\theta_0$ is needed to make it a unit vector in the spherical coordinate system). We obtain for the parallel transport vector

$$\hat{p}^{\theta} = \sin(\Omega\phi); \tag{26}$$

$$\hat{p}^{\phi} = \frac{1}{\sin\theta_0}\cos(\Omega\phi). \tag{27}$$

In contrast, for motion along a latitude line the tangent vector at all times has $t^{\theta} = 0$. Suppose the motion proceeds along a net longitudinal angle $\delta\phi$. Relative to the parallel transport vector, the tangent vector will have rotated by an angle $\delta\phi\cos(\theta_0)$. This rotation corresponds to the *geodesic curvature* of the path.

Recall that our aim here was to calculate how much tube two rotates as it moves. If this tube is rotated positively, then the twist induced will be negative (as in equation 3). Thus if footpoint 1 stays at the North pole and footpoint 2 travels through an angle $\delta\phi$ at co-latitude θ_0, then

$$\delta\mathcal{T}_1 = 0, \qquad \delta\mathcal{T}_2 = -\frac{\delta\phi}{2\pi}\cos(\theta_0). \tag{28}$$

The net helicity change due to the motion of tube 2 (including winding) is

$$\delta\mathcal{H} = \Phi^2\left(\delta\mathcal{T}_1 + \delta\mathcal{T}_2 + 2\delta H_{12}\right); \tag{29}$$

$$= \Phi^2\frac{\delta\phi}{2\pi}\left[0 - \cos(\theta_0) + 2\left(\frac{1}{2} + \frac{\cos(\theta_0)}{2}\right)\right]; \tag{30}$$

$$= \Phi^2\frac{\delta\phi}{2\pi}. \tag{31}$$

This coincides with the change in helicity we would have if we were working in a flat geometry.

4.3. Uniform rotation
Next we will consider what would happen if footpoint 1 rotates as well, and both tubes move around 2π. This could be thought of as one complete solid rotation of the sphere. The tube at the north pole will rotate by 2π causing a negative twist of $\mathcal{T}_1 = -1$. The lower tube will also rotate about by $2\pi\cos(\theta_0)$, and so twist by $\mathcal{T}_2 = -2\pi\cos(\theta_0)$. The tubes will also wind with each other according to equation 13. The change in helicity is

$$\delta\mathcal{H} = \Phi^2\left[-1-\cos(\theta_0)+2\left(\frac{1}{2}+\frac{\cos(\theta_0)}{2}\right)\right]; \quad (32)$$

$$= 0, \quad (33)$$

which is what is expected as, relative to each other, there is no winding going on.

5. Conclusions

The helicity of two flux tubes can be thought of as a sum of self and mutual helicities. If the tubes are located in all space or are restricted to a half space, then the self and mutual helicities are separately conserved during ideal motion. However, for tubes in the space outside (or inside) a sphere, only the total helicity is necessarily conserved. We have investigated the helicity flux through a spherical boundary, and how it separates into individual and mutual terms, i.e. the spins of individual footpoints and the orbiting of footpoints about each other. Here measures of rotation must be carefully defined. The spin of an individual footpoint as it follows a path can be measured by the geodesic curvature, calculating the rotation of the tangent vector relative to the parallel transport vector. As a consistency check, a uniform rotation of the sphere should provide zero net helicity flow. This provides a constraint on how to calculate the orbit term as well. These methods will readily generalize to non-spherical geometries.

References

Barenghi, C., Donnelly, R. & Vinen, W. 2001 **Quantized Vortex Dynamics and Superfluid Turbulence**. Springer Publishing, New York, USA.

Bates, A. & Maxwell, A. 2005 **DNA Topology**, 2nd edn. Oxford University Press, Oxford, UK.

Berger, M. 1986 Topological invariants of field lines rooted in planes. *Geophys. Astrophys. Fluid Dyn.* **34**, 265–281.

Berger, M. A. 1997 Magnetic helicity in a periodic domain. *J. Geophys Research* **102**, 2637–2644.

Berger, M. A. 2009 **Topological Magnetohydrodynamics and Astrophysics**. Springer, New York.

Berger, M. A. & Prior, C. 2006 The writhe of open and closed curves. *J. Phys. A* **39** (26), 8321–8348.

Epple, M. 1998 Orbits of asteroids, a braid, and the first link invariant. *Math. Intelligencer* **20**, 45–52.

Fuller, F. B. 1978 Decomposition of the linking number of a closed ribbon: a problem from molecular biology. *Proc. Nat. Acad. Sci. U.S.A.* **75** (8), 3557–3561.

Hannay, J. 1998 Cyclic rotations, contractibility, and gauss-bonnet. *J.Phys.A* **31**.

Moffatt, H. K. 1969 The degree of knottedness of tangled vortex lines. *Journal of Fluid Mechanics* **35**, 117–129.

Neukirch, S. & Starostin, E. L. 2008 Writhe formulas and antipodal points in plectonemic dna configurations. *Phys. Rev. E* **78**, 041912.

PRIEST, E. 1982 **Solar Magneto-Hydrodynamics**. D.Reidel Publishing Company, Dordrecht, Holland.

RICCA, R. & NIPOTI, B. 2011 Gauss' linking number revisited. *Journal of Knot Theory and Its Ramifications* **20** (10), 1325–1343.

RICCA, R. L. 1998 Applications of knot theory in fluid mechanics. In *Knot theory (Warsaw, 1995)*, *Banach Center Publ.*, vol. 42, pp. 321–346. Polish Acad. Sci., Warsaw.

WHITE, J. & BAUER, W. R. 1986 Calculation of the twist and the writhe for representative models of dna. *J. Mol Biol* **189** (2), 329–341.

A complete topological invariant for braided magnetic fields

A R Yeates[1] and G Hornig[2]

[1] Department of Mathematical Sciences, Durham University, Durham, DH1 3LE, UK
[2] Division of Mathematics, University of Dundee, Dundee, DD1 4HN, UK

E-mail: anthony.yeates@durham.ac.uk, gunnar@maths.dundee.ac.uk

Abstract. A topological flux function is introduced to quantify the topology of magnetic braids: non-zero line-tied magnetic fields whose field lines all connect between two boundaries. This scalar function is an ideal invariant defined on a cross-section of the magnetic field, whose integral over the cross-section yields the relative magnetic helicity. Recognising that the topological flux function is an action in the Hamiltonian formulation of the field line equations, a simple formula for its differential is obtained. We use this to prove that the topological flux function uniquely characterises the field line mapping and hence the magnetic topology. A simple example is presented.

1. Introduction

In this paper we present a complete topological invariant for so-called *magnetic braids*: magnetic fields in a flux tube with monotone guide field. Such magnetic fields arise, for example in coronal loops in the Sun's atmosphere (Reale, 2010), in toroidal fusion devices (Morrison, 2000), and in laboratory experiments (Van Compernolle & Gekelman, 2012). In each case, the magnetic field is embedded in a highly-conducting plasma so that the magnetic topology—or linking and connectivity of magnetic field lines—is approximately preserved. As such, topological invariants typically play a significant role in the dynamics of the plasma and its magnetic field (Woltjer, 1958; Taylor, 1986; Brown *et al.*, 1999; Yeates *et al.*, 2010; Candelaresi & Brandenburg, 2011).

Figure 1 shows our simply connected "flux tube" V, with a set (ρ, ψ, z) of orthogonal curvilinear coordinates satisfying $\nabla z \cdot (\nabla\rho \times \nabla\psi) = 0$. The magnetic field $\mathbf{B}$ is assumed to satisfy $B_z > 0$ everywhere in V, so that ∇z aligns with the "axis" of the flux tube. We take ψ to be an angular coordinate and ρ to be a radial coordinate in each plane of constant z. On the side boundaries of the tube, we require $\mathbf{B} \cdot \mathbf{n} = 0$, so that all magnetic field lines stretch from one end D_0 of the flux tube to the other end D_1. We can then define the *field line mapping* $F : D_0 \to D_1$ by $F(x_0) = f(x_0; 1)$, where $f(x_0; z)$ denotes the magnetic field line rooted at $x_0 \in D_0$.

Two magnetic braids $\mathbf{B}$, $\widetilde{\mathbf{B}}$ are *topologically equivalent* if and only if one can be reached from the other by an ideal evolution

$$\frac{\partial \mathbf{B}}{\partial t} = \nabla \times (\mathbf{v} \times \mathbf{B}) \tag{1}$$

Journal of Physics: Conference Series **544** (2014) 012002 doi:10.1088/1742-6596/544/1/012002

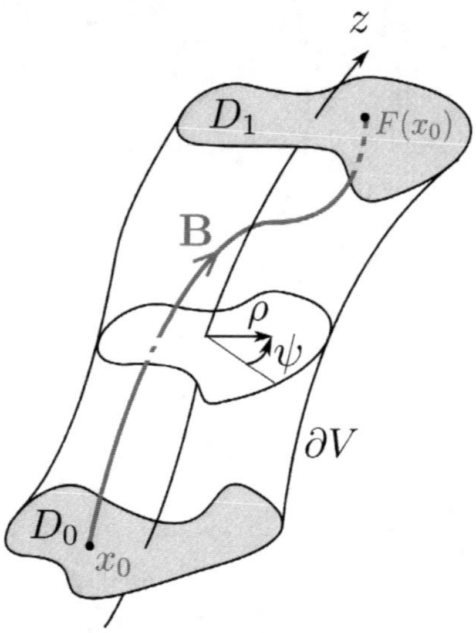

Figure 1. Coordinates and notation for a magnetic braid $\mathbf{B}$ in a flux tube V, with field line mapping $F : D_0 \to D_1$.

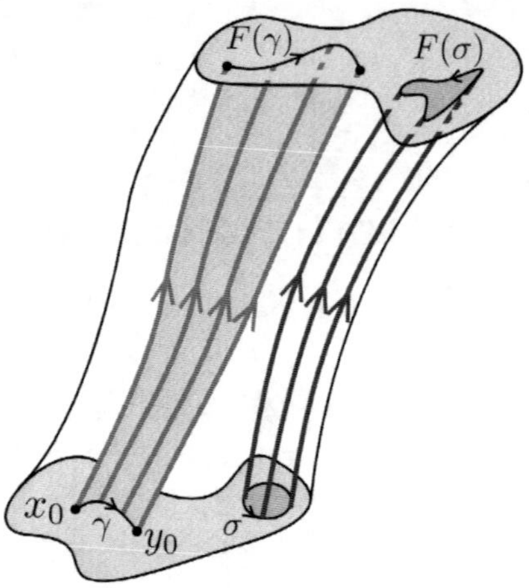

Figure 2. The physical meaning of Lemma 2 may be understood by integrating on the lower boundary D_0 along an open curve γ or a closed curve σ.

with $\mathbf{v}|_{\partial V} = 0$ throughout. We shall assume that their respective field line mappings match on the side boundary, $F|_{\partial D_0} = \widetilde{F}|_{\partial D_0}$, with the same winding number. In that case, $\mathbf{B}$ and $\widetilde{\mathbf{B}}$ are topologically equivalent if and only if $F = \widetilde{F}$. Notice that F has two components. Our main result is to prove that a single scalar function, which we call a "topological flux function" is both necessary and sufficient to determine the topology. This we state in the following theorem.

Theorem 1. *Let* $\mathbf{B}$, $\widetilde{\mathbf{B}}$ *be two magnetic braids on* V*, with respective field line mappings* F, $\widetilde{F}$ *that agree on* ∂D_0*, with the same winding number. Let* $\mathcal{A}$, $\widetilde{\mathcal{A}}$ *be their respective topological flux functions with the same reference field* $\mathbf{A}^{\text{ref}}$ *such that* $A_\rho^{\text{ref}} = 0$*. Then* $\mathcal{A} = \widetilde{\mathcal{A}}$ *if and only if* $F = \widetilde{F}$*.*

It is well-known (Berger & Field, 1984; Brown *et al.*, 1999) that having the same total magnetic helicity H_r (which will be defined below) is a necessary condition for topological equivalence but not a sufficient one (two magnetic braids can have the same H_r but different F). Indeed there are an infinity of other ideal invariants. For example, if $\theta_{x_0,y_0}(z)$ denotes the orientation of the line at height z between a pair of field lines $f(x_0; z)$ and $f(y_0; z)$, then their pairwise linking number

$$c_{x_0,x_1} = \frac{1}{2\pi}\int_0^1 \frac{\mathrm{d}\theta_{x_0,y_0}(z)}{\mathrm{d}z}\,\mathrm{d}z \tag{2}$$

is an invariant (Berger, 1986; Berger & Prior, 2006). So are analogous higher-order measures of the linking between triplets, 4-tuplets, etc., of field lines. Theorem 1 asserts that all of these ideal invariants are contained in $\mathcal{A}$ because, mathematically speaking, it is a *complete* invariant. It is the most economical description of field topology equivalence classes that is sought by Berger (1986). In future, the topological flux function should provide a useful tool not just to recognise but also to *quantify* changes in magnetic topology during non-ideal evolutions. For an initial application to measuring magnetic reconnection, see Yeates & Hornig (2011).

For the alternative geometry of a half-space, Berger (1988) introduces $\mathcal{A}$ as the helicity of an infinitesimal flux tube around a single field line. Indeed, Taylor (1986) points out that each closed field line yields such an invariant in a perfectly-conducting plasma. This paper aims to

Journal of Physics: Conference Series **544** (2014) 012002 doi:10.1088/1742-6596/544/1/012002

interpret these constraints in a different light. We show that Theorem 1 is a consequence of $\mathcal{A}$ being the *action* the Hamiltonian system of the magnetic field lines (Cary & Littlejohn, 1983). The Hamiltonian theory suggests an elegant expression for $\mathcal{A}$ in terms of differential forms that we exploit.

A less general form of Theorem 1 has recently appeared in Yeates & Hornig (2013). Here we present the results for a more general shape of flux tube with more general boundary conditions, and elaborate further on the Hamiltonian theory. We begin in Section 2 with the basic definition of $\mathcal{A}$, before discussing the Hamiltonian interpretation in Section 3. The proof of Theorem 1 is given in Section 4 and two explicit examples in Section 5.

2. Topological Flux Function

We define the *topological flux function* $\mathcal{A} : D_0 \to \mathbb{R}$ as the line integral of $\mathbf{A}$ along the magnetic field line rooted at $x_0 \in D_0$:

$$\mathcal{A}(x_0) = \int_{x_0}^{F(x_0)} \mathbf{A}\big(f(x_0; z)\big) \cdot \mathrm{d}\mathbf{l}. \tag{3}$$

As it stands, this definition is gauge dependent. Under a gauge transformation $\mathbf{A} \to \mathbf{A} + \nabla\chi$, it is easy to see that $\mathcal{A} \to \mathcal{A} + F^*\chi - \chi$, where the pull-back notation means $\big(F^*\chi\big)(x_0) = \chi\big(F(x_0)\big)$. We impose the gauge condition

$$\mathbf{n} \times \mathbf{A}|_{\partial V} = \mathbf{n} \times \mathbf{A}^{\mathrm{ref}}|_{\partial V}, \tag{4}$$

where $\mathbf{A}^{\mathrm{ref}}$ is the vector potential of a reference field $\mathbf{B}^{\mathrm{ref}}$ that matches $\mathbf{B} \cdot \mathbf{n}$ on ∂V but is otherwise arbitrary. This particular choice of gauge condition is motivated by the relative magnetic helicity (Section 2.1). Although this still leaves some freedom in χ, and consequently in $\mathcal{A}$, it is enough to make $\mathcal{A}$ an ideal invariant. To see this, assume an ideal evolution

$$\frac{\partial \mathbf{A}}{\partial t} = \mathbf{v} \times \mathbf{B} + \nabla\Phi. \tag{5}$$

The boundary conditions of $\mathbf{v}|_{\partial V} = 0$ and fixed $\mathbf{n} \times \mathbf{A}|_{\partial V}$ mean that $\mathbf{n} \times \nabla\Phi|_{\partial V} = 0$. Then, using the formula for the rate of change of a line integral over a moving domain (e.g., Frankel, 1997), we find

$$\frac{\mathrm{d}\mathcal{A}}{\mathrm{d}t} = \frac{\mathrm{d}}{\mathrm{d}t}\int_{x_0}^{F(x_0)} \mathbf{A} \cdot \mathrm{d}\mathbf{l} \tag{6}$$

$$= \int_{x_0}^{F(x_0)} \left(\frac{\partial \mathbf{A}}{\partial t} - \mathbf{v} \times \nabla \times \mathbf{A} + \nabla(\mathbf{v} \cdot \mathbf{A})\right) \cdot \mathrm{d}\mathbf{l} \tag{7}$$

$$= \int_{x_0}^{F(x_0)} \nabla(\Phi + \mathbf{v} \cdot \mathbf{A}) \cdot \mathrm{d}\mathbf{l} = 0. \tag{8}$$

2.1. Relation to Magnetic Helicity

Berger (1988) defines $\mathcal{A}$ as a "field line helicity", or limiting magnetic helicity in an infinitesimal flux tube around a single magnetic field line. Since our domain is magnetically open, we define a gauge invariant relative helicity (Berger & Field, 1984; Finn & Antonsen, 1985) relative to the reference field $\mathbf{B}^{\mathrm{ref}} = \nabla \times \mathbf{A}^{\mathrm{ref}}$. Assuming (4), the relative helicity may be written

$$H_r = \int_V (\mathbf{A} + \mathbf{A}^{\mathrm{ref}}) \cdot (\mathbf{B} - \mathbf{B}^{\mathrm{ref}})\, \mathrm{d}^3x = \int_V \mathbf{A} \cdot \mathbf{B}\, d^3x - \int_V \mathbf{A}^{\mathrm{ref}} \cdot \mathbf{B}^{\mathrm{ref}}\, \mathrm{d}^3x. \tag{9}$$

The last term is a constant (call it H^{ref}) which depends only on the choice of reference field. For the first term, let x_0 be the footpoint of the field line through $x = (\rho, \psi, z)$, so that $x = f(x_0; z)$. Changing coordinates gives

$$H_r - H^{\text{ref}} = \int_V \mathbf{A}(x) \cdot \mathbf{B}(x)\, \mathrm{d}^3x, \tag{10}$$

$$= \int_V \mathbf{A}\big(f(x_0; z)\big) \cdot \mathbf{B}\big(f(x_0; z)\big) \frac{B_z(x_0)}{B_z\big(f(x_0; z)\big)} \mathrm{d}^2x_0 \mathrm{d}z, \tag{11}$$

$$= \int_{D_0} \mathcal{A}(x_0) B_z(x_0)\, \mathrm{d}^2x_0. \tag{12}$$

Thus we see that $\mathcal{A}$ represents the helicity per field line.

3. Hamiltonian Interpretation

We will show in this section that the differential of $\mathcal{A}$ has the following succinct formula in terms of differential forms (Frankel, 1997), where $h_\rho \equiv ||\partial(\mathbf{r})/\partial\rho||$, $h_\psi \equiv ||\partial(\mathbf{r})/\partial\psi||$ are the coordinate scale factors.

Lemma 2. *Let* $\mathbf{B}$ *be a magnetic braid with field line mapping* F *and topological flux function* $\mathcal{A}$. *Then*

$$\mathrm{d}\mathcal{A} = F^*\alpha - \alpha,$$

where $\alpha = A_\rho^{\text{ref}} h_\rho\, \mathrm{d}\rho + A_\psi^{\text{ref}} h_\psi\, \mathrm{d}\psi$ *is the 1-form associated to* $\mathbf{A}^{\text{ref}}$ *on* D_0 *and (perhaps differently) on* D_1.

This formula will be used to prove Theorem 1 in Section 4. It may be understood physically by integrating along a curve $\gamma \in D_0$, so that

$$\int_\gamma \mathrm{d}\mathcal{A} = \int_\gamma F^*\alpha - \int_\gamma \alpha. \tag{13}$$

In vector notation, equation (13) reads

$$\mathcal{A}(y_0) - \mathcal{A}(x_0) = \int_{F(\gamma)} \mathbf{A} \cdot \mathrm{d}\mathbf{l} - \int_\gamma \mathbf{A} \cdot \mathrm{d}\mathbf{l}, \tag{14}$$

where $x_0, y_0 \in D_0$ are the start and end points of γ (see Figure 2). If $y_0 \neq x_0$ then γ is an open curve, and (14) expresses the fact that the field lines rooted in γ form a flux surface with $\oint \mathbf{A} \cdot \mathrm{d}\mathbf{l} = 0$. This is the vertical shaded surface in Figure 2. On the other hand, if $y_0 = x_0$ then the curve is closed (the curve σ in Figure 2), in which case (14) expresses conservation of (vertical) magnetic flux in the corresponding flux tube.

3.1. $\mathcal{A}$ as an Action

It is insightful to derive Lemma 2 by interpreting $\mathcal{A}$ as an action in a Hamiltonian system. Suppose we parametrise a magnetic field line in terms of its length l as $\mathbf{x}(l)$, and think of the topological flux function as a functional,

$$\mathcal{A}(x_0) = \int_{x_0}^{F(x_0)} \mathbf{A}(\mathbf{x}) \cdot \frac{\mathrm{d}\mathbf{x}}{\mathrm{d}l}\, \mathrm{d}l. \tag{15}$$

Then Cary & Littlejohn (1983) point out that extremising $\mathcal{A}$ for given $\mathbf{A}$ and fixed end-points gives the path of the field line $\mathbf{x}(l)$ through the domain. To see this, note that the Euler-Lagrange equations are

$$\frac{\partial L}{\partial x_j} - \frac{\mathrm{d}}{\mathrm{d}l}\left(\frac{\partial L}{\partial(\partial x_j/\partial l)}\right) = 0. \tag{16}$$

In our case, the Lagrangian is $L = \mathbf{A}\cdot(\mathrm{d}\mathbf{x}/\mathrm{d}l)$, so

$$\frac{\partial A_i}{\partial x_j}\frac{\partial x_i}{\partial l} - \frac{\partial x_i}{\partial l}\frac{\partial A_j}{\partial x_i} = 0, \tag{17}$$

which is equivalent to

$$(\nabla\times\mathbf{A})\times\frac{\mathrm{d}\mathbf{x}}{\mathrm{d}l} = 0. \tag{18}$$

Hence $\mathrm{d}\mathbf{x}/\mathrm{d}l$ is everywhere parallel to $\mathbf{B}$, and $\mathbf{x}(l)$ is a magnetic field line.

In fact, it is well known that the equations of the magnetic field lines are a Hamiltonian system. To show this, we follow Cary & Littlejohn (1983) and change the gauge of $\mathbf{A}$ so as to write the action explicitly in the canonical form of a Hamiltonian system. A gauge transformation $\mathbf{A}\to\mathbf{A}+\nabla\chi$ changes the integrand (the Lagrangian) to $L+\mathrm{d}\chi/\mathrm{d}l$ and also changes the integral (the action $\mathcal{A}$), but leaves the Euler-Lagrange equations unchanged. If we set $A_\rho = 0$ everywhere in space, then $\mathbf{A}\cdot\mathrm{d}\mathbf{x} = A_\psi h_\psi\,\mathrm{d}\psi + A_z h_z\,\mathrm{d}z$. Making the identifications $p\leftrightarrow A_\psi(\rho,\psi,z)h_\psi, q\leftrightarrow\psi, t\leftrightarrow z, H\leftrightarrow -A_z(\rho,\psi,z)h_z$, our action becomes

$$\mathcal{A} = \int_{x_0}^{F(x_0)}\left(p\frac{\mathrm{d}q}{\mathrm{d}t} - H(p,q,t)\right)\mathrm{d}t, \tag{19}$$

which is a 1 degree-of-freedom Hamiltonian system in canonical form. The generalised coordinate is ψ, the generalised momentum is $A_\psi h_\psi$, and the Hamiltonian is $-A_z h_z$. Time is the z-direction, so our Hamiltonian is, in general, time dependent.

The gauge where $A_\rho = 0$ everywhere may be called a *canonical gauge* since it renders the Hamiltonian system in canonical form. For Theorem 1 it is sufficient that $\mathbf{n}\times\mathbf{A}|_{\partial V}$ be in canonical gauge, so in practice we restrict the gauge of $\mathbf{A}^{\mathrm{ref}}$ to be canonical, i.e. $A^{\mathrm{ref}}_\rho = 0$. The gauge $A_\rho = 0$ is equivalent to writing $\mathbf{B}$ in the form $\mathbf{B} = \nabla(A_\psi h_\psi)\times\nabla\psi + \nabla(A_z h_z)\times\nabla z$ (Boozer, 1983; Yoshida, 1994).

3.2. Proof of Lemma 2

Given that $\mathcal{A}$ is the action in a Hamiltonian system, we can use known properties of Hamiltonian systems. Let f_z be the "height-z" mapping of our field line system, defined by $f_z(x_0) = f(x_0; z)$. This mapping must preserve magnetic flux. In terms of differential forms, $f_z^*\omega - \omega = 0$, where $\omega = B_z h_\rho h_\psi\,\mathrm{d}\rho\wedge\mathrm{d}\psi$. Every Hamiltonian system preserves such a symplectic form ω. But notice that

$$\omega = \left(\frac{\partial}{\partial\rho}(A_\psi h_\psi) - \frac{\partial}{\partial\psi}(A_\rho h_\rho)\right)\mathrm{d}\rho\wedge\mathrm{d}\psi = \mathrm{d}\alpha \tag{20}$$

where $\alpha = A_\psi h_\psi\,\mathrm{d}\psi + A_\rho h_\rho\,\mathrm{d}\rho$. The primitive α is called the canonical/Liouville 1-form.

Assume now that $\mathbf{A}$ is in canonical gauge as above, so that $\alpha = A_\psi h_\psi\,\mathrm{d}\psi$. Then Lemma 2 follows from the following argument (Haro, 1998). In canonical coordinates, Hamilton's equations have the form

$$\dot{p} = -\frac{\partial H}{\partial q}, \quad \dot{q} = \frac{\partial H}{\partial p}. \tag{21}$$

If we define the Hamiltonian vector field $\mathbf{x}(t)\equiv(x^q, x^p) = (\dot{q},\dot{p})$, then Hamilton's equations may be succinctly written (Marsden & Ratiu, 1994) in terms of α as

$$i_{\mathbf{x}}\,\mathrm{d}\alpha = -\mathrm{d}H. \tag{22}$$

By the definition of the Lie derivative (Theorem 4.3.1 of Marsden & Ratiu, 1994), using Cartan's magic formula, and the fact that differentials and pull-backs commute, we have

$$\frac{d}{ds}(f_s^*\alpha) = f_s^*\mathcal{L}_{\mathbf{x}(s)}\alpha = f_s^*\left(i_{\mathbf{x}(s)}\,\mathrm{d}\alpha + \mathrm{d}i_{\mathbf{x}(s)}\alpha\right) = \mathrm{d}f_s^*\left(i_{\mathbf{x}(s)}\alpha - H\right). \tag{23}$$

Journal of Physics: Conference Series **544** (2014) 012002 doi:10.1088/1742-6596/544/1/012002

Integrating both sides from $s = 0$ to $s = z$ yields the desired formula

$$f_z^*\alpha - \alpha = \mathrm{d}\Big(\int_0^z f_s^*\big(i_{\mathbf{x}(s)}\alpha - H\big)\,\mathrm{d}s\Big). \tag{24}$$

Taking $z = 1$ in this expression and recognising (19) gives

$$F^*\alpha - \alpha = \mathrm{d}\Big(\int_0^1 f_z^*(p\dot{q} - H)\,\mathrm{d}z\Big) = \mathrm{d}\Big(\int_0^1 (p\dot{q} - H)\circ f_z\,\mathrm{d}z\Big) = \mathrm{d}\mathcal{A}, \tag{25}$$

which is Lemma 2.

To show that Lemma 2 holds even in non-canonical gauge, suppose that $F^*\alpha - \alpha = \mathrm{d}\mathcal{A}$ for some $\mathbf{A}$. Under the change of gauge $\mathbf{A} \to \mathbf{A} + \nabla\chi$, we find $\alpha \to \alpha + \mathrm{d}\chi$, so

$$F^*\alpha - \alpha \quad \to \quad F^*\alpha - \alpha + \mathrm{d}(F^*\chi - \chi) = \mathrm{d}(\mathcal{A} + \mathcal{F}^*\chi - \chi). \tag{26}$$

But this is precisely $\mathrm{d}\mathcal{A}$ in the new gauge (Section 2), so if the Lemma holds in one gauge it holds in any gauge.

4. Proof of Theorem 1

Lemma 2 allows for an elegant proof of Theorem 1.

Necessity

We already know that $\mathcal{A}$ is a *necessary* condition for topological equivalence, because it is an ideal invariant. But we can also see this from Lemma 2. Assuming $\widetilde{F} = F$, it follows that

$$\mathrm{d}\widetilde{\mathcal{A}} = \widetilde{F}^*\alpha - \alpha = F^*\alpha - \alpha = \mathrm{d}\mathcal{A}, \tag{27}$$

so that $\widetilde{\mathcal{A}}$ and $\mathcal{A}$ differ by at most an overall constant. But since $\widetilde{F}|_{\partial D_0} = F_{\partial D_0}$ (with the same winding number) we must have $\widetilde{\mathcal{A}} = \mathcal{A}$ (see Figure 3).

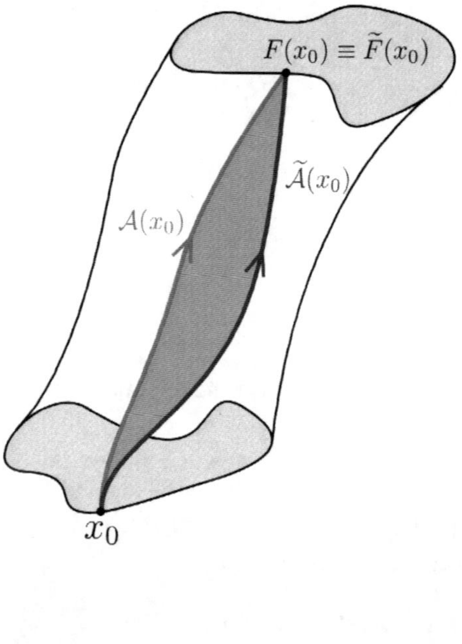

Figure 3. Illustration of why $\widetilde{F} = F$ implies $\widetilde{\mathcal{A}} = \mathcal{A}$ when the winding numbers are the same. Both braids have the same $\mathbf{n} \times \mathbf{A}$ on ∂V, although they may have different field lines on this boundary linking x_0 and $F(x_0)$ (as shown). But the flux through the loop must vanish, so $\widetilde{\mathcal{A}} = \mathcal{A}$ by Stokes' Theorem.

Sufficiency
To prove the converse—that $\mathcal{A}$ is a *sufficient* condition for topological equivalence—assume that $\mathcal{A} = \widetilde{\mathcal{A}}$ and define the mapping $G = \widetilde{F} \circ F^{-1}$. Then, using Lemma 2,

$$\begin{aligned} G^*\alpha - \alpha &= (F^{-1})^* \circ \widetilde{F}^*\alpha - \alpha && (28)\\ &= (F^{-1})^*(\alpha + \mathrm{d}\mathcal{A}) - \alpha && (29)\\ &= (F^{-1})^*\alpha - \alpha + (F^{-1})^*\mathrm{d}\mathcal{A} && (30)\\ &= (F^{-1})^*\alpha - (F^{-1})^* \circ F^*\alpha + (F^{-1})^*\mathrm{d}\mathcal{A} && (31)\\ &= (F^{-1})^*(\alpha - F^*\alpha) + (F^{-1})^*\mathrm{d}\mathcal{A} && (32)\\ &= -(F^{-1})^*\mathrm{d}\mathcal{A} + (F^{-1})^*\mathrm{d}\mathcal{A} && (33)\\ &= 0. && (34) \end{aligned}$$

Now we determine the possible mappings $G : D_1 \to D_1$ satisfying $G^*\alpha = \alpha$, which Haro (1998, 2000) calls "actionmorphisms". If α is in the canonical gauge $A_\rho^{\mathrm{ref}} = 0$, then the possible mappings G that preserve α are known (Marsden & Ratiu, 1994, Proposition 6.3.2) to take the form of cotangent lifts

$$G(q,p) = \left(T(q), \frac{p}{T'(q)}\right) \qquad (35)$$

where T is a diffeomorphism. Since ∂D_0 is not a line of constant $q(\equiv \psi)$, it follows from our assumption $\widetilde{F}|_{\partial D_0} = F|_{\partial D_0}$ that $T(q) = q$ and hence that $G_q = q$, $G_p = p$. We see that $G_\psi = \psi$. Recalling that $p = A_\psi^{\mathrm{ref}} h_\psi$, we note in canonical gauge that $B_z > 0$ implies that $\partial p/\partial\rho > 0$. It follows that the coordinate transformation from (ρ, ψ) to (p, q) has non-zero Jacobian, and hence that $G_\rho = \rho$. So $G = \mathrm{id}$ and therefore $\widetilde{F} = F$. □

Remarks

(i) If α is not in the canonical gauge, then G need not take the form of a cotangent lift. Indeed, the nature of the actionmorphisms appears to depend on the gauge of α. For suppose that $G^*\alpha - \alpha = 0$ for some α. Then after a gauge transformation $\alpha \to \alpha + \mathrm{d}\chi$ we find

$$G^*(\alpha + \mathrm{d}\chi) - (\alpha + \mathrm{d}\chi) = \mathrm{d}(G^*\chi - \chi) \neq 0 \qquad (36)$$

so that G is no longer an actionmorphism in the new gauge.

(ii) Notice that the property of whether or not two magnetic braids are topologically equivalent does not depend on the choice of reference field. This choice is arbitrary; all that matters is that the *same* reference field is used to compute both $\mathcal{A}$ and $\widetilde{\mathcal{A}}$. On the other hand, the nature of differences in $\mathcal{A}$ between two inequivalent braids may depend on the choice of reference: this is an issue for future investigation.

5. Examples: Magnetic Braids on a Cylinder

We conclude with two simple examples on the cylinder $V = \{(r, \phi, z) : 0 \leq r \leq 4, -8 \leq z \leq 8\}$. The first serves to illustrate the theory with a simple magnetic braid where $\mathcal{A}$ may be calculated analytically, and the second to illustrate an important topological interpretation of $\mathcal{A}$ introduced by Yeates & Hornig (2013) for magnetic braids in the cylinder.

5.1. Centred Flux Ring

Take standard cylindrical coordinates $\rho = r$ and $\psi = \phi$, so that the scale factors are $h_\rho = h_z = 1$, $h_\psi = r$. Consider the magnetic braid

$$\mathbf{B} = \mathbf{e}_z + \sqrt{2}r\, e^{-r^2/2 - z^2/4}\mathbf{e}_\phi \qquad (37)$$

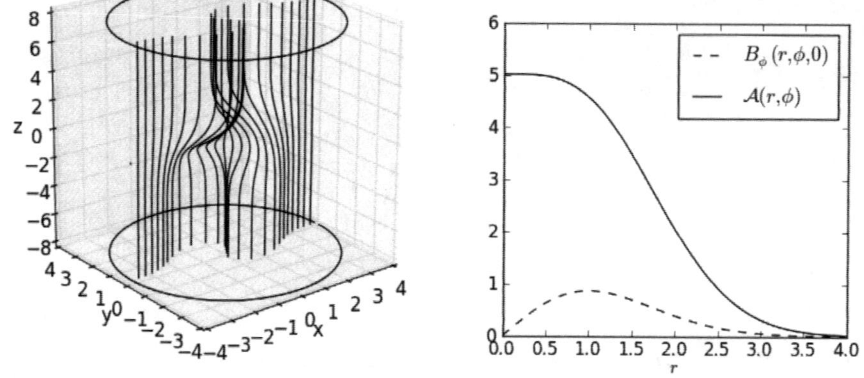

Figure 4. The example magnetic field from Section 5.1, showing field lines (*left*) and profiles of $B_\phi(z=0)$ and $\mathcal{A}$ as a function of r (*right*).

which consists of a localised toroidal "flux ring" in a uniform background field (Wilmot-Smith *et al.*, 2009). For this field the field lines are

$$f_r(r,\phi;z) = r, \qquad f_\phi(r,\phi;z) = \sqrt{2\pi}\, e^{-r^2/a^2}\Big(\mathrm{erf}(z/2) - \mathrm{erf}(z_0/2)\Big) + \phi, \tag{38}$$

and taking $z_0 \to -\infty$, $z \to \infty$ in (37), the field line mapping is

$$F_r(r,\phi) = r, \qquad F_\phi(r,\phi) = 2\sqrt{2\pi}\, e^{-r^2/2} + \phi. \tag{39}$$

A vector potential in the appropriate gauge is $\mathbf{A} = (r/2)\mathbf{e}_\phi + \sqrt{2}\exp(-r^2/2 - z^2/4)\mathbf{e}_z$. Since $A_z \approx 0$ on all boundaries to high accuracy, the corresponding reference is $\mathbf{A}^{\mathrm{ref}} = (r/2)\mathbf{e}_\phi$, giving $\mathbf{B}^{\mathrm{ref}} = \mathbf{e}_z$. One can then compute the topological flux function analytically, finding

$$\mathcal{A}(r,\phi) = \sqrt{2\pi}(r^2+2)\, e^{-r^2/2}. \tag{40}$$

Notice that $\partial\mathcal{A}/\partial r = -\sqrt{2\pi}r^3\, e^{-r^2/2}$ and $\partial\mathcal{A}/\partial\phi = 0$. In this case $\alpha = (r^2/2)\,\mathrm{d}\phi$ and we can verify that

$$\Big(F^*\alpha\Big)(r,\phi) = \frac{F_r^2}{2}\left(\frac{\partial F_\phi}{\partial r}\,\mathrm{d}r + \frac{\partial F_\phi}{\partial \phi}\,\mathrm{d}\phi\right) = \frac{r^2}{2}\left(-2\sqrt{2\pi}r\, e^{-r^2/2}\,\mathrm{d}r + \mathrm{d}\phi\right) \tag{41}$$

so that indeed $F^*\alpha - \alpha = \mathrm{d}\mathcal{A}$.

5.2. Displaced Flux Ring

Suppose now that the flux ring is displaced from the origin and centred at $(r=1, \phi=0)$. Then the new vector potential is

$$\mathbf{A} = \frac{r}{2}\mathbf{e}_\phi + \sqrt{2}\exp\left(-\frac{(r\cos\phi - 1)^2 + (r\sin\phi)^2}{2} - \frac{z^2}{4}\right)\mathbf{e}_z. \tag{42}$$

This is not topologically equivalent to the original example, so $\mathcal{A}$ must differ. One might expect that the $\mathcal{A}$ distribution would simply be translated, but Figure 5(a) shows that this is not the

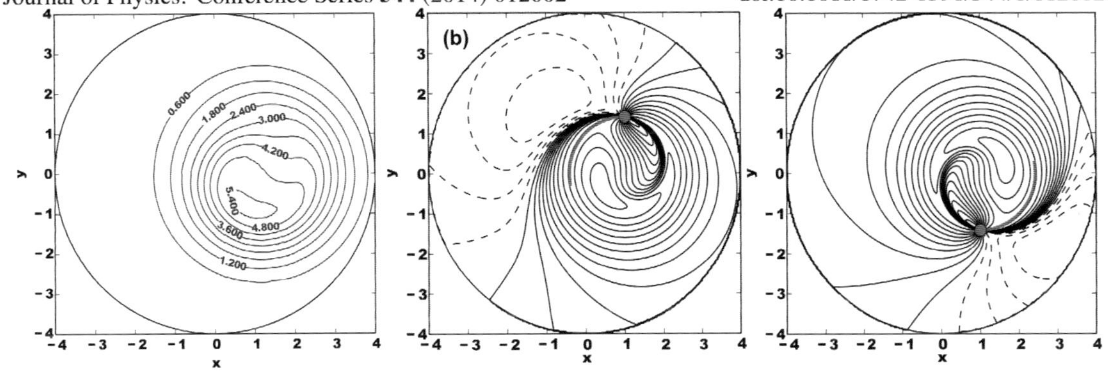

Figure 5. Asymmetry when the flux ring is displaced from the origin (Section 5.2). Panel (a) shows contours of $\mathcal{A}$. Panels (b) and (c) show contours of the pairwise linking number c_{x_0,y_0} as a function of y_0, for $x_0 = (\sqrt{3}, 55°)$ (in b) and $x_0 = (\sqrt{3}, -55°)$ (in c). The red circles show x_0 and the thick red line shows the projection of the field line $f(x_0; z)$. Solid (dashed) contours show positive (negative) values of c_{x_0,y_0}. All three panels are computed numerically.

case: the profile of $\mathcal{A}$ within the disturbed region is no longer symmetric about the ring centre. This reflects the fact that $\mathcal{A}$ is a *global* quantity: it counts not just the horizontal flux of the ring itself, but how it is linked with the reference field $\mathbf{B}^{\mathrm{ref}} = \mathbf{e}_z$. In (42), the reference field is asymmetrical with respect to the position of the flux ring.

This may be understood using the following alternative formula for $\mathcal{A}$ on the cylinder. Yeates & Hornig (2013) show that $\mathcal{A}(x_0)$ is the *average pairwise linking number* between the field line $f(x_0; z)$ and all other field lines $f(y_0; z)$. The linking number in this context (Berger, 1988) is the integral

$$c_{x_0,y_0} = \frac{1}{2\pi}\int_{-8}^{8} \frac{\mathrm{d}\theta_{x_0,y_0}(z)}{\mathrm{d}z}\,\mathrm{d}z, \tag{43}$$

where $\theta_{x_0,y_0}(z)$ is the orientation of the line connecting $f(x_0; z)$ and $f(y_0; z)$ in the plane at height z. It is proved by Yeates & Hornig (2013) that

$$\mathcal{A}(x_0) = \int_{D_0} c_{x_0,y_0} B_z(y_0)\,\mathrm{d}^2 y_0. \tag{44}$$

As an aside, notice that equation (44) gives a formula to compute $\mathcal{A}$ directly from $\mathbf{B}$ without needing to know $\mathbf{A}$.

The asymmetry in Figure 5(a) arises due to differing c_{x_0,y_0} for different field lines $f(x_0; z)$ through the flux ring. To see this, consider two field lines $f(x_0; z)$ passing through the ring, but starting on opposite sides (Figures 5b and 5c). The contours show c_{x_0,y_0} as a function of y_0 for each of these x_0. For y_0 within the ring itself, the c_{x_0,y_0} contours are the same in each case up to rotation, so these points make the same contribution to the integral (44). But for most y_0 outside the ring, c_{x_0,y_0} has the opposite sign in Figure 5(b) compared to Figure 5(c), because of the field line direction $f_y(x_0; z) < 0$ in Figure 5(b) while $f_y(x_0; z) > 0$ in Figure 5(c). Furthermore, since c_{x_0,y_0} is opposite in sign on the left and right of the ring, and since there is a larger area to the left of the ring, the net contributions to (44) from y_0 outside the ring differ in each case. This explains the asymmetry. In Section 5.1, the area to the left and right of the ring was the same, hence the symmetric $\mathcal{A}$ profile.

Acknowledgments

The work was supported by STFC Grant ST/G002436/1 to the University of Dundee. We thank J. B. Taylor, J.-J. Aly and P. Boyland for useful suggestions.

Journal of Physics: Conference Series **544** (2014) 012002 doi:10.1088/1742-6596/544/1/012002

References

Berger, M. A. 1986 Topological invariants of field lines rooted to planes. *Geophys. Astrophys. Fluid Dyn.* **34**, 256–281.

Berger, M. A. 1988 An energy formula for nonlinear force-free magnetic fields. *Astron. Astrophys.* **201**, 355–361.

Berger, M. A. & Field, G. B. 1984 The topological properties of magnetic helicity. *J. Fluid Mech.* **147**, 133–148.

Berger, M. A. & Prior, C. 2006 The writhe of open and closed curves. *J. Phys. A: Math. Theor.* **39**, 8321–8348.

Boozer, A. H. 1983 Evaluation of the structure of ergodic fields. *Phys. Fluids* **26**, 1288–1291.

Brown, M. R., Canfield, R. C. & Pevtsov, A. A. (ed.) 1999 **Magnetic Helicity in Space and Laboratory Plasmas**, Washington. AGU.

Candelaresi, S. & Brandenburg, A. 2011 Decay of helical and nonhelical magnetic knots. *Phys. Rev. E* **84**, 01646.

Cary, J. R. & Littlejohn, R. G. 1983 Noncanonical hamiltonian mechanics and its application to magnetic field line flow. *Annal. Phys.* **151**, 1–34.

Finn, J. H. & Antonsen, T. M. 1985 Magnetic helicity: What is it and what is it good for? *Comments Plasma Phys. Contr. Fusion* **9**, 111–126.

Frankel, T. 1997 **The Geometry of Physics**. CUP, Cambridge.

Haro, A. 1998 The primitive function of an exact symplectomorphism. PhD thesis, Universitat de Barcelona, Barcelona.

Haro, A. 2000 The primitive function of an exact symplectomorphism. *Nonlinearity* **13**, 1483–1500.

Marsden, J. E. & Ratiu, T. S. 1994 **Introduction to Mechanics and Symmetry**. Springer-Verlag, New York.

Morrison, P. J. 2000 Magnetic field lines, hamiltonian dynamics, and nontwist systems. *Phys. Plasmas* **7**, 2279–2289.

Reale, F. 2010 Coronal loops: Observations and modeling of confined plasma. *Living Rev. Solar Phys.* **7**, 5.

Taylor, J. B. 1986 Relaxation and magnetic reconnection in plasmas. *Rev. Mod. Phys.* **58**, 741–763.

Van Compernolle, B. & Gekelman, W. 2012 Morphology and dynamics of three interacting kink-unstable flux ropes in a laboratory magnetoplasma. *Physics of Plasmas* **19** (10), 102102.

Wilmot-Smith, A. L., Hornig, G. & Pontin, D. I. 2009 Magnetic braiding and parallel electric fields. *Astrophys. J.* **696**, 1339–1347.

Woltjer, L. 1958 A theorem on force-free magnetic fields. *Proc. Nat. Acad. Sci.* **44**, 489–491.

Yeates, A. R. & Hornig, G. 2011 A generalized flux function for three-dimensional magnetic reconnection. *Phys. Plasmas* **18**, 102118.

Yeates, A. R. & Hornig, G. 2013 Unique topological characterization of braided magnetic fields. *Phys. Plasmas* **20**, 012102.

Yeates, A. R., Hornig, G. & Wilmot-Smith, A. L. 2010 Topological constraints on magnetic relaxation. *Phys. Rev. Lett.* **105**, 085002.

Yoshida, Z. 1994 A remark on the hamiltonian form of the magnetic-field-line equations. *Phys. Plasmas* **1**, 208–209.

Journal of Physics: Conference Series **544** (2014) 012003 doi:10.1088/1742-6596/544/1/012003

Lower and upper bounds on the energy of braided magnetic fields

J.-J. Aly

AIM, Service d'Astrophysique, CE Saclay, 91191 Gif sur Yvette Cedex, France

E-mail: jean-jacques.aly@cea.fr

Abstract. Let $\mathcal{B}$ be a class of topologically equivalent regular magnetic fields occupying a cylindrical domain of axis Oz and produced by ideal MHD deformations of the uniform field $B_0\hat{\mathbf{z}}$, and let $\mathcal{B}_{ff} \subset \mathcal{B}$ be constituted of the fields of $\mathcal{B}$ which are force-free. We derive topological lower and upper bounds on the energy of the fields in $\mathcal{B}$ and in $\mathcal{B}_{ff}$, respectively. We also establish new formulae for the field line helicity and the relative magnetic helicity of the fields in $\mathcal{B}$, and give a sufficient criterion of ideal MHD linear stability for the fields in $\mathcal{B}_{ff}$.

1. Introduction

In the simple model of solar coronal loop proposed by Parker (Parker (1994), and references therein), the magnetic field $\mathbf{B}$ is taken to occupy a cylindrical domain D of axis Oz, filled up with a low beta perfectly conducting plasma, and to satisfy at any given time two conditions: it is obtained from the uniform vertical field $\mathbf{B}_0$ by an ideal MHD deformation produced by continuous footpoints displacements on the lower and upper parts of the boundary of D, and it is in a force-free equilibrium state. One is thus immediately confronted with a nonstandard problem: is it really possible to find a force-free configuration having an arbitrarily prescribed topology? Parker argued that such a configuration does exist, but that it is singular when it is forced to be fully 3D by the braiding imposed to the lines: in that case, it needs to contain infinitely thin current sheets, i.e., surfaces across which $\mathbf{B}$ is discontinuous. If a small resistivity is accounted for, current sheets are replaced by thin current layers in which intense magnetic energy dissipation eventually occurs. This results into plasma heating and Parker suggested that the corona maintains its high temperature by this mechanism. Actually, Parker's conclusion about the necessary appearance of current sheets in the ideal situation has been criticized by many authors, and strong arguments have been advanced in favor of the existence of smooth topologically constrained 3D equilibria at least when the field deformation is not too large. An alternative scenario for generating current sheets has also been proposed by Ng & Bhattacharjee (1998). These authors argue that a progressively deformed smooth equilibrium reaches at some stage an unstable configuration from which it needs to relax to a necessarily singular lower energy state. The debate is still going on, and the reader may get a flavor of it by looking at some recent references (e.g., Aly & Amari (2010); Huang *et al.* (2010); Janse *et al.* (2010)).

The aim of this Communication is to present some new general properties of Parker's model. We consider the set of all the regular magnetic fields which can be obtained by a continuous ideal deformation of $\mathbf{B}_0$. We divide it into classes $\mathcal{B}$ of topologically equivalent fields and take the point of view – opposite to Parker's conclusion – that each of them contains at least one

Journal of Physics: Conference Series **544** (2014) 012003 doi:10.1088/1742-6596/544/1/012003

force-free configuration. A part of the paper is devoted to deriving new formulae for the field line helicity and the relative magnetic helicity of the fields in $\mathcal{B}$, and computing lower bounds on their energy. The force-free fields in $\mathcal{B}$ are considered in another part, and we obtain for them an energy upper bound and a sufficient criterion of ideal MHD linear stability. Most of the results reported here have been obtained years ago, but they have never been published before. Proofs are only briefly sketched. Details will be given in forthcoming papers.

2. Assumptions

Hereafter, (x, y, z) are Cartesian coordinates and D is the cylinder $S_0 \times]0, h[\subset \mathbb{R}^3$ of base $S_0 \subset \{(x, y, 0)\}$, axis Oz, and height h. The boundary ∂D of D is constituted of S_0, $S_h = S_0 + h\hat{\mathbf{z}}$, and a vertical surface S. We denote as $\hat{\mathbf{n}}$ the exterior unit normal to ∂D and as $|Y|$ the "natural" measure of the set Y (e.g., $|D|$ is the volume of D). If $\mathbf{X}$ is a vector field, we set $\mathbf{X} = \mathbf{X}_\perp + X_z\hat{\mathbf{z}}$ ($\mathbf{r} = \mathbf{R} + z\hat{\mathbf{z}}$ for the position vector), and $\boldsymbol{\nabla}\mathbf{X} = \sum \partial_i X_j \hat{\mathbf{e}}_i \otimes \hat{\mathbf{e}}_j$ $(i, j = x, y, z)$.

We take D to contain a regular magnetic field $\mathbf{B}$ embedded in a low beta perfectly conducting plasma. $\mathbf{B}$ is assumed to have been produced by a continuous deformation, constrained by the frozen-in law, of the uniform field $\mathbf{B}_0 = B_0\hat{\mathbf{z}}$, with $B_0 > 0$ a constant. This deformation transports an element of plasma from its initial position $\mathbf{r}_0 \in D$ (resp., S_0, S_h, S) to the new position $\overline{\mathbf{r}}(\mathbf{r}_0) = (\overline{\mathbf{R}}, \overline{z})(\mathbf{r}_0) \in D$ (resp., S_0, S_h, S). The mapping $\overline{\mathbf{r}}$ is one-to-one and orientation preserving, whence $J = \det \mathbf{F} > 0$, where $\mathbf{F} = \boldsymbol{\nabla}_0\overline{\mathbf{r}}$ and $\boldsymbol{\nabla}_0$ is the gradient with respect to $\mathbf{r}_0$. The horizontal positions of the footpoints of $\mathbf{B}_0$ on S_0 and S_h are thus transported from $\mathbf{R}_0$ to $\overline{\mathbf{R}}(\mathbf{R}_0, 0)$ and $\overline{\mathbf{R}}(\mathbf{R}_0, h)$, respectively, while the magnetic lines on S acquire their new positions by slipping tangentially to that surface. As it is well known, $\mathbf{B}$ is explicitly given by

$$\mathbf{B}[\overline{\mathbf{r}}(\mathbf{r}_0)] = \frac{\mathbf{B}_0 \cdot \mathbf{F}(\mathbf{r}_0)}{J(\mathbf{r}_0)} = \frac{B_0}{J(\mathbf{r}_0)}\frac{\partial \overline{\mathbf{r}}}{\partial z_0}(\mathbf{r}_0). \tag{1}$$

We say that two fields $\mathbf{B}$ and $\mathbf{B}'$ produced by the mappings $\overline{\mathbf{r}}$ and $\overline{\mathbf{r}}'$, respectively, are *topologically equivalent* if they can be continuously transformed into each other by an ideal MHD deformation keeping fixed the positions of the footpoints. This implies in particular that $\overline{\mathbf{r}}(\mathbf{r}_0) = \overline{\mathbf{r}}'(\mathbf{r}_0)$ for $\mathbf{r}_0 \in S_0 \cup S_h$. We denote as $\mathcal{B}$ a class of equivalent fields and as $\mathcal{B}_{ff}$ the subset of $\mathcal{B}$ containing the fields that are force-free and then satisfy

$$\boldsymbol{\nabla} \times \mathbf{B} = \alpha\mathbf{B} \tag{2}$$

for some function α. As $\boldsymbol{\nabla} \cdot \mathbf{B} = 0$, $\mathbf{B} \cdot \boldsymbol{\nabla}\alpha = 0$ and α is constant along any magnetic field line. Our assumption here is that $\mathcal{B}_{ff}$ is nonempty – at least for a large class of $\mathcal{B}$. A field $\mathbf{B}$ belonging to $\mathcal{B}_{ff}$ can be possibly constructed as the final state of one of the two following processes: (i) the first one is an MHD viscous evolution with zero resistivity in which the footpoints are fixed on S_0 and S_h. The initial state is selected in $\mathcal{B}$ and then the field stays in that set at all subsequent times (Moffatt, 1985). (ii) The second one is a quasi-static evolution during which the field passes through a sequence of force-free configurations, with the initial field being the uniform $\mathbf{B}_0$. This evolution is driven by imposing the footpoints on S_0 and S_h to move horizontally at the adequately prescribed velocity $\mathbf{v}(\mathbf{R}, 0, t)$ and $\mathbf{v}(\mathbf{R}, h, t)$, respectively.

One of our essential goal hereafter is to establish lower bounds on the energy W of the fields of $\mathcal{B}$, and upper bounds on that of the fields in $\mathcal{B}_{ff}$. W is given by the various expressions

$$W = \int_D \frac{B^2}{8\pi}\,\mathrm{d}v = W_0 + w = W_0 + w_\perp + w_z = \int_D \omega_0\,\mathrm{d}v_0, \tag{3}$$

where we have set $\mathbf{B} = B_0\hat{\mathbf{z}} + \mathbf{b} = (B_0 + b_z)\hat{\mathbf{z}} + \mathbf{b}_\perp$ and

$$W_0 = \frac{B_0^2|D|}{8\pi}, \quad w_k = \int_D \frac{b_k^2}{8\pi}\,\mathrm{d}v \text{ for } k = ., \perp, z, \quad \text{and } \omega_0 = \frac{B_0^2}{8\pi J}\left|\frac{\partial\overline{\mathbf{r}}}{\partial z_0}\right|^2, \tag{4}$$

and used that $\int_{S_z} b_z\,\mathrm{d}s = 0$ by flux conservation (S_z is the cross section of D at height z).

3. Some considerations on the topology of the fields in $\mathcal{B}$

3.1. Basic notions and characterization of the topology

We associate to the arbitrary field $\mathbf{B} \in \mathcal{B}$ four functions which play an important role hereafter:

- *Boundary flux distribution.* It is just the vertical component B_z of $\mathbf{B}$ on S_0 and S_h.
- *Magnetic mapping.* It is the one-to-one mapping $\mathbf{M} : S_0 \to S_h$ which associates to the lower endpoint $\mathbf{r} = \mathbf{R}$ of the line $\mathcal{L} = \mathcal{L}(\mathbf{R})$ on S_0 its upper endpoint $\mathbf{M}(\mathbf{R}) = \mathbf{R}^h(\mathbf{R}) + h\hat{\mathbf{z}}$ on S_h.
- *Winding number $n(\mathbf{R})$ of the line $\mathcal{L}(\mathbf{R})$ drawn on S.* It is the algebraic number of turns,

$$n(\mathbf{R}) = \frac{1}{|\partial S_0|} \int_{\mathcal{L}(\mathbf{R})} \mathrm{d}q = \frac{1}{|\partial S_0|} \int_{\mathcal{L}(\mathbf{R})} \frac{B_q}{B} \,\mathrm{d}l \,\in\, \mathbb{R}, \tag{5}$$

  made by $\mathcal{L}(\mathbf{R})$ around S ((q, z) are orthonormal coordinates on S, with $\boldsymbol{\nabla}_s q = \hat{\mathbf{z}} \times \hat{\mathbf{n}}$).
- *Relative twist $\phi(\mathbf{R}, \mathbf{R}')$ of the pair of magnetic lines $(\mathcal{L}(\mathbf{R}), \mathcal{L}(\mathbf{R}'))$.* When all the lines of $\mathbf{B}$ are pointing upwards at all the points of D ($B_z > 0$ in D), $\phi(\mathbf{R}, \mathbf{R}')$ is defined to be the total angle by which the horizontal vector connecting $\mathcal{L}(\mathbf{R})$ to $\mathcal{L}(\mathbf{R}')$ at height z rotates when moving vertically from $z = 0$ to $z = h$. We refer to Berger & Prior (2006) for the general definition of ϕ, which accounts for the possibility of having $B_z < 0$ in some part of D.

These functions can be checked to not depend on the particular field $\mathbf{B}$ which is considered in $\mathcal{B}$, and they can be taken to be properties of the class $\mathcal{B}$ itself.

Finally, we claim that the topology of the fields in $\mathcal{B}$ can be fully characterized by the magnetic mapping, $\mathbf{M}$, and the winding number, $n_1 = n(\mathbf{R}_1)$, of a particular line $\mathcal{L}(\mathbf{R}_1)$ drawn on S. Roughly speaking, we can chose for $\mathcal{L}(\mathbf{R}_1)$ a simple curve on S starting at $\mathbf{R}_1 \in \partial S_0$ and ending on ∂S_h after making n_1 turns around S, and then use continuity for connecting together in a topologically unique way the points $\mathbf{R}$ and $\mathbf{M}(\mathbf{R})$ by pairwise disjoint simple curves.

3.2. Relative magnetic helicity

We now define the *relative magnetic helicity*, H, of a field $\mathbf{B} \in \mathcal{B}$ (Berger & Field, 1984). This important quantity is known to stay invariant when a field is submitted to an ideal MHD deformation keeping fixed the footpoints and it thus takes the same value for all the fields in $\mathcal{B}$.

Let us introduce the unique potential field $\mathbf{B}_\pi$ ($\nabla \times \mathbf{B}_\pi = 0$) with the same boundary flux distribution as the fields in $\mathcal{B}$ ($\mathbf{B}_\pi \notin \mathcal{B}$ in general), and its unique potential vector $\mathbf{C}_\pi$ ($\mathbf{B}_\pi = \nabla \times \mathbf{C}_\pi$) satisfying $\nabla \cdot \mathbf{C}_\pi = 0$ in D and $\hat{\mathbf{n}} \cdot \mathbf{C}_\pi = 0$ on ∂D. Next choose a vector potential $\mathbf{C}$ for $\mathbf{B}$ ($\mathbf{B} = \nabla \times \mathbf{C}$) such that $\hat{\mathbf{n}} \times (\mathbf{C} - \mathbf{C}_\pi) = 0$ on ∂D (this defines $\mathbf{C}$ up to an additive $\boldsymbol{\nabla} f$, with $f =$ const. on ∂D). Then H is given by (Berger, 1988)

$$H = \int_D \mathbf{C} \cdot \mathbf{B} \,\mathrm{d}v = \int_{S_0} \lambda B_z \,\mathrm{d}s, \qquad \text{where} \quad \lambda(\mathbf{R}) = \int_{\mathcal{L}(\mathbf{R})} \mathbf{C} \cdot \mathrm{d}\boldsymbol{l}. \tag{6}$$

The function λ (which is independent of the gauge chosen for $\mathbf{C}$) is called either *field line helicity* (Berger, 1988) or *topological flux function* (Yeates & Hornig, 2013). It can be given an alternative expression by proceeding as follows. We first note that $\mathbf{B}$ and then also $\mathbf{B}_\pi$ have equal fluxes through, respectively, the closed oriented curve $\mathcal{C}_0 \subset S_0$ and its image by the magnetic mapping, $\mathcal{C}_h = \mathbf{M}(\mathcal{C}_0) \subset S_h$. This implies $\int_{\mathcal{C}_0} \mathbf{C}_{\pi\perp} \cdot \mathrm{d}\boldsymbol{l} = \int_{\mathcal{C}_h} \mathbf{C}_{\pi\perp} \cdot \mathrm{d}\boldsymbol{l}$ by Stokes theorem, whence (by setting $F^h(\mathbf{R}) = F(\mathbf{R}^h(\mathbf{R}), h)$)

$$\forall \mathcal{C}_0 : \int_{\mathcal{C}_0} \mathrm{d}\boldsymbol{l} \cdot (\mathbf{C}_{\pi\perp} - \boldsymbol{\nabla}_\perp \mathbf{R}^h \cdot \mathbf{C}^h_{\pi\perp}) = 0 \quad \Rightarrow \quad \exists \psi(\mathbf{R}) : \; \mathbf{C}_{\pi\perp} - \boldsymbol{\nabla}_\perp \mathbf{R}^h \cdot \mathbf{C}^h_{\pi\perp} = -\boldsymbol{\nabla}_\perp \psi. \tag{7}$$

Next we choose an open curve $\mathcal{C}_0 \subset S_0$ connecting $\mathbf{R}_1$ to $\mathbf{R}$, and note that the flux of $\mathbf{B}$ through the contour $\mathcal{C}_1 = \mathcal{C}_0 \cup \mathcal{L}(\mathbf{R}) \cup -\mathcal{C}_h \cup -\mathcal{L}(\mathbf{R}_1)$ ($\mathcal{C}_h = \mathbf{M}(\mathcal{C}_0) \subset S_h$) vanishes. Hence

$\lambda(\mathbf{R}_1) - \psi(\mathbf{R}_1) = \lambda(\mathbf{R}) - \psi(\mathbf{R})$ by Stokes theorem and the previous result. As ψ is defined up to an additive constant, we can impose $\psi(\mathbf{R}_1) = \lambda(\mathbf{R}_1)$ at some specific point $\mathbf{R}_1$ and we get $\lambda(\mathbf{R}) = \psi(\mathbf{R})$ for all $\mathbf{R} \in S_0$. Choosing $\mathbf{R}_1 \in \partial S_0$, we obtain similarly

$$\lambda(\mathbf{R}_1) = \psi(\mathbf{R}_1) = \int_0^h C_{\pi z}(\mathbf{R}_1, z)\,\mathrm{d}z + \int_{\mathcal{C}^h(\mathbf{R}_1)} C_{\pi q}\,\mathrm{d}q \tag{8}$$

by considering the contour $\mathcal{C}_2 = \mathcal{C}_v(\mathbf{R}_1) \cup \mathcal{C}^h(\mathbf{R}_1) \cup -\mathcal{L}(\mathbf{R}_1)$, with $\mathcal{C}_v(\mathbf{R}_1)$ the vertical segment on S above $\mathbf{R}_1$ and $\mathcal{C}^h(\mathbf{R}_1)$ the curve connecting $(\mathbf{R}_1, h)$ to $(\mathbf{R}^h(\mathbf{R}_1), h)$ after running $n(\mathbf{R}_1)$ times along ∂S_h. Whence

$$\lambda(\mathbf{R}) = \psi(\mathbf{R}) = \int_0^h C_{\pi z}(\mathbf{R}_1, z)\,\mathrm{d}z + \int_{\mathcal{C}^h(\mathbf{R}_1)} C_{\pi q}\,\mathrm{d}q - \int_{\mathcal{C}(\mathbf{R}_1,\mathbf{R})} \mathrm{d}\boldsymbol{l} \cdot (\mathbf{C}_{\pi\perp} - \boldsymbol{\nabla}_\perp \mathbf{R}^h \cdot \mathbf{C}^h_{\pi\perp}), \tag{9}$$

where $\mathcal{C}(\mathbf{R}_1, \mathbf{R})$ connects $\mathbf{R}_1$ to $\mathbf{R}$. This shows that all the fields $\mathbf{B} \in \mathcal{B}$ have the same line helicity λ. Some other properties of λ (with λ possibly computed from a potential vector $\mathbf{A} \neq \mathbf{C}$) have been discussed by Yeates & Hornig (2013) (see also Yeates & Hornig, these Proceedings).

Using Eq. (9) in Eq. (6) leads to a new formula in which H is directly expressed in terms of the field topology. We thus recover the fact that H has the same value for all $\mathbf{B} \in \mathcal{B}$. The helicity formula can be transformed by integrating by parts, and one gets for instance

$$H = \int_{S_0} [\mathbf{C}_{\pi\perp} \times (\boldsymbol{\nabla}_\perp \mathbf{R}^h \cdot \mathbf{C}^h_{\pi\perp})] \cdot \hat{\mathbf{z}}\,\mathrm{d}s + n\Phi_0^2 \tag{10}$$

in the simple case where $B_z(\mathbf{R}, 0) = B_z(\mathbf{R}, h)$ and all the lines on S have the same winding number $n \in \mathbb{Z}$ ($\Phi_0 = B_0|S_0|$ is the magnetic flux through S_0).

4. Some properties of the fields in $\mathcal{B}_{ff}$

4.1. Some remarks on the force-free equations

Eq. (2) can be given the alternative form

$$\boldsymbol{\nabla} \cdot \mathbf{T} = 0, \quad \text{where } \mathbf{T} = -(B^2/8\pi)\boldsymbol{\delta} + \mathbf{B} \otimes \mathbf{B}/4\pi \tag{11}$$

($\boldsymbol{\delta}$ denotes the unit tensor, i.e., $(\boldsymbol{\delta})_{ij} = \delta_{ij}$ = Kronecker symbol). By using Eq. (1) and the standard relation $\boldsymbol{\nabla}_0 \cdot \mathbf{Cof\,F} = 0$ satisfied by the cofactor of $\mathbf{F}$, one may also write

$$\boldsymbol{\nabla}_0 \cdot \mathbf{T}_0 = 0, \quad \text{where } \mathbf{T}_0 = \mathbf{Cof\,F} \cdot \mathbf{T} = -\frac{B_0^2}{8\pi J^2}\left|\frac{\partial \mathbf{r}}{\partial z_0}\right|^2 \mathbf{Cof\,F} + \frac{B_0^2}{4\pi J}\hat{\mathbf{z}} \otimes \frac{\partial \mathbf{r}}{\partial z_0}. \tag{12}$$

Let us make three important points about these equations:

- Let us use the equation of a magnetic line $\mathcal{L}$ of $\mathbf{B}$, $\mathrm{d}\mathbf{r}/\mathrm{d}l = \mathbf{B}/B$, in Eq. (11) ($l$ is an arclength along $\mathcal{L}$). We thus obtain

$$\frac{\mathrm{d}}{\mathrm{d}l}\left(B\frac{\mathrm{d}\mathbf{r}}{\mathrm{d}l}\right) = \boldsymbol{\nabla} B, \tag{13}$$

  which is identical to the equation of propagation of a light ray in a medium with refractive index $n = B$. Hence the magnetic lines of $\mathbf{B}$ also obey a Fermat principle: with B given, the line $\mathcal{L}(\mathbf{R})$ makes stationary the functional $\int_{\mathcal{C}(\mathbf{R})} B\,\mathrm{d}l$ defined on the set of curves $\mathcal{C}(\mathbf{R})$ connecting $\mathbf{R}$ and $\mathbf{M}(\mathbf{R})$. Alternatively, one can say that $\mathcal{L}(\mathbf{R})$ is a geodesic line of the conformal metric $B^2\mathrm{d}l^2$. The ray equation above was previously written by Parker (1994) for the potential field $\mathbf{B}_\pi = \boldsymbol{\nabla}\psi_\pi$. In that case, the lines are orthogonal to the equipotential surfaces $\{\psi_\pi = \tau\}$ as the light rays are orthogonal to the wave fronts. But no family of normal surfaces does exist for the lines of $\mathbf{B} \in \mathcal{B}_{ff}$, as such a field has nonzero *abnormality* $\mathbf{B} \cdot \boldsymbol{\nabla} \times \mathbf{B}/B^2 = \alpha$.

Journal of Physics: Conference Series **544** (2014) 012003 doi:10.1088/1742-6596/544/1/012003

- Eq. (12) is formally similar to the equation of equilibrium determining the deformation of an elastic body with respect to a configuration of reference, with $\mathbf{T}_0 = \mathbf{T}_0(\mathbf{F})$ corresponding to the *first Piola-Kirchhoff stress tensor*. The analogy is even with the equilibrium of an *hyperelastic body*, as $\mathbf{T}_0$ is related to the *stored energy density* ω_0 (see Eq. (4)) by $(T_0)_{ij} = \partial\omega_0/\partial(\partial\overline{x}_j/\partial x_{0i})$ (see, e.g., Ciarlet (1988)). This remark opens the possibility of transferring some results from the well-developed theory of elasticity to the theory of topologically constrained MHD equilibria. Such equilibria have, however, some unusual properties from the point of view of elasticity. In particular, there are finite displacements that do not change the energy (those for which each plasma element is moved along a line of $\mathbf{B}$; this degeneracy disappears when plasma pressure is taken into account).
- The structure of the force-free fields in $\mathcal{B}_{ff}$ is often studied in the reduced MHD approximation, in which one has $B_z \equiv B_0$ and $\nabla_\perp \cdot \mathbf{b}_\perp = 0$ in D while Eq. (11) reduces to $B_0\partial_z\mathbf{b}_\perp + \mathbf{b}_\perp \cdot \nabla_\perp\mathbf{b}_\perp + \nabla f = 0$, for some function f (van Ballegooijen, 1985; Parker, 1994; Ng & Bhattacharjee, 1998). Under the substitutions $z/\mathbf{B}_0 \to t$, $\mathbf{b}_\perp \to \mathbf{v}$, and $f \to p/\rho$, these equations become identical to the Euler equations for a 2D incompressible fluid of velocity $\mathbf{v}$, pressure p, and uniform mass density ρ. Those are usually solved in S_0, say, under two different types of boundary conditions. Either one seeks for a solution having a prescribed vorticity $\omega = \hat{\mathbf{z}} \cdot \nabla \times \mathbf{v}$ (or velocity $\mathbf{v}$) at $t = 0$ (initial value problem, most often considered). Or one works in the Lagrangian framework, in which the unknown is the mapping $\overline{\mathbf{R}}^t(\mathbf{R}_0)$ giving the position at time t of the fluid element located at $\mathbf{R}_0$ at $t = 0$, and one seeks for a solution with a prescribed $\overline{\mathbf{R}}^T$ at some T. Such a solution $\overline{\mathbf{R}}^t$ can be interpreted as a geodesic line in the (infinite dimensional) set of area preserving diffeomorphisms of S_0 (Arnold & Khesin, 1998). When reformulated in our MHD context, these problems amount to determine a field $\mathbf{B}$ having, respectively, a given α on S_0 and a prescribed magnetic mapping. The second problem coincides with ours here (but for the imposition of the winding number n_1), and we can benefit of the many results obtained so far in the hydrodynamical framework. For instance, we may state the existence of a unique regular field minimizing the energy as long as the imposed deformation is not too large.

4.2. A consequence of current conservation

For a force-free field $\mathbf{B}$ in D, the total vertical electric current $I(z)$ through S_z is independent of z (there is no current flowing through S), and we obtain with help from Ampère's law

$$\int_{\partial S_z} b_q \,\mathrm{d}q = \frac{4\pi}{c} I(0). \tag{14}$$

On the other hand, the total currents flowing through, respectively, the arbitrary closed contour $\mathcal{C}_0 \subset S_0$ and its magnetic image $\mathcal{C}_h = \mathbf{M}(\mathcal{C}_0) \subset S_h$ are equal, while no current flows through the contours $\mathcal{C}_1$ and $\mathcal{C}_2$ defined in Sect. 3.2. Proceeding as therein (with $\mathbf{B}$ substituted for $\mathbf{C}$), we thus obtain

$$\exists\chi(\mathbf{R}): \ \mathbf{b}_\perp(\mathbf{R}, 0) - (\nabla_\perp\mathbf{R}^h \cdot \mathbf{b}_\perp^h)(\mathbf{R}) = -\nabla_\perp\chi(\mathbf{R}), \tag{15}$$

$$\int_{\mathcal{L}(\mathbf{R})} B \,\mathrm{d}l = \chi(\mathbf{R}) + B_0 h, \tag{16}$$

$$\chi(\mathbf{R}) = \int_0^h b_z(\mathbf{R}, z)\,\mathrm{d}z + \int_{\mathcal{C}^h(\mathbf{R})} b_q \,\mathrm{d}q \quad \text{for } \mathbf{R} \in \partial S_0. \tag{17}$$

The term $B_0 h$ in Eq. (16) just fixes the gauge of χ.

Using Eq. (16) in the expression $W = \int_{S_0}(\int_{\mathcal{L}} B\,\mathrm{d}l)B_z\,\mathrm{d}s/8\pi$ of the energy leads to

$$w = \frac{1}{8\pi}\int_{S_0} \chi B_z \,\mathrm{d}s. \tag{18}$$

Journal of Physics: Conference Series **544** (2014) 012003 doi:10.1088/1742-6596/544/1/012003

4.3. Global equilibrium and virial theorem

Integrating the z-component of Eq. (11) between S_0 and S_z and using Gauss theorem leads to

$$\int_{S_z} (B_\perp^2 - B_z^2)\,\mathrm{d}s = \int_{S_0} (B_\perp^2 - B_z^2)\,\mathrm{d}s, \tag{19}$$

from which we get immediately

$$w_\perp - w_z = \frac{h}{8\pi}\int_{S_0} (b_\perp^2 - b_z^2)\,\mathrm{d}s = \frac{h}{8\pi}\int_{S_h} (b_\perp^2 - b_z^2)\,\mathrm{d}s. \tag{20}$$

Similarly, multiplying Eq. (11) scalarly by $\mathbf{r}$, integrating the result by parts in D, and using once more Gauss theorem, we obtain the well-known scalar virial relation

$$W = \int_{\partial D} \left(\frac{B^2}{8\pi}\hat{\mathbf{n}}\cdot\mathbf{r} - \frac{(\hat{\mathbf{n}}\cdot\mathbf{B})(\mathbf{B}\cdot\mathbf{r})}{4\pi}\right)\mathrm{d}s. \tag{21}$$

Another interesting virial-type relation (analogous to the one derived by Green (1973) for homogeneous hyperelastic bodies) can be established by starting from Eq. (12), multiplying it scalarly by $\mathbf{r}_0\cdot\mathbf{F}-\bar{\mathbf{r}}$, and integrating the result with respect to $\mathbf{r}_0$ over D. By using in particular the relation between $\mathbf{T}_0$ and ω_0 quoted above, one gets after some algebra

$$3W = \int_{\partial D} \hat{\mathbf{n}}_0\cdot[\mathbf{T}_0\cdot(\bar{\mathbf{r}} - \mathbf{r}_0\cdot\mathbf{F}) + \omega_0\mathbf{r}_0]\,\mathrm{d}s_0. \tag{22}$$

4.4. A new energy formula

Using the decomposition $\mathbf{B} = \mathbf{B}_0 + \mathbf{b}$ and Eq. (20) in Eq. (21) and combining the resulting expression with Eqs. (17), (18), and (15), one gets after some algebra the new energy formula containing information of topological nature

$$\begin{aligned} 3w + w_\perp - w_z &= -\int_S \frac{b^2}{8\pi}\mathbf{R}\cdot\hat{\mathbf{n}}\,\mathrm{d}s + \frac{B_0}{4\pi}\int_{\partial S_h} \mathbf{R}\cdot\hat{\mathbf{n}}\left(\int_{\mathcal{C}^h(\mathbf{R})} b_q'\,\mathrm{d}q'\right)\mathrm{d}q \\ &+ \int_{S_0} (\mathbf{R}^h - \mathbf{R}\cdot\nabla_\perp\mathbf{R}^h)\cdot\frac{\mathbf{b}_\perp^h B_z}{4\pi}\,\mathrm{d}s + \int_{S_0}(2\chi + \mathbf{R}\cdot\nabla_\perp\chi)\frac{b_z}{4\pi}\,\mathrm{d}s. \end{aligned} \tag{23}$$

It is also possible to arrive at this relation, a special case of which was previously derived in Aly (2005), by starting from Eq. (22).

5. Lower bounds on the energy of the fields in $\mathcal{B}$

5.1. A nontopological lower bound

As it is well known, the energy W of an arbitrary field $\mathbf{B}\in\mathcal{B}$ is always larger than the energy W_π of the potential field $\mathbf{B}_\pi$ introduced in Sect. 3.2, whence $W_\pi - W_0 = w_\pi \le w$. Of course, one has $\mathbf{B}_\pi = \mathbf{B}_0$ and $w_\pi = 0$ when the flux distribution is uniform on S_0 and S_h, i.e., $B_z\,|_{S_0,S_h}\equiv B_0$. When $B_z\,|_{S_0,S_h}\not\equiv B_0$, the lower bounds given below are relevant only when they exceed w_π.

5.2. A first topological lower bound

The magnetic energy w admits the lower bound

$$\frac{w_1}{W_0} = \left[\int_{S_0}\left(1 + \left|\frac{\xi}{h}\right|^2\right)^{1/2}\frac{\mathrm{d}s_0}{|S_0|}\right]^2 - 1 \le \frac{w}{W_0}, \tag{24}$$

where $\boldsymbol{\xi}(\mathbf{R}_0) = \overline{\mathbf{R}}(\mathbf{R}_0, h) - \overline{\mathbf{R}}(\mathbf{R}_0, 0)$ represents the horizontal separation between the two footpoints of the line of $\mathbf{B}$ emerging at $\mathbf{R}_0$ in the initial uniform configuration. Eq. (24) is obtained by writing $\boldsymbol{\xi}(\mathbf{R}_0) + h\hat{\mathbf{z}}$ as the integral of $\partial\overline{\mathbf{R}}(\mathbf{R}_0, z_0)/\partial z_0$ with respect to z_0, applying Schwartz inequality to that integral, injecting the result into the last expression of W in Eq. (3), applying once more Schwartz inequality, and using the relation $\int_D J\, dv_0 = |D|$.

This lower bound is interesting when the deformation of the field is sufficiently small. But it certainly strongly underestimates w when $\mathbf{B}$ is much deformed. For instance, $\underline{w}_1$ keeps the same value if we add to the field an overall twist of $2\pi n$, with $|n| \in \mathbb{N}$ arbitrarily large.

5.3. A second topological lower bound

Our second lower bound does not suffer the previous limitation but it is derived under the particular assumption that $B_z = B_0 + b_z \geq B_1 > 0$ in D. This bound is a generalization of the one derived by Berger (1993) under the much more severe restriction $B_z \equiv B_0$ in D. Under our assumption the line $\mathcal{L}(\mathbf{R})$ can be represented by a function $\mathbf{R}^z(\mathbf{R})$, while the relative twist of the two lines $\mathcal{L}(\mathbf{R})$ and $\mathcal{L}(\mathbf{R}')$ (see Sect. 3.1) can be expressed as

$$\phi(\mathbf{R}, \mathbf{R}') = \int_0^h \frac{\partial \varphi^z}{\partial z}(\mathbf{R}, \mathbf{R}')\, dz. \tag{25}$$

$\varphi^z(\mathbf{R}, \mathbf{R}')$ is the angle of $\mathbf{X}^z(\mathbf{R}, \mathbf{R}') = \mathbf{R}^z(\mathbf{R}') - \mathbf{R}^z(\mathbf{R})$ and $\hat{\mathbf{x}}$, and one has with $\hat{\mathbf{u}} = \hat{\mathbf{z}} \times \mathbf{X}^z/X^z$

$$\frac{\partial \varphi^z}{\partial z}(\mathbf{R}, \mathbf{R}') = \left[\left(\frac{\mathbf{b}_\perp}{B_z}\right)(\mathbf{R}^z(\mathbf{R}'), z) - \left(\frac{\mathbf{b}_\perp}{B_z}\right)(\mathbf{R}^z(\mathbf{R}), z)\right] \cdot \left(\frac{\hat{\mathbf{u}}}{X^z}\right)(\mathbf{R}, \mathbf{R}'). \tag{26}$$

To derive a lower bound on w, we start from the quantity

$$w^* = \frac{1}{16\pi R_0} \int_{S_0 \times S_0} |\phi(\mathbf{R}, \mathbf{R}')| B_z(\mathbf{R}, 0) B_z(\mathbf{R}', 0)\, ds\, ds' \tag{27}$$

which takes the same value for all the $\mathbf{B} \in \mathcal{B}$ ($2R_0$ is the diameter of S_0). Using Eqs. (25)-(26) in Eq. (27) and applying the triangle inequality leads after some simple algebra to

$$8\pi R_0 w^* \leq \int_D B_0 b_\perp \left(\int_{S_z} \frac{|\hat{\mathbf{u}} \cdot \hat{\mathbf{b}}_\perp|}{X}\, ds'\right) dv + \int_D b_\perp \left(\int_{S_z} \frac{b'_z}{X}\, ds'\right) dv, \tag{28}$$

where $X = |\mathbf{R}' - \mathbf{R}|$ and $b'_z = b_z(\mathbf{R}', z)$. The first integral in the right-hand side is similar to the one appearing in Berger (1993) and we proceed essentially as therein to bound it from above. To bound the second integral, we use Schwartz inequality and the easily obtained inequality $\int_{S_0}[\int_{S_0}(f(\mathbf{R}')/|\mathbf{R}' - \mathbf{R}|)\, ds']^2\, ds \leq (\nu R_0)^2 \int_{S_0} |f(\mathbf{R}')|^2\, ds'$ valid for any function f ($\nu = \nu(S_0) \leq 2\pi$ is a geometric constant). We thus get eventually

$$\frac{\underline{w}_2}{W_0} = \min_{s \geq 0}\left[s + \frac{(w^*/W_0)^2}{(\mu + \nu\sqrt{s})^2}\right] \leq \frac{w}{W_0}, \tag{29}$$

where $\mu = \mu(S_0)$ is another geometric constant. When S_0 is a disk, $\nu = 2\pi$ and $\pi(\mu R_0)^2 = \int_{S_0} m^2\, ds$, with $m = m(\mathbf{R}) = 2[\arcsin t/t + (1 - t^2)^{1/2}]$, $t = R/R_0$.

$\underline{w}_2/B_0$ can be replaced by the slightly less accurate but often more convenient lower bound

$$\frac{\underline{w}_3}{W_0} = \max\left\{\left(\frac{w^*}{\mu W_0}\right)^2 \left[\left(1 + \left(\frac{2\nu w^*}{\mu^2 W_0}\right)^2\right)^{1/2} + 1\right]^{-1}, \frac{\sqrt{2}w^*}{\nu W_0} - \frac{\mu^2}{\nu^2}\right\} \leq \frac{w}{W_0}. \tag{30}$$

These bounds are still valid if we substitute the crossing number $C(\mathbf{R}, \mathbf{R}') = \int_0^h |\partial_z \varphi^z|\, dz$ of a pair of lines for $|\phi(\mathbf{R}, \mathbf{R}')|$ in the definition of w^*. The formulae obtained that way naturally reduce to Berger's when $B_z \equiv B_0$ on S_0 and S_h, and w^* is small.

Journal of Physics: Conference Series **544** (2014) 012003 doi:10.1088/1742-6596/544/1/012003

6. Upper bound on the energy of the fields in $\mathcal{B}_{ff}$

We next present an upper bound on the energy w of any force-free field $\mathbf{B} \in \mathcal{B}_{ff}$. We make here the following additional assumptions on $\mathcal{B}_{ff}$ and S_0: (i) $B_z \equiv B_0$ and then $b_z \equiv 0$ on S_0 and S_h; (ii) all the lines drawn on S have the same winding number $n(\mathbf{R}) = n$, with $n \in \mathbb{Z}$ but possibly in the case where S_0 is a disk; and (iii) the domain S_0 is strictly star-shaped, i.e., $\hat{\mathbf{n}} \cdot \mathbf{R} \geq R_1 > 0$ on ∂S_0 for an adequately chosen origin of the frame. Eq. (23) thus reduces to

$$\begin{aligned} 4w - 2w_z &= \left[\frac{B_0}{4\pi} \int_{\partial S_h} \mathbf{R} \cdot \hat{\mathbf{n}} \left(\int_{C^h(\mathbf{R})} b'_q \, \mathrm{d}q' \right) \mathrm{d}q - \int_S \frac{b_q^2}{8\pi} \mathbf{R} \cdot \hat{\mathbf{n}} \, \mathrm{d}s \right] \\ &+ \frac{B_0}{4\pi} \int_{S_0} (\mathbf{R}^h - \mathbf{R} \cdot \boldsymbol{\nabla}_\perp \mathbf{R}^h) \cdot \mathbf{b}_\perp^h \, \mathrm{d}s - \int_S \frac{b_z^2}{8\pi} \mathbf{R} \cdot \hat{\mathbf{n}} \, \mathrm{d}s. \end{aligned} \tag{31}$$

Taking into account the nonpositivity of the last term in the right-hand side of Eq. (31) and bounding the other terms with help from Eqs. (14) and (20), Schwartz inequality, and the elementary inequality $2xy \leq x^2 + y^2$, we obtain eventually

$$\frac{4w - 2w_z}{W_0} \leq 2N + 2L \sqrt{\frac{w - 2w_z}{W_0}}, \tag{32}$$

where $w - 2w_z \geq 0$ by Eq. (20) and

$$L^2 = \frac{1}{|S_0| h^2} \int_{S_0} |\mathbf{R} \cdot \boldsymbol{\nabla}_\perp \mathbf{R}^h - \mathbf{R}^h|^2 \, \mathrm{d}s \quad \text{and} \quad N = \frac{4\pi\delta n^2 |S_0|}{h^2}. \tag{33}$$

The factor δ is purely geometric ($\delta = \delta(S_0)$). When S_0 is a disk, one has $\delta = 1$ and L vanishes if the boundary motions are pure rigid rotations.

A simple discussion of inequality (32) thus allows to conclude that

$$\frac{w}{W_0} \leq \frac{\overline{w}}{W_0} = \begin{cases} \left(L^2 + 4N + \sqrt{L^2(L^2 + 8N)} \right) / 8 \leq (L^2 + 2N)/3 & \text{when } N \leq L^2, \\ (L^2 + 2N)/3 & \text{when } L^2 < N, \end{cases} \tag{34}$$

$$\frac{w_z}{W_0} \leq \frac{1}{24}(L^2 + 8N). \tag{35}$$

A similar upper bound on w may be established in the framework of reduced MHD (with then $w_z = 0$), and for arbitrary flux distributions on the lower and upper boundaries, in which case $\overline{w}$ is given by an expression much more complex than the one above. We also mention that an alternative energy bound can be derived in a simple way when $\mathbf{B}$ is the final state of a quasi-static evolution during which the field passes through a sequence of regular force-free configurations $\mathbf{B}(t)$, $0 \leq t \leq T$, with $\mathbf{B}(0) = \mathbf{B}_0$. An equation of evolution can be obtained for the energy $w(t)$ by starting from the ideal induction equation $\partial_t \mathbf{B}(t) = \boldsymbol{\nabla} \times (\mathbf{v} \times \mathbf{B}(t))$ and the equilibrium equation (2) satisfied by $\mathbf{B}(t)$, and one can bound its right-hand side by applying Schwartz inequality and Eq. (20). One gets

$$4\pi \frac{\mathrm{d}w(t)}{\mathrm{d}t} = \left(\int_{S_h} - \int_{S_0} \right) (\mathbf{v} \cdot \mathbf{b}_\perp B_z)(t) \, \mathrm{d}s \leq G_1(t) \sqrt{8\pi w(t) + G_2(t)}, \tag{36}$$

where the functions G_k can be computed from $\mathbf{v}$, and an upper bound follows by time integration from 0 to T. This bound does not suppose the assumptions (i)-(iii) to be verified, but it has the defect of depending on the specific velocity field that is applied on the boundary to produce $\mathbf{B}$ (of course, there are an infinity of such fields and it seems difficult to determine an optimal one among them).

7. A stability criterion for the fields in $\mathcal{B}_{ff}$

We consider finally the important problem of the ideal stability properties of a force-free magnetic field $\mathbf{B} \in \mathcal{B}_{ff}$ (we recall that $\alpha = \alpha(\mathcal{L}) = \mathbf{B} \cdot \nabla \times \mathbf{B}/B^2$). At the linear level, stability can be determined by using the energy principle (Bernstein *et al.*, 1958): an arbitrary displacement $\boldsymbol{\xi}(\mathbf{r})$, satisfying $\boldsymbol{\xi} = 0$ on $S_0 \cup S_h$ (line-tying condition) and $\hat{\mathbf{n}} \cdot \boldsymbol{\xi} = 0$ on S, is applied to the plasma elements in D and one looks at the sign of the resulting second variation of the energy

$$\delta^2 W[\boldsymbol{\xi}] = \frac{1}{8\pi} \int_D [|\nabla \times \mathbf{A}|^2 - \alpha \mathbf{A} \cdot \nabla \times \mathbf{A}] \, \mathrm{d}v, \quad \text{where } \mathbf{A} = \boldsymbol{\xi} \times \mathbf{B}. \tag{37}$$

$\mathbf{B}$ is linearly stable if $\delta^2 W[\boldsymbol{\xi}] \geq 0$ for every $\boldsymbol{\xi}$, while it is unstable if there exists a $\boldsymbol{\xi}$ such that $\delta^2 W[\boldsymbol{\xi}] < 0$. Following Aly (1990), one can reexpress $\delta^2 W$ as a functional of the field perturbation, $\delta\mathbf{b} = \nabla \times \mathbf{A}$, with the admissible $\delta\mathbf{b}$ being constrained to satisfy $\nabla \cdot \delta\mathbf{b} = 0$ in D, $\hat{\mathbf{n}} \cdot \delta\mathbf{b} = 0$ on ∂D, and $\int_{\mathcal{S}} \hat{\mathbf{n}} \cdot \delta\mathbf{b} \, \mathrm{d}s = 0$ for any surface $\mathcal{S} \subset D$ bounded by a magnetic line $\mathcal{L}$ and a curve on ∂D connecting the footpoints of $\mathcal{L}$. The method is as follows. One introduces the Euler representation $\mathbf{B} = \nabla U \times \nabla V$ of the equilibrium field, writes $\mathbf{A} = \lambda \nabla U + \mu \nabla V$, and expresses the coefficients λ and μ as integrals of $\delta\mathbf{b} \cdot \nabla V/B$ and $\delta\mathbf{b} \cdot \nabla U/B$, respectively, along a magnetic line. Whence $\mathbf{A} = \mathbf{A}[\delta\mathbf{b}]$, and $\delta^2 W = \delta^2 W[\delta\mathbf{b}]$ indeed.

It can be shown that $\delta^2 W[\delta\mathbf{b}] \geq 0$ for any $\delta\mathbf{b} \not\equiv 0$ if one has on every line $\mathcal{L}$

$$|\alpha(\mathcal{L})| \leq \left(\int_{\mathcal{L}} \frac{|\nabla U|^2}{B} \, \mathrm{d}l \int_{\mathcal{L}} \frac{|\nabla V|^2}{B} \, \mathrm{d}l \right)^{-1/2}. \tag{38}$$

$\mathbf{B} \in \mathcal{B}_{ff}$ is then linearly stable if $|\alpha(\mathcal{L})|$ is not larger than some fraction of $|\mathcal{L}|^{-1}$. This result is quite general, but it is not fully satisfying as the stability criterion is expressed in terms of α rather than in terms of topological quantities characterizing $\mathcal{B}$. Another sufficient condition of stability, also expressed in terms of α, can be derived by the method devised in Aly (1990a).

8. Conclusion

In this Communication, we have reported some new general results on the model introduced long ago by Parker (1994) for studying a basic mechanism of plasma heating in confined magnetic structures. On the one hand, we have derived new formulae for the field line helicity and the relative helicity of the fields belonging to a class $\mathcal{B}$ of topologically equivalent regular fields, and established lower bounds on their energy. In particular, we have generalized Berger's bound (Berger, 1993) to the case of a field with a nonuniform z-component. On the other hand, we have considered the force-free fields contained in $\mathcal{B}$, assuming the existence of at least one. We have made some new points about the analogies existing between the force-free equations and those of geometrical optics, elastostatics, and 2D time-dependent hydrodynamics, established upper bounds on their energy, and proposed a sufficient criterion of ideal MHD linear stability.

We note that studying the relations between the energy of a magnetic field and its topology is also quite important for constructing models of solar large scale eruptive processes. These phenomena exhibit huge expansions of the magnetic structures and one needs for studying them to introduce models in which the force-free coronal magnetic field is not confined as in Parker's, but rather occupies an unbounded domain (e.g., the exterior of a sphere, or a half-space). By developing considerations similar to those in this paper, it turns out to be possible to establish basic inequalities which imply in particular that the radial scale of the field eventually increases very fast (at least as $\exp{(t/T)^2}$) when energy is continuously transferred from the subphotospheric layers into the corona (Aly & Amari (2007), and references therein).

Finally we recall that the problem of the constraints imposed by topology on magnetic energy has also been addressed in the framework of knot theory. In that case, the fields one considers are not tied to a boundary as in Parker's model, but are confined within a tube constructed about a closed curve of a given topological type (or a set of tubes associated to a link). One of the basic issues is then to find the (non force-free!) configurations accessible by ideal MHD incompressible motions which minimize energy, and much efforts have been done to establish topological energy lower bounds on the energy minimum (Arnold & Khesin (1998), and references therein).

References

Aly, J.-J. 1990 Quasi-static evolution of a force-free magnetic field. *Comp. Phys. Comm.* **59**, 13–20.

Aly, J.-J. 1990a The stability of a line-tied force-free magnetic field in an unbounded region of space. *Phys. Fluids B* **2**, 1928–1935.

Aly, J.-J. 2005 A uniqueness result for a simple force-free equilibrium submitted to a topological constraint. *Astron. Astrophys.* **429**, 15–18.

Aly, J.-J. & Amari, T. 2007 Structure and evolution of the solar coronal magnetic field. *Geophys. Astrophys. Fluid Dyn.* **101**, 249–287.

Aly, J.-J. & Amari, T. 2010 Does the compression or the expansion of a simple topology potential magnetic field lead to the development of current sheets? *Astrophys. J.* **709**, L99–L102.

Arnold, V. & Khesin, B. 1998 **Topological Methods in Hydrodynamics.** New York, NY: Springer.

Berger, M. A. 1988 An energy formula for nonlinear force-free magnetic fields. *Astron. Astrophys* **201**, 355–361.

Berger, M. A. 1993 Energy-crossing number relations for braided magnetic fields. *Phys. Rev. Let.* **70**, 705–708.

Berger, M. A. & Field, G. B. 1984 *J. Fluid Mech.* **35**, 147.

Berger, M. A. & Prior, C. 2006 The writhe of open and closed curves. *J. Phys. A: Math. Gen.* **39**, 8321–8348.

Bernstein, I., Frieman, E., Kruskal, M. & Kulsrud, R. 1958 An energy principle for hydromagnetic stability problems. *Proc. R. Soc. Lond., Ser. A* **244**, 17–40.

Ciarlet, P. G. 1988 **Mathematical Elasticity. Volume I: Three-dimensional Elasticity.** Amsterdam: North- Holland.

Green, A. E. 1973 On some general formulae in finite elastostatics. *Arch. Ration. Mech. Anal.* **50**, 73–80.

Huang, Y.-M., Bhattacharjee, A. & Zweibel, E. G. 2010 Effects of line-tying on magnetohydrodynamic instabilities and current sheet formation. *Phys. Plasmas* **17** (5), 055707.

Janse, A. M., Low, B. C. & Parker, E. N. 2010 Topological complexity and tangential discontinuity in magnetic fields. *Phys. Plasmas* **17** (9), 092901.

Moffatt, H. K. 1985 *J. Fluid Mech.* **159**, 359-378.

Ng, C. S. & Bhattacharjee, A. 1998 Nonequilibrium and current sheet formation in line-tied magnetic fields. *Phys. Plasmas* **5** (11), 4028–4040.

Parker, E. N. 1994 **Spontaneous Current Sheets in Magnetic Fields, with Applications to Stellar X-rays**. Oxford: Oxford University Press.

van Ballegooijen, A. A. 1985 Electric currents in the solar corona and the existence of magnetostatic equilibrium. *Astrophys. J.* **298**, 421–430.

Yeates, A. R. & Hornig, G. 2013 Unique topological characterization of braided magnetic fields. *Phys. Plasmas* **20**, 012102.

Journal of Physics: Conference Series **544** (2014) 012004 doi:10.1088/1742-6596/544/1/012004

Magnetic field generated by current filaments

Y. Kimura

Graduate School of Mathematics, Nagoya University
Furo-cho, Chikusa-ku, Nagoya 464-8602, Japan

E-mail: kimura@math.nagoya-u.ac.jp

Abstract. We investigate the magnetic field generated by two straight current filaments using the analogy between steady MHD and Euler flows. Using the Biot-Savart law, we present a dynamical system describing the extension of magnetic lines around the current filaments. It is demonstrated that, if two current filaments are non-parallel, a magnetic line starting near one current goes to infinity by the drifting effect of the other.

1. Introduction

It is well-known that a practically important analogy resides between steady Euler and MHD flows. If we denote incompressible velocity and vorticity as $\mathbf{u}$ and $\boldsymbol{\omega}$ for steady Euler flows, and a magnetic field (or the magnetic flux density) and an electric current distribution as $\mathbf{B}$ and $\mathbf{j}$ for steady MHD flows, the following analogy for the two sets holds exactly :[1]

$$\nabla \times \mathbf{u} = \boldsymbol{\omega} \qquad \nabla \cdot \mathbf{u} = 0 \tag{1}$$

$$\nabla \times \mathbf{B} = \mathbf{j} \qquad \nabla \cdot \mathbf{B} = 0 \tag{2}$$

From the Biot-Savart law, which is equivalent to take the inverse of rotation, we have

$$\mathbf{u}(\mathbf{x}) = \frac{1}{4\pi} \int \frac{\boldsymbol{\omega}(\mathbf{x}') \times (\mathbf{x} - \mathbf{x}')}{|\mathbf{x} - \mathbf{x}'|^3} \mathrm{d}V', \tag{3}$$

for $\mathbf{u}(\mathbf{x})$, and

$$\mathbf{B}(\mathbf{x}) = \frac{1}{4\pi} \int \frac{\mathbf{j}(\mathbf{x}') \times (\mathbf{x} - \mathbf{x}')}{|\mathbf{x} - \mathbf{x}'|^3} \mathrm{d}V', \tag{4}$$

for $\mathbf{B}(\mathbf{x})$.

We should note that this analogy breaks if we deal with a time-dependent problem, because corresponding dynamics, the Euler equation for $\mathbf{u}$ and the magnetic induction equations for $\mathbf{B}$, involve $\mathbf{u}$ and $\mathbf{B}$ in a different manner. As long as we stay in steady problems, however, we can extract useful information from one side and apply to the other. The content of this article make full use of this analogy.

Using the Biot-Savart law (3), we have investigated the particle motion around a vortex soliton (Kimura & Koikari (2004)). It is found that particles near the loop part of a vortex soliton are trapped and move inside a knotted torus around the loop of the vortex soliton.

[1] We have omitted the permeability index μ_0 which would appear when S. I. units are used.

To understand this torus formation, a simple toy model, which replaced a vortex soliton with a couple of vortex filaments tangent to the soliton filament was introduced and named the chopsticks model. A variety of particle motions including chaotic ones are found inside the torus, and the discovery of such convoluted particle motions even for a simple setting of vortex filaments is the direct motivation to study possible complication of the magnetic field structure around a simple configuration of electric current filaments. For similar motivations, reader is referred to, for example, Gascon & Peralta-Salas (2005) and Aguirre *et al.* (2008) and the references therein.

2. Formulation

Let us first calculate the magnetic field generated by a single electric current filament of a (steady) current J which passes $\mathbf{x}_0$ with the unit tangent vector $\mathbf{t}$ (Figure 1). A direct integration of (4) results

$$\mathbf{B}(\mathbf{x}) = \frac{J}{2\pi}\frac{\mathbf{t}\times(\mathbf{x}-\mathbf{x}_0)}{|\mathbf{t}\times(\mathbf{x}-\mathbf{x}_0)|^2}\,. \tag{5}$$

Now a couple of current filaments, J_1 and J_2, are placed as in Figure 2. We may assume that J_1 is on the z-axis, and J_2 is on a plane parallel to the yz-plane with a distance of d, and is tilted by the angle θ from the z direction. For the formula (5), we have $\mathbf{t} = (0, 0, 1)$, $\mathbf{x}_0 = (0, 0, 0)$ for J_1 and $\mathbf{t} = (0, \sin\theta, \cos\theta)$, $\mathbf{x}_0 = (d, 0, 0)$ for J_2. The total magnetic field is a superposition of the contributions from each current filament. To see how a given magnetic line extends in the 3D space, we start at the initial point and connect segments of the magnetic vector field "nicely". For the present case, it is equivalent to integrate the following dynamical system

$$\frac{\mathrm{d}}{\mathrm{dt}}\begin{pmatrix} x \\ y \\ z \end{pmatrix} = \frac{j_1}{x^2+y^2}\begin{pmatrix} -y \\ x \\ 0 \end{pmatrix} + \frac{j_2}{(z\sin\theta - y\cos\theta)^2 + (x-d)^2}\begin{pmatrix} z\sin\theta - y\cos\theta \\ (x-d)\cos\theta \\ -(x-d)\sin\theta \end{pmatrix}, \tag{6}$$

where we have replaced $J_1/2\pi \to j_1$ and $J_2/2\pi \to j_2$.

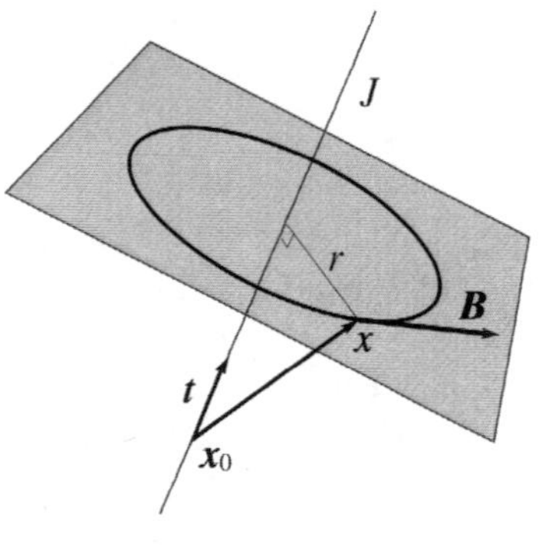

Figure 1. A Single electric current filament J and the magnetic field B generated by J.

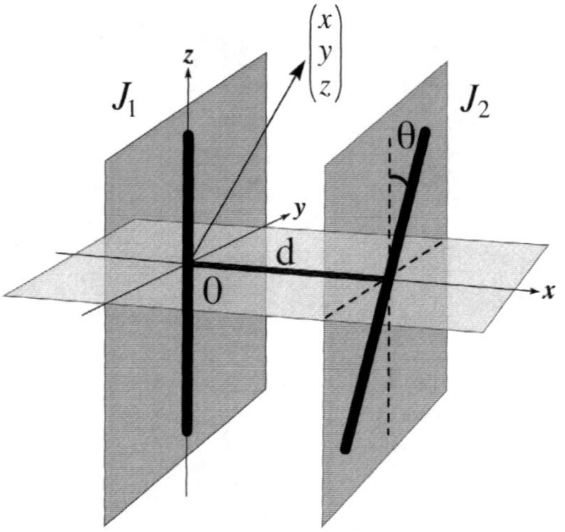

Figure 2. The configuration of a couple of electric current filaments J_1 and J_2.

We should note that the right hand side of (6) can be written with a vector potential as follows,

$$\frac{d}{dt}\begin{pmatrix} x \\ y \\ z \end{pmatrix} = \nabla\times\left[\frac{-j_1}{2}\log(x^2+y^2)\begin{pmatrix} 0 \\ 0 \\ 1 \end{pmatrix} + \frac{-j_2}{2}\log\left\{(z\sin\theta - y\cos\theta)^2 + (x-d)^2\right\}\begin{pmatrix} 0 \\ \sin\theta \\ \cos\theta \end{pmatrix}\right]. \tag{7}$$

3. Comparison with the vortex chopsticks model

With the vortex chopsticks model, we obtained the following expression corresponding to (7),

$$\begin{aligned} \frac{d}{dt}\begin{pmatrix} x \\ y \\ z \end{pmatrix} &= \nabla\times\left[-\frac{\Gamma_1}{2\pi}\log|\mathbf{t}_1\times(\mathbf{x}-\mathbf{x}_1)|\ \mathbf{t}_1 - \frac{\Gamma_2}{2\pi}\log|\mathbf{t}_2\times(\mathbf{x}-\mathbf{x}_2)|\ \mathbf{t}_2\right] \\ &\quad +\nabla\times\left[\frac{1}{2}(\nu^2+\tau^2)G(x^2+y^2)\hat{z}\right] + \nabla\times\nabla\times\left[\tau G(x^2+y^2)\hat{z}\right], \end{aligned} \tag{8}$$

where Γ_1, Γ_2 are the circulation of vortex filaments. In the above equation, the first two terms on the right hand side correspond directly to the right hand side of (7), but there are additional terms. These terms attribute to the steady rotational and translational motion of a vortex soliton, which can be parametrized by a half of the maximum curvature ν, a constant torsion τ and the local induction constant G of a vortex soliton(Hasimoto (1972), Kida (1981)). In Kimura & Koikari (2004), a wide variety of combinations of these parameters are examined in integrating (8) to calculate the volume of the torus as a fluid mass transported by the vortex chopsticks. While examining the parameter dependence on the volume of the torus, we noticed a tendency that the volume decrease drastically with decreasing value of the torsion parameter τ, and in the limit when $\tau \to 0$, we predict that the volume of torus also goes to 0. On the other hand, as a finite volume remains even for zero ν (with nonzero τ), we may assume that the last term in (8), which corresponds to a uniform translation of the system, is essential for a finite volume torus to exist. So if we use the analogy between a vortex filament and a current filament, we may expect that the translation of the system, which is understood as an anti-drifting motion, is also essential for the existence of closed magnetic lines or their stability, if any. To corroborate this, we will investigate (6) for some special cases.

For simplicity, if we assume that the initial point (x, y, z) is much closer to J_1 than J_2 in Figure 2, *i.e.* $\sqrt{x^2+y^2}/d \ll 1$, then (6) is approximated as

$$\frac{\mathrm{d}}{\mathrm{dt}}\begin{pmatrix} x \\ y \\ z \end{pmatrix} = \frac{j_1}{x^2+y^2}\begin{pmatrix} -y \\ x \\ 0 \end{pmatrix} + \frac{j_2/d}{((z/d)\sin\theta)^2+1}\begin{pmatrix} (z/d)\sin\theta \\ -\cos\theta \\ \sin\theta \end{pmatrix}. \tag{9}$$

From the above equation, we see that a trajectory is near a circle around J_1 which is perturbed by the effect of J_2. We are concerned with how the effect of perturbation remains if d becomes larger. To see this we check the development of the z-component of the above equation. The z-component is

$$\frac{\mathrm{dz}}{\mathrm{dt}} = \frac{(j_2/d)\sin\theta}{(z^2/d^2)\sin^2\theta+1},$$

and if we denote $a = \sin^2\theta/d^2$ and $b = (j_2/d)\sin\theta$, and assume the initial condition $z(0) = 0$, we have an initial value problem of

$$dz/dt = b/(az^2+1), \quad z(0) = 0, \tag{10}$$

Journal of Physics: Conference Series **544** (2014) 012004 doi:10.1088/1742-6596/544/1/012004

which can be integrated easily to give

$$\frac{a}{3}z^3 + z = bt \qquad \rightarrow \qquad z \sim \left(\frac{3b}{a}t\right)^{1/3} = \left(\frac{3\,d\,j_2}{\sin\theta}\,t\right)^{1/3}. \tag{11}$$

The speed depends on the parameters, but $z(t)$ goes to plus or minus infinity as $t \to \infty$ for any values of θ except 0 and π. From (9) the equation for the square distance of trajectory from J_1, $r^2(t) = x(t)^2 + y(t)^2$, is given by

$$\frac{dr^2}{dt} = \frac{2j_2\,d}{(z^2/d^2)\sin^2\theta + 1}\left((xz/d)\sin\theta - y\cos\theta\right). \tag{12}$$

Using (11), the right hand side of this equation is evaluated as $\sim t^{-1/3}$ for $t \to \infty$ which provides the asymptotic behavior $r^2 \sim t^{2/3}$.

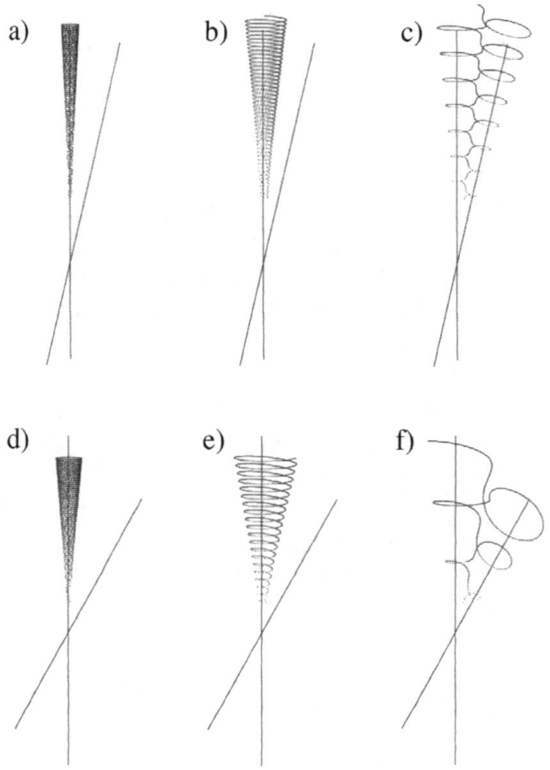

Figure 3. Simulation results of (6). Parameters are set $j_1 = j_2 = 1$ and $d = 1$, and $\theta = \pi/6$ for $a), b), c)$ and $\theta = \pi/3$ for $d), e), f)$. The initial conditions are $(0.1, 0.333, 0)$ for $a), d)$, $(0.3, 0.333, 0)$ for $b), e)$, $(0.5, 0.333, 0)$ for $c), f)$.

Figure 3 shows some plots of the trajectories (magnetic lines) obtained by integrating (6). The parameters are set $j_1 = j_2 = 1$ and $d = 1$, and $\theta = \pi/6$ for $a), b), c)$ and $\theta = \pi/3$ for $d), e), f)$. All the initial conditions are set on the $z = 0$ plane with $y = 0.0333$, and the x component is 0.1 for $a), d)$, 0.3 for $b), e)$ and 0.5 for $c), f)$. If the initial positions are close to J_1, the trajectories spiral out along J_1 ($a), b), d), e)$). On the other hand, if they are close to

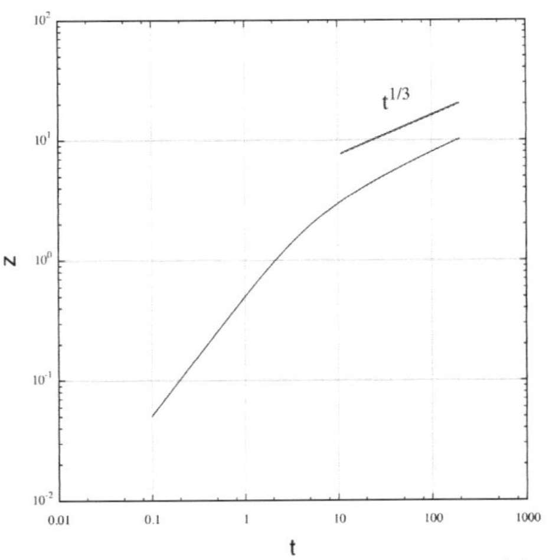

Figure 4. The time development of $z(t)$ for Figure 3a). Asymptotically, it approach the line of $t^{1/3}$.

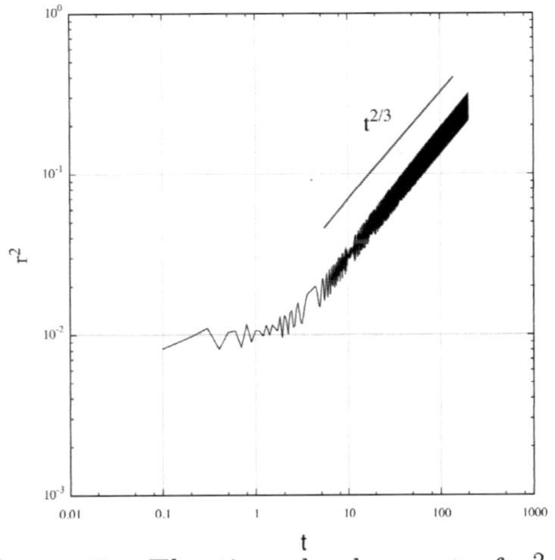

Figure 5. The time development of $r^2 = x(t)^2 + y(t)^2$ for Figure 3a). With oscillations, it approach the line of $t^{2/3}$ asymptotically.

the (approximated) hyperbolic point $(0.5, 0, 0)$, the trajectories meander between J_1 and J_2, but still the z component tends to go to infinity.

In Figure 4 and Figure 5, the time development of $z(t)$ and of $r^2(t)$ are plotted respectively for the case of Figure 3a). As estimated, they show the asymptotic dependence of $t^{1/3}$ and $t^{2/3}$, respectively.

We have demonstrated that a magnetic line which passes a point close to one of the two straight current filaments in a skewed configuration goes to infinity by the drifting effect of the other. Although this observation implies that such a magnetic line can not be closed, it still remains an open problem whether any magnetic lines not close to either current filament could make a closed loop or not. To answer to this question, a thorough investigation of (6) is necessary.

Acknowledgments
Comments and references from Professor Peralta-Salas(ICMAT) are gratefully acknowledged.

References

Aguirre, J., Giné, J. & Peralta-Salas, D. 2008 Integrability of magnetic field created by current distributions. *Nonlinearity* **21**, 51–69.

Gascon, F. & Peralta-Salas, D. 2005 Some properties of magnetic fields generated by symmetric configurations of wires. *Physica D* **206**, 109–120.

Hasimoto, H. 1972 A soliton on a vortex filament. *Journal of Fluid Mechanics* pp. 477–485.

Kida, K. 1981 A vortex filament moving without change of form. *Journal of Fluid Mechanics* pp. 397–409.

Kimura, Y. & Koikari, S. 2004 Particle transport by a vortex soliton. *Journal of Fluid Mechanics* **510**, 201–218.

Flux in QCS IOP Publishing
Journal of Physics: Conference Series **544** (2014) 012005 doi:10.1088/1742-6596/544/1/012005

Multiple breathers on a vortex filament

H. Salman[1]

[1] School of Mathematics, University of East Anglia, Norwich Research Park, Norwich, NR4 7TJ, UK

E-mail: H.Salman@uea.ac.uk

Abstract. In this paper we investigate the correspondence between the Da Rios-Betchov equation, which appears in the three-dimensional motion of a vortex filament, and the nonlinear Schrödinger equation. Using this correspondence we map a set of solutions corresponding to breathers in the nonlinear Schrödinger equation to waves propagating along a vortex filament. The work presented generalizes the recently derived family of vortex configurations associated with these breather solutions to a wider class of configurations that are associated with combination homoclinic/heteroclinic orbits of the 1D self-focussing nonlinear Schrödinger equation. We show that by considering these solutions of the governing nonlinear Schrödinger equation, highly nontrivial vortex filament configurations can be obtained that are associated with a pair of breather excitations. These configurations can lead to loop-like excitations emerging from an otherwise weakly perturbed helical vortex. The results presented further demonstrate the rich class of solutions that are supported by the Da Rios-Betchov equation that is recovered within the local induction approximation for the motion of a vortex filament.

1. Introduction

The vorticity representation of the equations of motion of a fluid is appealing in that it provides a direct understanding of the dynamics in terms of the rotational component of the velocity field associated with coherent vortex structures. In the limit of an ideal, inviscid fluid, fluid elements with non-zero vorticity evolve according to the Helmholtz laws of vortex motion which permits a geometric interpretation of the evolution of the vorticity field. This geometric interpretation is most succinctly illustrated in the case of a vortex filament which consists of vorticity of infinite strength concentrated along the length of the filament. In such a case, deformations or excitations created along the filament evolve in a non-trivial way due to the self-induced velocity produced by the instantaneous configuration of the filament. While the vortex filament concept is an attractive mathematical idealisation in the context of classical fluids, vortex filaments are physically real entities in superfluids. In this respect, understanding the fundamental type of excitations that can survive along these filaments becomes even more relevant as such excitations turn out to be the main degrees of freedom remaining in a superfluid in the ultra low temperature regime.

The study of excitations on a vortex dates back to the work of Kelvin (1880), who considered low amplitude perturbations of a straight vortex. These linear perturbations, now referred to as Kelvin waves, have been invoked by Vinen (2000); Kozik & Svistunov (2009); L'vov & Nazarenko (2010) to explain the dissipation mechanism in superfluids at extremely low temperatures. Kelvin waves can be recovered by truncating the full equations that govern the motion of a filament at leading order. In this approximation, the dynamics are governed by the local shape of the filament and is, therefore, referred to as the local induction approximation (LIA). This approximation removes the nonlocal contributions arising from the Biot-Savart law by expressing the velocity field induced in terms of the local curvature. The governing equation of motion of a vortex then reduces to the Da Rios-Betchov equation (Rios, 1906; Betchov, 1965; Ricca, 1996). While Kelvin waves are the main low amplitude excitations, it is also important to consider larger amplitude excitations propagating along the filament. Such excitations can still be studied within the framework of the Da Rios-Betchov equation. The first analytical study in this direction was carried out by Hasimoto (1972) who showed that the LIA can be mapped onto the nonlinear Schrödinger equation (NLS) of the self-focusing type. This connection was profound since the NLS equation is known to be integrable and admits soliton solutions. The correspondence between the two implied that these solitons represent nonlinear excitations and have a corresponding geometric interpretation in the context of a vortex filament. Hasimoto presented an explicit expression for these soliton excitations propagating along a vortex and classified them into three different categories, a hump soliton, a cusp soliton, and a loop soliton depending on the relative magnitude of the curvature and torsion of the filament. The discovery of Hasimoto's solitons lead to a wealth of results on solitons on vortices (Konno *et al.*, 1991; Konno & Ichikawa, 1992; Leibovich & Ma, 1983). In particluar, we refer to the work of Hopfinger & Browand (1982), who observed these structures experimentally. This further illustrated that while the LIA neglects the full dynamics containd in the Biot-Savart law, the approximation can, nevertheless, be used to infer excitations that arise on real vortices. Subsequent work has lead to a number of generalisations of Hasimoto's work. This included a formulation by Cieśliński (1992) for two soliton solutions propagating along the filament as well as N-soliton solutions on a vortex (Fukumoto & Miyazaki, 1986; Levi *et al.*, 1983; Maksimović *et al.*, 2003).

Despite the progress made, there is much less work focused on the integrability of the Da Rios-Betchov equation of a vortex filament in comparison to the NLS equation. This is particularly true of integrable solutions of the NLS equation known as breathers. A breather is a nonlinear wave in which energy concentrates in a localized and oscillatory manner. Unlike solitons, breathers appear unsteady in any frame of reference. Breather solutions of the NLS equation were first predicted by Kuznetsov (1977) and Ma (1979) as a temporally periodic modulation, and by Akhmediev *et al.* (1987) as a spatially periodic modulation of a plane wave solution. A limiting case of these two solutions arises in the form of the Peregrine soliton (Peregrine, 1983) which is recovered either when the spacial period of the modulation of the Akhmediev breather tends to infinity or when the temporal period of the Ma-Kuznetsov breather tends to infinity (Dysthe & Trulsen, 1999). In contrast to solitons, which can arise both for the self-focusing and self-defocusing forms of the NLS equation and correspond to bright and dark solitons, respectively, breathers only occur in the former case. We note that the difference between the two equations arises from the relative signs of the nonlinear and dispersion terms. Since the Hasimoto transformation leads to an NLS equation of the self-focusing type, we would anticipate breather like excitations to arise in the case of the Da Rios-Betchov equations. It is, therefore, surprising that essentially no work has been carried out to investigate breathers in the case of vortex motion within the LIA. An exception to this is the work of Umeki (2010) who derived a set of equations permitting exact and explicit solutions to be obtained for the motion of the tangent vector of the vortex filament within the LIA. Umeki's work was motivated by the observation that the breather solutions obtained for the NLS equation are commonly presented in a form in which they are homoclinic to a spatially uniform plane wave. In the context of a vortex filament, this would imply that the excitations arise on an n-ply wound circular vortex which is somewhat unphysical. Umeki, therefore, worked directly with the Da Rios-Betchov equation obtaining an explicit solution for the tangent vector of the vortex filament. The vortex could then be reconstructed by direct numerical integration from this solution of the tangent vector. Very recently, Salman (2013) generalised the breather solutions by establishing a connection between

Journal of Physics: Conference Series **544** (2014) 012005 doi:10.1088/1742-6596/544/1/012005

the breather solutions of the NLS equation and solutions of the Da Rios-Betchov equation. This approach has the advantage that it allows results already derived and established for the NLS equation to be mapped onto novel vortex configurations corresponding to these localised excitations. Using this mapping, Salman was able to illustrate how the Akhmediev breather solution maps onto loop like excitations of a periodic helical vortex. Moreover, he illustrated that these solutions persist even under the full dynamics of the Biot-Savart law where the loop-like excitations can ultimately induce a self-reconnection on the vortices. He also demonstrated these vortex excitations arise in a microscopic model of a superfluid provided by the mean field description of the Gross-Pitaevskii equation (Gross, 1961; Pitaevskii, 1961). The mechanism identified by Salman is distinct from other mechanisms previously discussed (e.g. Kursa *et al.* (2011)) for generating vortex loops. In particular, it is consistent with the aforementioned theories of the superfluid turbulence regimes identified.

Given the generality of the formulation presented in Salman (2013), we will extend the results to a more general class of breather solutions. In particular, we will consider how so-called combination homoclinic/heteroclinic solutions of the NLS, as derived by Ablowitz & Herbst (1990), map onto vortex configurations of the NLS equation. We point out that while Ablowitz & Herbst refer to these solutions as combination homoclinic orbits, Akhmediev & Korneev (1986) previously illustrated that the phase shifts that arise in such solutions are more appropriately described in terms of heteroclinic orbits. We will, therefore, refer to these solutions as homoclinic/heteroclinic orbits with the understanding that they are, in fact, associated with the same form of solution. Given that we have already established that in Salman (2013), the results of the LIA extend to more realistic models, we will focus exclusively on analysing our results for the Da Rios-Betchov equation.

2. The Da Rios-Betchov Equation and Hasimoto's Transformation

We begin by recalling that a vortex filament with circulation Γ induces a velocity according to the Biot-Savart law given by

$$\mathbf{v}(\mathbf{r}) = \frac{\Gamma}{4\pi}\int_{\mathcal{C}(t)} \frac{d\mathbf{s}\times(\mathbf{r}-\mathbf{s})}{|\mathbf{r}-\mathbf{s}|^3}, \tag{1}$$

where $\mathbf{r}$ is the position vector for any point in the fluid, and the vector $\mathbf{s}$ runs along the vortex filament parameterised by the initial arclength ξ. The integral is evaluated along the curve $\mathcal{C}(t)$ that varies with time. This equation shows that the velocity induced on a superfluid vortex depends on the instantaneous configuration of a vortex line where non-local contributions arise from the terms under the integral. To simplify the problem, it can be shown Schwarz (1988); Donnelly (1991) that, at leading order, the vortex velocity depends on the local radius of curvature R and the vortex core radius a_o with

$$\dot{\mathbf{s}} = (\Gamma/4\pi)\ln(R/a_o)\mathbf{s}'\times\mathbf{s}'' \equiv \beta(\mathbf{s}'\times\mathbf{s}''), \tag{2}$$

where dot denotes partial differentiation with respect to time with ξ held fixed, and primes denote partial differentiation with respect to arclength ξ with time t held fixed. Equation (2) corresponds to the Da Rios-Betchov equation of motion of a vortex filament. By introducing the Seret-Frenet equations given by (Pismen, 1999),

$$\mathbf{r}' = \mathbf{t}, \quad \mathbf{t}' = \kappa\mathbf{n}, \quad \mathbf{n}' = \tau\mathbf{b} - \kappa\mathbf{t}, \quad \mathbf{b}' = -\tau\mathbf{n}, \tag{3}$$

where $\mathbf{t}, \mathbf{n}, \mathbf{b}$ are a right-handed system of mutually perpendicular unit vectors corresponding to the tangent, the principal normal and binormal directions, respectively, Eq. (2) can be written as $\dot{\mathbf{r}} = \beta\kappa\mathbf{b}$. In general, although β can vary since it depends on the radius of curvature R, it will be assumed to be constant. Here, $\kappa(\xi,t)$ and $\tau(\xi,t)$ denote the local curvature and torsion of the filament, respectively. In order to transform Eq. (2) into a form corresponding to the NLS equation, we introduce the variables

$$\mathbf{N} = (\mathbf{n} + i\mathbf{b}) \exp\left(i \int_{\xi_p}^{\xi} \tau(\tilde{\xi}) d\tilde{\xi}\right), \quad \phi = \kappa \exp\left(i \int_{\xi_p}^{\xi} \tau(\tilde{\xi}) d\tilde{\xi}\right). \tag{4}$$

These two quantities are the intrinsic variables of the vortex and completely determine the configuration of the filament up to an arbitrary orientation in physical space.

Following the procedure set out in Hasimoto (1972), Eq. (2) can then be mapped to

$$\beta^{-1}(i\phi_t) = -\phi_{\xi\xi} - \frac{1}{2}|\phi|^2\phi. \tag{5}$$

We observe that the resulting equation is a self-focussing NLS equation and can, therefore, admit breather solutions. The breather solutions of this equation will be presented in the next section. In order to reconstruct the vortex from the function $\phi(\xi,t)$, the orientation and position of the vortex is specified at a particular instant in time. A procedure is then required whereby the location and orientation of the vortex must be recovered from the equation of motion self-consistently. To accomplish this, we revert back to the Seret-Frenet equations (3) and consider the time derivatives of the tangent vector $\mathbf{t}$ and the complex vector $\mathbf{N}$ to obtain

$$\dot{\mathbf{t}} = \frac{1}{2}\left(\phi'\overline{\mathbf{N}} - \overline{\phi}'\mathbf{N}\right), \quad \dot{\mathbf{N}} = \frac{i}{2}\left(\kappa^2\mathbf{N} - 2\phi'\mathbf{t}\right). \tag{6}$$

Here, bars denote complex conjugate quantities. To simplify the prescription of the initial tangent, normal, and binormal vectors, we will choose a point on the vortex filament that coincides with the point $\xi = \xi_p(t)$ such that $\kappa'(\xi_p) = 0$. Therefore, given the initial condition at time t_i for the vectors $\mathbf{t}(\xi_p(t_i))$, $\mathbf{n}(\xi_p(t_i))$, $\mathbf{b}(\xi_p(t_i))$, we must evaluate these vectors at some later time t. This would provide the initial conditions to determine the constant of integration of the Seret-Frenet system of equations. To obtain the desired equations for the change of $\mathbf{t}$ and $\mathbf{N}$ moving along the point $\xi_p(t)$, we simply consider the rate of change $\frac{D}{Dt} = \frac{\partial}{\partial t}\big|_s + 2\tau_o \frac{\partial}{\partial s}\big|_t$ where τ_o is a constant torsion associated with the unperturbed helical vortex filament as derived in the following section. The equations describing the rate of change of the three vectors of the Seret-Frenet system can then be cast in the form

$$\frac{D}{Dt}\begin{pmatrix} \sqrt{2}\mathbf{t} \\ \mathbf{N} \\ \overline{\mathbf{N}} \end{pmatrix} = \begin{pmatrix} \mathbf{0} & -\overline{h}\mathbf{I} & -h\mathbf{I} \\ h\mathbf{I} & \frac{i\kappa^2}{2}\mathbf{I} & \mathbf{0} \\ \overline{h}\mathbf{I} & \mathbf{0} & -\frac{i\kappa^2}{2}\mathbf{I} \end{pmatrix} \begin{pmatrix} \sqrt{2}\mathbf{t} \\ \mathbf{N} \\ \overline{\mathbf{N}} \end{pmatrix}, \tag{7}$$

where, $h(s) = -\left(i\phi'/\sqrt{2} + \sqrt{2}\tau_o\phi\right)$, and $\mathbf{0}$ and $\mathbf{I}$ are 3×3 zero and identity matrices, respectively. We have written the governing equations in skew-Hermitian form which arises as a consequence of the underlying symplectic structure of the LIA. The algorithm prescribed above permits complete reconstruction of the vortex. We will now consider the class of breather solutions that are associated with the function ϕ.

3. Breathers of the Nonlinear Schrödinger Equation

The main breather solutions covered in the literature come in two different types; the Kuznetsov-Ma breather (Kuznetsov, 1977; Ma, 1979) which has a temporally periodic solution, and the Akhmediev breather which is spatially periodic. In this work, we will be primarily interested in vortices that have a periodic structure along their arclength. For this reason, we will focus on the Akhmediev breather solution, which will turn out to be more relevant for our considerations. The Akhmediev breather solution is typically cast in a form that corresponds to an orbit that is homoclinic/heteroclinic to a spatially uniform plane wave fixed point and is given by (Dysthe & Trulsen, 1999)

$$\phi_A(\xi,t) = \mathrm{e}^{i\beta t/2}\frac{\cosh(\Omega\beta t/4 - 2i\varphi) - \cos(\varphi)\cos(k\xi/2)}{\cosh(\Omega\beta t/4) - \cos(\varphi)\cos(k\xi/2)},$$
$$k = 2\sin(\varphi), \qquad \Omega = 2\sin(2\varphi). \tag{8}$$

When interpreting this form of the solution in terms of a vortex, the basic state of the solution has no torsion and so would not correspond to vortices that are of general interest in classical fluids or superfluids. In order to generalise the above solution to a helical vortex, what is needed is an expression for a homoclinic/heteroclinic orbit to a spatially travelling plane wave solution. We can obtain such a solution by making use of two transformations satisfied by the NLS equation. Specifically, we note that Eq. (5) is invariant under the Galilean (gauge) and scaling transformations given respectively by

$$t \to t, \quad \xi \to \xi - 2\tau_o t, \quad \phi \to \phi\exp[i(\tau_o\xi - \tau_o^2 t)],$$
$$t \to \kappa_o^2 t, \quad \xi \to \kappa_o\xi, \quad \phi \to \kappa_o\phi(\kappa_o\xi, \kappa_o^2 t). \tag{9}$$

We note that under the above Galilean transformation, the spatially uniform plane wave solution maps onto a solution that corresponds to an unperturbed helical vortex with curvature κ_o and torsion τ_o. The breather solution we seek can, therefore, be recovered from Eq. (8) following these transformations to obtain

$$\phi_A(\xi,t) = \kappa_o \mathrm{e}^{i\kappa_o^2\beta t/2 + i\tau_o\xi + i(\kappa_o^2-\tau_o^2)\beta t}\frac{\cosh(\Omega\kappa_o^2\beta t/4 - 2i\varphi) - \cos(\varphi)\cos(k\kappa_o\xi/2 - k\tau_o\kappa_o\beta t)}{\cosh(\Omega\kappa_o^2\beta t/4) - \cos(\varphi)\cos(k\kappa_o\xi/2 - k\tau_o\kappa_o\beta t)}. \tag{10}$$

This breather solution was considereded in Salman (2013) to study nonlinear excitations in the Da Rios-Betchov equation, the Biot-Savart law, and the Gross-Pitaevskii model of a superfluid. We note that a particularly important solution of the Akhmediev breather arises in the limit $k \to 0$. In this case, we obtain the analogue of the Peregrine soliton given by

$$\phi_p = \lim_{k\to 0} q_A \;=\; \kappa_o \mathrm{e}^{i\tau_o\xi + i(\kappa_o^2-\tau_o^2)\beta t}\left\{1 - \frac{4(1 + i\kappa_o^2\beta t)}{1 + 4(\kappa_o\xi/2 - \tau_o\kappa_o\beta t)^2 + i(\kappa_o^2\beta t)^2}\right\}. \tag{11}$$

This solution has the maximum possible amplitude of ϕ (equivalently maximum curvature for the vortex) among the family of breather solutions given by Eq. (10).

The above Akhmediev breather solution was subsequently derived by Ablowitz and Herbst using the direct bilinear method of Hirota (1978, 1982). They expressed the solution in the form $\phi(\xi,t) = G(\xi,t)/F(\xi,t)$. By invoking the transformation given in Eq. (9), the expressions for G and F given in Ablowitz & Herbst (1990) transform to

$$\begin{aligned} G(\xi,t) &= \kappa_o\exp(i\kappa_o^2\beta t/2 + i\tau_o\xi + i(\kappa_o^2 - \tau_o^2)\beta t)\times \\ &\quad \left\{1 + 2\cos(\eta)\exp(\Omega\kappa_o^2\beta t/4 + 2i\phi + \gamma) + \frac{\exp(\Omega\kappa_o^2\beta t/2 + 4i\phi + 2\gamma)}{(\cos^2\phi)}\right\}, \\ F(\xi,t) &= \left\{1 + 2\cos(\eta)\exp(\Omega\kappa_o^2\beta t/4 + \gamma) + \frac{\exp(\Omega\kappa_o^2\beta t/2 + 2\gamma)}{(\cos^2\phi)}\right\}, \end{aligned} \tag{12}$$

where $\eta = k\kappa_o\xi/2 - k\tau_o\kappa_o\beta t$, $\Omega = \pm k\sqrt{4-k^2}$ and γ is an arbitrary phase. We note that by setting $\gamma = \log(-\cos(\phi))$ and choosing the negative root of Ω, we recover the expression given in Eq. (10). We, therefore, see that the form given in Eq. (12) is equivalent to the breather solution discussed above.

Using this form of solution, we can now also consider a pair of breather solutions also referred to in Ablowitz & Herbst (1990) as combination homoclinic orbits. In this case, the solution is given by

$$
\begin{aligned}
\phi(\xi,t) &= \kappa_o e^{(i\kappa_o^2\beta t/2 + i\tau_o\xi + i(\kappa_o^2-\tau_o^2)\beta t)}\frac{G(\xi,t)}{F(\xi,t)} \\
G(\xi,t) &= 1 + 2\exp(-2i\phi_1)\cos(\eta_1)\exp(\Omega_1\kappa_o^2\beta t/4+\gamma_1) + A_{12}\exp(4i\phi_1)\exp(\Omega_1\kappa_o^2\beta t/2+2\gamma_1) \\
&\quad +2\exp(-2i\phi_3)\cos(\eta_3)\exp(\Omega_3\kappa_o^2\beta t/4+\gamma_3) + A_{34}\exp(4i\phi_3)\exp(\Omega_3\kappa_o^2\beta t/2+2\gamma_3) \\
&\quad +2\exp[-2i(\phi_1+\phi_3)]\left[A_{13}\cos(\eta_1+\eta_3)+A_{23}\cos(\eta_1-\eta_3)\right] \\
&\qquad \times\exp[(\Omega_1+\Omega_3)\kappa_o^2\beta t/4+(\gamma_1+\gamma_3)] \\
&\quad +2A_{13}A_{23}A_{34}\exp[-i(2\phi_1+4\phi_3)]\cos(\eta_1)\exp[(\Omega_1+2\Omega_3)\kappa_o^2\beta t/4+(\gamma_1+2\gamma_3)] \\
&\quad +2A_{12}A_{13}A_{23}\exp[-i(4\phi_1+2\phi_3)]\cos(\eta_3)\exp[(2\Omega_1+\Omega_3)\kappa_o^2\beta t/4+(2\gamma_1+\gamma_3)] \\
&\quad +A_{12}A_{13}^2A_{23}^2A_{34}\exp[-4i(\phi_1+\phi_3)]\exp[2(\Omega_1+\Omega_3)\kappa_o^2\beta t/4+2(\gamma_1+\gamma_3)]
\end{aligned}
$$

$$
\begin{aligned}
F(\xi,t) &= 1 + 2\cos(\eta_1)\exp(\Omega_1\kappa_o^2\beta t/4+\gamma_1) + A_{12}\exp(4i\phi_1)\exp(\Omega_1\kappa_o^2\beta t/2+2\gamma_1) \\
&\quad +2\cos(\eta_3)\exp(\Omega_3\kappa_o^2\beta t/4+\gamma_3) + A_{34}\exp(4i\phi_3)\exp(\Omega_3\kappa_o^2\beta t/2+2\gamma_3) \\
&\quad +2\left[A_{13}\cos(\eta_1+\eta_3)+A_{23}\cos(\eta_1-\eta_3)\right] \\
&\qquad \times\exp[(\Omega_1+\Omega_3)\kappa_o^2\beta t/4+(\gamma_1+\gamma_3)] \\
&\quad +2A_{13}A_{23}A_{34}\cos(\eta_1)\exp[(\Omega_1+2\Omega_3)\kappa_o^2\beta t/4+(\gamma_1+2\gamma_3)] \\
&\quad +2A_{12}A_{13}A_{23}\cos(\eta_3)\exp[(2\Omega_1+\Omega_3)\kappa_o^2\beta t/4+(2\gamma_1+\gamma_3)] \\
&\quad +A_{12}A_{13}^2A_{23}^2A_{34}\exp[2(\Omega_1+\Omega_3)\kappa_o^2\beta t/4+2(\gamma_1+\gamma_3)]
\end{aligned} \tag{13}
$$

where

$$
\begin{aligned}
A_{jk} &= \left(\frac{\sin\frac{1}{2}(\Phi_j-\Phi_k)}{\sin\frac{1}{2}(\Phi_j+\Phi_k)}\right)^2, \quad \eta_i = k_i\kappa_o\xi/2 - k_i\tau_o\kappa_o\beta t, \quad k_i = 2\sin\Phi_i, \\
\Omega_i &= k_i\sqrt{4-k_i^2}, \quad \Phi_2 = \Phi_1+\pi, \quad \Phi_4 = \Phi_3+\pi.
\end{aligned} \tag{14}
$$

In the next section we will present examples of breather excitations on vortices constructed from the above breather solutions. However, we point out that a general expression for constructing an even more complex set of breather solutions was presented in Ablowitz & Herbst (1990). Therefore, in principle, the method presented here could be applied to these breathers also.

4. Results

We begin by illustrating the breather solution given by Eq. (10). We have set $\kappa_o = r/(r^2+h^2)$, and $\tau_o = h/(r^2+h^2)$ where $r = 10$, $h = L/(2\pi n_H)$, $n_H = 6$ and $L = 128$, and set $\beta = 4\pi$. Figure (1a) shows how the curvature along the filament varies with time. Initially, the curvature is constant along the filament and corresponds to a helical vortex. As $t \to 0$, we observe the emergence of large deviations from the ambient level. The curvature eventually relaxes back to the background value as $t \to \infty$. Now looking at the plot of the torsion, presented in Fig. (1b), we notice that torsion is essentially a nonzero constant at early times. However, peaked values of the torsion (both positive and negative) emerge as $t \to 0$ which eventually relax back to τ_o for large times. We note that the values of these peaks are dependent on the resolution of the grid used to generate these surface plots of τ as a function of (ξ, t) which in this case was set to

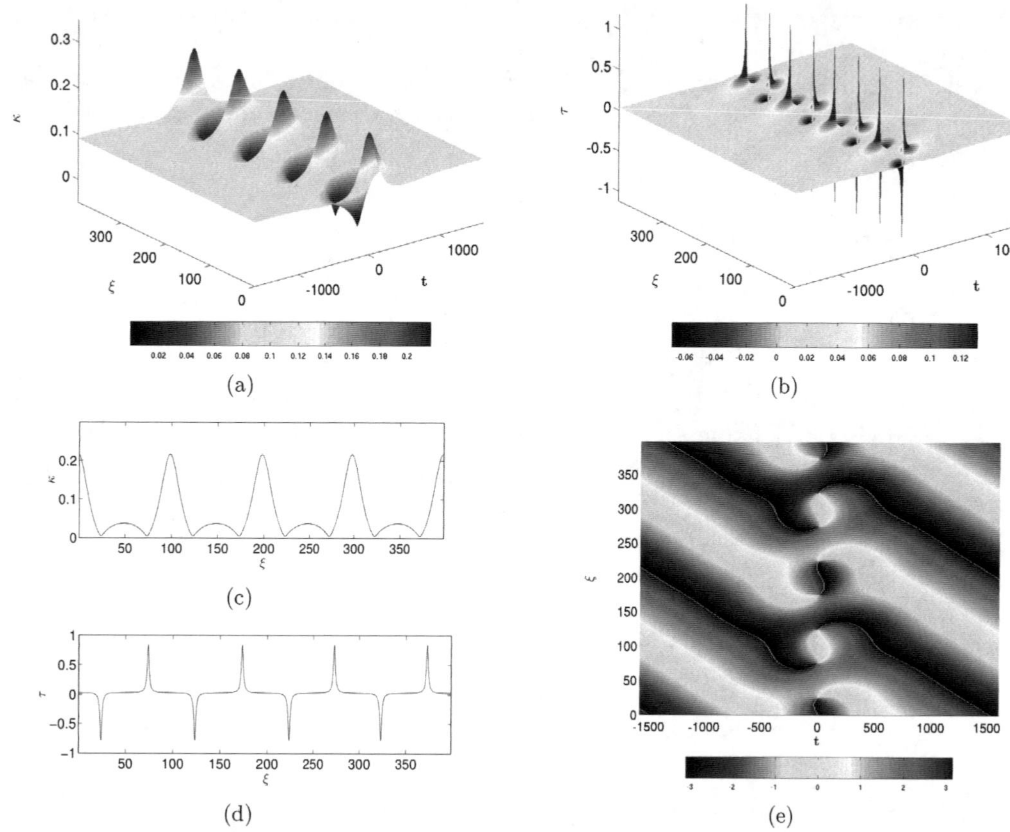

Figure 1. Akhmediev breather solution of the NLS equation for a travelling plane wave solution; (a) Temporal evoluation of amplitude of wavefunction, κ; (b) Temporal evoluation of gradient of phase of wavefunction, τ; (c) Variation of $\kappa(\xi)$ at $t = 0$; (d) Variation of $\tau(\xi)$ at $t = -0.1\beta$; (e) Temporal evolution of phase of wavefunction.

$(\Delta\xi, \Delta t) = (0.01, 10)$. This indicates that a singular behaviour in the value of τ is reached as $t \to 0$ is approached. To understand this behaviour more fully, we have presented plots in Figs. (1c,d) of κ and τ as a function of ξ at $t = 0$ and $t = -0.1\beta$, respectively.

We note that at $t = 0$, the curvature vanishes at eight points along the length of the filament. Comparing with the plot of τ, we observe that these correspond to peaked structures which, in fact, approach delta functions as $t \to 0$. The singular behaviour of τ is more evident in the phase plot of ϕ_A presented in Fig. (1e) as a function of arclength ξ and time t. The figure clearly illustrates that the phase becomes ill defined at the points where $\kappa = 0$. Therefore, constructing the filament at $t = 0$ becomes problematic. To uncover the physical implication of this singular behaviour in the solution, we will now present the configuration of the vortex filament associated with these functions of $\kappa(\xi, t)$ and $\tau(\xi, t)$. Given our choice of parameters, we expect that in the limit $t \to \pm\infty$, we will have a helical vortex which can be parameterised as

$$x = x_o + r\cos\left(\frac{2\pi n_H \xi}{AL}\right), \quad y = y_o + r\sin\left(\frac{2\pi n_H \xi}{AL}\right), \quad z = \frac{\xi}{A},$$

where $A = \left[1 + (2\pi n_H/L)^2 r^2\right]^{1/2}$. The helical vortex contains 6 waves extending along the z-direction over an interval of length L and with an amplitude, r. The vortex can be clearly seen in Fig. (2) which deforms as the breather solution is approached as $t \to 0$. As can be

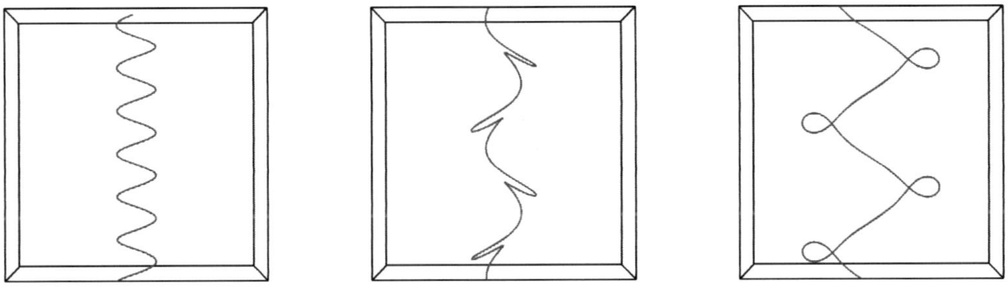

(a) $t = -800\beta$ (b) $t = -20\beta$ (c) $t = -0.1\beta$

Figure 2. Configuration of vortex filament corresponding to the Akhmediev breather at different times. Horizontal and vertical directions correspond to x and z coordinates respectively and depth corresponds to the y coordinate.

seen, the breather is spatially periodic with a wavenumber of the spacial modulation given by $k = 4\pi n_B/(L\kappa_o)$ with $n_B = 4$. Near $t = 0$, we see loop like excitations. Upon more careful inspection, it becomes clear that the points corresponding to $\kappa = 0$ correspond to the severe twisting that the vortex experiences near $t = 0$. This provides the physical interpretation for the lack of a well defined torsion in this case. The breather solutions we have obtained for the vortex, therefore, provide a mechanism for generating vortex loops from a vortex. Indeed, this was confirmed in Salman (2013) with numerical simulations based on the Biot-Savart law and the Gross-Pitaevskii equation.

We pointed out in Section (3) that a particularly important limit of a breather is recovered in the limit $k \to 0$ and corresponds to the so-called Peregrine soliton given by Eq. (11). In the context of rogue waves in the ocean, the Peregrine soliton corresponds to a wave of maximum amplitude that can arise from the family of these breather solutions. In the context of vortices, this would imply breather excitations with maximum curvature emerging from the weakly perturbed helical vortex. We illustrate this case in Fig. (3a) where we clearly see a localised excitation in the function $\kappa(\xi, t)$ emerging as $t \to 0$. The corresponding plots for κ at $t = 0$ and τ at $t = -0.1\beta$ are also included in Figs. (3b,c), respectively. We note that the maximum value of κ at $t = 0$, is larger than the corresponding value presented for the breather in Fig. (3c) as expected for the Peregrine soliton Dysthe & Trulsen (1999). The corresponding vortex configuration presented in Fig. (4) now admits a single loop-like excitation since we are dealing with the limiting case where the wavenumber of the breathing mode tends to zero and so the spatial periodicity of the solution is lost.

We will now consider the multiple breather solutions presented in Eqs. (12) and (13). This case contains a larger number of parameters as the breathers can have different modulation wavenumbers, and can be offset with respect to one another in time. We will consider a particular example corresponding to $\gamma_1 = \gamma_2 = 0$ which displays quite complex spatio-temporal behaviour in the breathing modes. All other parameters are the same as the Akhmediev breather. In Fig. (5c), we present a surface plot of the curvature as a function of arclength and time. The solution now clearly exhibits much more complex behaviour than the Akhmediev breather solution. In particular, we note that the peaks that form emerge at different times and in different locations. This is also clearly reflected in the phase plot presented in Fig. (5d) that reveals the points where the curvature vanishes and where multiple self-reconnection events would be expected. To illustrate the non-trivial property of the vortex configurations associated with such breather solutions, we have plotted in Fig. (6) the vortex filament configurations corresponding to these breathers. Given our parameters, we point out that the initial vortex configuration at early times is the same as that presented in Fig. (2a). However, significant differences emerge at later

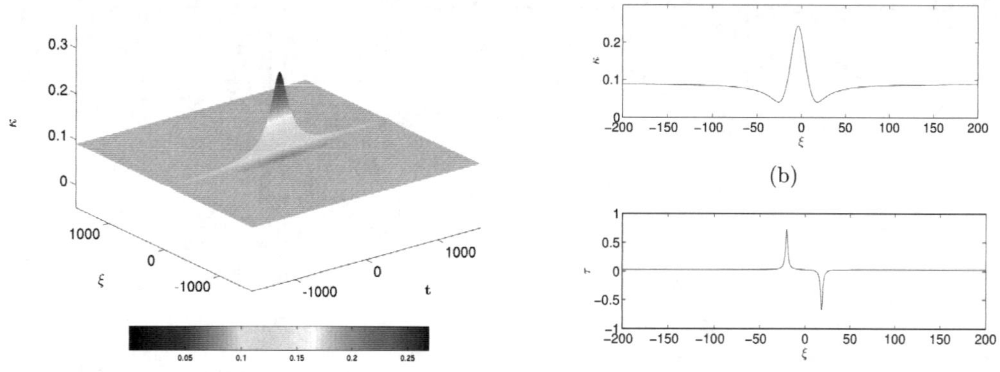

(a) (b) (c)

Figure 3. Peregrine solution of the NLS equation for a travelling plane wave solution; (a) Temporal evoluation of amplitude of wavefunction, κ; (b) Variation of $\kappa(\xi)$ at $t = 0$; (c) Variation of $\tau(\xi)$ at $t = -0.1\beta$.

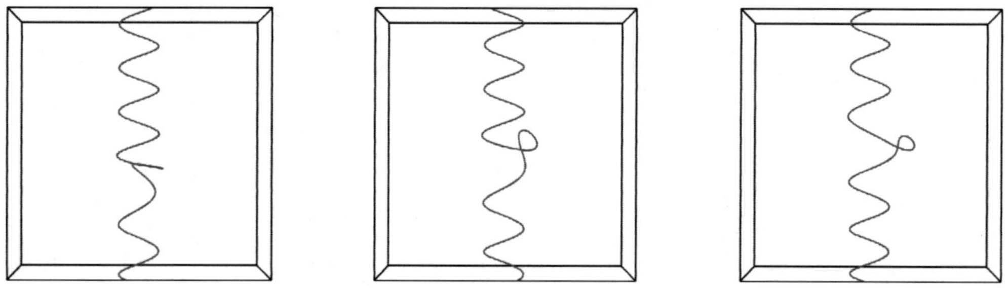

(a) $t = -20\beta$ (b) $t = -10\beta$ (c) $t = -0.1\beta$

Figure 4. Configuration of vortex filament corresponding to Peregrine soliton at different times. Horizontal and vertical directions correspond to x and z coordinates respectively and depth corresponds to the y coordinate.

times in that the vortex loops that emerge lack the more regular structures observed with the Akhmediev breather. Moreover, different sized loop-like excitations emerge at different times indicating a more irregular and complex evolution of the vortex with time. While reconnections, do not arise in our simple LIA description presented here, based on our previous work, we would expect the dynamics observed from this simple model to generalize to models based on the Biot-Savart law or the Gross-Pitaevskii equation.

5. Conclusions

We have considered breather solutions of the Da Rios-Betchov equation which governs the motion of a vortex filament in the localised induction approximation. Using the correspondence between this equation and the self-focusing NLS equation that emerges upon application of Hasimoto's transformation, we have presented a method for mapping breather solutions of the NLS equation to the vortex filament. This approach has the benefit that it permits direct application of previously derived breather solutions of the nonlinear Schrödinger (NLS) equation to reconstruct the corresponding vortex configurations. To demonstrate our approach, we have considered three particular solutions of the NLS equation. This included the Akhmediev breather for a spatially traveling plane wave, the Peregrine soliton that is recovered when the period of the spacial modulation tends to infinity, and finally a pair of breather solutions that is recovered from

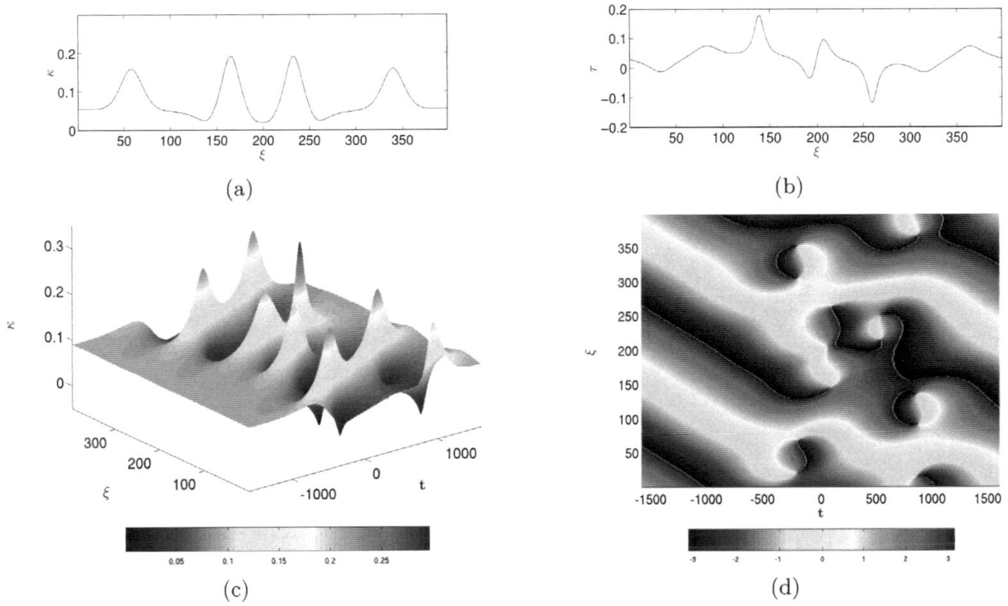

Figure 5. Multiple breather solution of the NLS equation for a travelling plane wave solution; (a) Temporal evoluation of amplitude of wavefunction, κ; (b) Temporal evoluation of gradient of phase of wavefunction, τ; (c) Variation of $\kappa(\xi)$ at $t = 0$; (d) Variation of $\tau(\xi)$ at $t = -0.1\beta$; (e) Temporal evolution of phase of wavefunction.

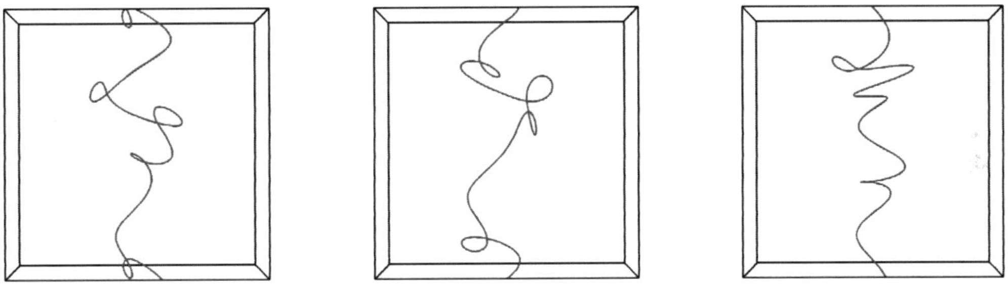

(a) $t = -15\beta$ (b) $t = 0$ (c) $t = 40\beta$

Figure 6. Configuration of vortex filament corresponding to a multiple breather at different times. Horizontal and vertical directions correspond to x and z coordinates respectively and depth corresponds to the y coordinate.

a more general class of solutions permitting the construction of multiple breather excitations on a vortex filament. Our results clearly demonstrated that such excitations propagate along the filament giving rise to loop-like structures. Since our analysis was based exclusively on the localised induction approximation, no self-reconnections and change of topology arises in this model. However, the recent work presented in Salman (2013) clearly illustrates that when these excitations are considered in more realistic models of a vortex, such as the Biot-Savart law, the emergent loops lead to self-recrossings and the subsequent emission of vortex rings. This change in topology of the vortex that occurs in the nonlinear stages of the instability identified here can play an impotant role in how vortices release the excess energy stored within them. Such considerations can have important implications in a number of different physical contexts with the energy cascade in superfluid turbulence of liquid ^{4}He being among one of the main potential applications of the new theoretical results presented here.

References

Ablowitz, M.J. & Herbst, B.M. 1990 On homoclinic structure and numerically induced chaos for the nonlinear schrödinger equation. *SIAM Journal of Applied Mathematics* **50**, 339.

Akhmediev, N.N., Eleonskii, V.M. & Kulagin, N.E. 1987 Exact first-order solutions of the nonlinear schrödinger equation. *Translation from Teoreticheskaya i Matematicheskaya Fizika* **72**, 183.

Akhmediev, N.N. & Korneev, V.I. 1986 Modulation instability and periodic solutions of the nonlinear schrödinger equation. *Translation from Teoreticheskaya i Matematicheskaya Fizika* **69**, 189.

Betchov, R. 1965 On the curvature and torsion of an isolated vortex filament. *Journal of Fluid Mechanics* **22**, 471–479.

Cieśliński, J. 1992 Two solitons on a thin vortex filament. *Physics Letters A* **171**, 323.

Donnelly, R.J. 1991 **Quantized Vortices in Helium II**. Cambridge University Press.

Dysthe, K.B. & Trulsen, K. 1999 Note on breather type solutions of the nls as models for freak-waves. *Physica Scripta* p. 48.

Fukumoto, Y. & Miyazaki, T. 1986 N-solitons on a curved vortex filament. *Journal of Physical Society of Japan* **55**, 4152–4155.

Gross, E.P. 1961 Structure of a quantized vortex in boson systems. *Nuovo Cimento* **20**, 454–457.

Hasimoto, H. 1972 A soliton on a vortex filament. *Journal Fluid Mechanics* p. 477.

Hirota, R. 1978 Nonlinear partial difference equations. iv. bcklund transformation for the discrete-time toda equation. *Journal of Physical Society of Japan* **45**, 321–332.

Hirota, R. 1982 Bilinearisation of soliton equations. *Journal of Physical Society of Japan* **51**, 323–331.

Hopfinger, E.J. & Browand, F.K. 1982 Vortex solitary waves in a rotating, turbulent flow. *Nature* **295**, 393.

Kelvin, W.T. 1880 Vibrations of a columnar vortex. *Phil. Mag. Ser.* **5** (10), 155–168.

Konno, K. & Ichikawa, Y.H. 1992 Solitons on a vortex filament with axial flow. *Chaos, Solitons & Fractals* **2**, 237–250.

Konno, K., Mituhashi, M. & Ichikawa, Y.H. 1991 Soliton on thin vortex filament. *Chaos, Solitons & Fractals* **1**, 55–65.

Kozik, E.V. & Svistunov, B.V. 2009 Theory of decay of superfluid turbulence in the low-temperature limit. *Journal of Low Temperature Physics* **156**, 215.

Kursa, M., Bajer, K. & Lipniacki, T. 2011 Cascade of vortex loops initiated by a single reconnection of quantum vortices. *Physical Review B* **83**, 014515.

Kuznetsov, E. 1977 Solitons in a parametrically unstable plasma. *Soviet Physics Doklady* **22**, 507.

Leibovich, S. & Ma, H.Y. 1983 Soliton propagation on vortex cores and the hasimoto soliton. *Physics of Fluids* **26**, 3173.

Levi, D., Sym, A. & Wojciechowski, S. 1983 N-solitons on a vortex filament. *Physics Letters A* **94**, 408.

L'vov, V.S. & Nazarenko, S.A. 2010 Spectrum of kelvin-wave turbulence in superfluids. *Pisma v Zhurnal Èksperimental'noi i Teoreticheskoi Fiziki* **91**, 464–470.

Journal of Physics: Conference Series **544** (2014) 012005 doi:10.1088/1742-6596/544/1/012005

MA, H.Y. 1979 The perturbed plane-wave solutions of the cubic schrdinger equation. *Studies in Applied Mathematics* **60**, 43.

MAKSIMOVIĆ, A., LUGOMER, S. & MICHIELI, I. 2003 Multisolitons on a vortex filaments: the origin of axial tangling. *Journal of Fluids and Structures* **17**, 317–330.

PEREGRINE, D.H. 1983 Water waves, nonlinear schrdinger equations and their solutions. *J. Austral. Math. Soc. Ser. B* **25**, 16.

PISMEN, L.M. 1999 **Vortices in Nonlinear Fields: From Liquid Crystals to Superfluids, from Nonequilibrium Patterns to Cosmic Strings**. Oxford University Press.

PITAEVSKII, L.P. 1961 Vortex lines in an imperfect bose gas. *Soviet Physics JETP* **13**, 451–454.

RICCA, R.L. 1996 The contibutions of Da Rios and Levi-Civita to asymptotic potential theory and vortex filament dynamics. *Fluid Dynamics Research* **18**, 245–268.

RIOS, L. D. 1906 Sul moto d'un liquido indefinite con un filetto vorticoso di forma qualunque. *Rend. Circ. Mat. Palermo* **22**, 117.

SALMAN, H. 2013 Breathers on quantized superfluid vortices. *Physical Review Letters* **111**, 165301.

SCHWARZ, K.W. 1988 Three-dimensional vortex dynamics in superfluid ^{4}He: Homogeneous superfluid turbulence. *Phys. Rev. B* **38**, 2398.

UMEKI, M. 2010 A locally induced homoclinic motion of a vortex filament. *Theoretical and Computational Fluid Dynamics* **24**, 383.

VINEN, W.F. 2000 Classical character of turbulence in a quantum liquid. *Physical Review B* **61**, 1410.

Journal of Physics: Conference Series **544** (2014) 012006 doi:10.1088/1742-6596/544/1/012006

Topological constraints in magnetic field relaxation

S. Candelaresi[1,2]

[1] NORDITA, KTH Royal Institute of Technology and Stockholm University, Roslagstullsbacken 23, SE-10691 Stockholm, Sweden

[2] Department of Astronomy, AlbaNova University Center, Stockholm University, SE-10691 Stockholm, Sweden

E-mail: iomsn@physto.se, iomsn1@gmail.com

Abstract.
Stability and reconnection of magnetic fields play a fundamental role in natural and man-made plasma. In these applications the field's topology determines the stability of the magnetic field. Here I will describe the importance of one topology quantifier, the magnetic helicity, which impedes any free decay of the magnetic energy. Further constraints come from the fixed point index which hinders the field to relax into the Taylor state.

1. Introduction

Examples of topologically intriguing magnetic fields can be found both in nature and in laboratory plasma. During solar eruptions, ejections of hot plasma often appear in the shape of pig tail like loops or sigmoidal structures (Canfield *et al.*, 1999). These shapes are the product of strong magnetic fields which force the charged particles to move along the field lines, which themselves have a loop like shape. In laboratory plasma topologically non-trivial magnetic fields are used to shield the walls of the apparatus from the hot plasma. In order to improve stability, the field is often twisted (Taylor, 1986) and can form even more sophisticated helical shapes, such as those in the Large Helical Device in Toki, Japan (Motojima O. et al., 2005).

During their evolution, astrophysical magnetic fields can break up, reconnect or simply diffuse, which does not only alter their geometry but in some cases also their topology. Two magnetic vector fields are topologically different if one cannot be transformed into the other without magnetic reconnection, i.e. without breaking field lines and connecting them in a different manner. Examples are magnetic flux tubes of the shape of knots and links (Fig. 1–Fig. 3).

2. Magnetic Helicity

Magnetic helicity $H = \int \boldsymbol{A} \cdot \boldsymbol{B} \, \mathrm{d}V$, where $\boldsymbol{B} = \boldsymbol{\nabla} \times \boldsymbol{A}$ is the magnetic field expressed as its vector potential $\boldsymbol{A}$, is the most widely used measure for qualifying and quantifying the field's topology. It measures the number of mutual linkage of two magnetic field lines (Moffatt, 1969). This popularity arises from both its universal applicability, i.e. it is useful for delimited and diffusive fields, and from its conservation in ideal MHD.

In astrophysics and plasma physics one is often interested in the stability of the field. Magnetic helicity is approximately conserved in such systems, suggesting that helical systems are more stable. Arnold (1974) captured this characteristic in the realizability condition, which expresses

Figure 1. Hopf link consisting of magnetic flux tubes.

Figure 2. Volume rendering of the initial magnetic energy for the Borromean rings configuration.

Figure 3. Volume rendering of the initial magnetic energy for the 8_{18} knot.

a lower limit for the magnetic energy in presence of magnetic helicity

$$E(k) \geq k|H(k)|/(2\mu_0), \tag{1}$$

with the spectral magnetic energy $E(k)$, magnetic helicity $H(k)$, the wave number k and the vacuum permeability μ_0. It should be noted that, although magnetic helicity is generally gauge dependent, the realizability condition is not, which preserves its physical relevance.

Helical field configurations like the Hopf link (see Fig. 1) consist of field lines which are mutually interlocked or self interlinked. Such linkage obviously obstructs the evolution of the field as long as magnetic reconnection is not very strong (Moffatt, 1985). This resistance against free evolution is captured in the realizability condition (1). There are, however, field configurations of interlinked field lines which are not helical, like the Borromean rings (Fig. 2) and the 8_{18} knot (Fig. 3). Are such magnetic links and knots restricted in their evolution because of their linkage, or are they free, because the realizability condition does not apply?

In a resistive and viscous plasma it is easy to simulate the decay of magnetic knots and links (Del Sordo *et al.*, 2010). Three initial magnetic field configurations are considered, each comprising three flux rings (Fig. 4). The first and second configurations are linked, the first of which is helical. The third un-linked setup servers for comparison. Initially all three setups show a very similar decay in magnetic energy (Fig. 5). After about 10 Alfvenic times, however, they decay distinctively. For the helical setup we observe a rather slow decay in magnetic energy, which for late times, behaves approximately as a power law of $t^{-1/3}$. For the two non-helical configurations this power law is steeper and of the order of $t^{-3/2}$. Unlike it might be assumed, the actual linking of the rings does not affect the dynamics considerably. Magnetic helicity, on the other hand, restricts the relaxation such that magnetic energy decays only slowly. The two non-helical cases decay equally fast and the significance of the realizability condition is confirmed.

3. Beyond magnetic helicity

Other than the magnetic helicity there exist third and forth order topological invariants which are non-zero for the Borromean rings (Ruzmaikin & Akhmetiev, 1994) . Whether such invariants affect the field's evolution is tested in comparison with the triple rings (Fig. 6) (Candelaresi & Brandenburg, 2011). Reconnection causes the Borromean rings to reshape into two twisted tori (Fig. 7). Those tori are helical, and so the realizability condition has an effect locally, which can be observed in the slowed down decay of the magnetic energy (Fig. 6). Any higher order topological invariant is not needed to explain the intermediate decay, but cannot be excluded either.

Figure 4. Iso surfaces for the initial magnetic field for the triple ring configurations. The left configuration is helical, while the other two are non-helical. Taken from (Del Sordo *et al.*, 2010).

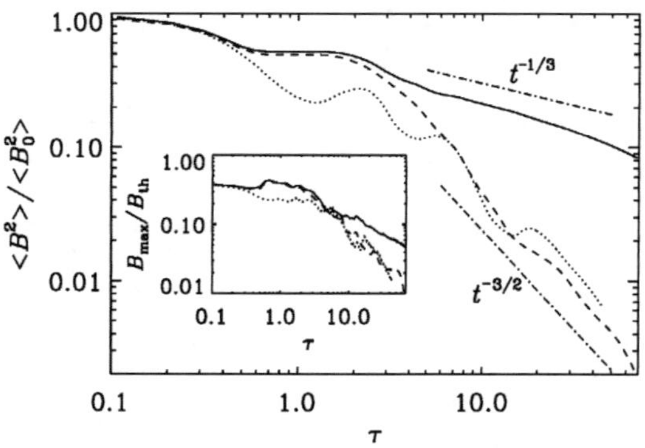

Figure 5. Normalized magnetic energy in time for the helical triple rings (solid line), the non-helical interlocked rings (dashed line) and un-linked rings (dotted line). The linking of field lines is not able to slow down the decay of magnetic energy. On the other hand, the presence of magnetic helicity poses a constraint in the field's relaxation. Taken from (Del Sordo *et al.*, 2010).

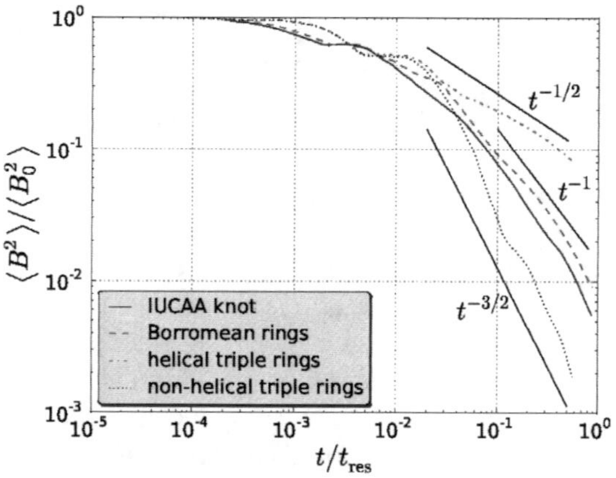

Figure 6. Comparison of the magnetic energy evolution for the interlocked triple rings, the Borromean rings and the IUCAA knot (8_{18} knot). The non-helical Borrmean rings and the IUCAA knot show an intermediate decay behavior, of which the first one can be explained by the occurrence of separated helical structures. Taken from (Candelaresi & Brandenburg, 2011).

Perhaps the most successful application of topological invariants other than the magnetic helicity is the fixed point index (Brown, 1971). Yeates *et al.* (2010) showed that its conservation imposes an additional restriction in the field's evolution such that the relaxed state reaches higher energies than proposed by Taylor (1974). Any application of the fixed point index on particularly non-helical knots and links in the form of braids would give considerable new insights in their relaxation dynamics.

4. Conclusions

We have shown that magnetic helicity, rather than the topology, is the most important parameter in determining the evolution of magnetic links and knots. Nevertheless, a non-helical initial field can spontaneously transform into separate helical fields which are restricted in their evolution (see Borromean rings). Combining the fixed point index with various topologically non-trivial

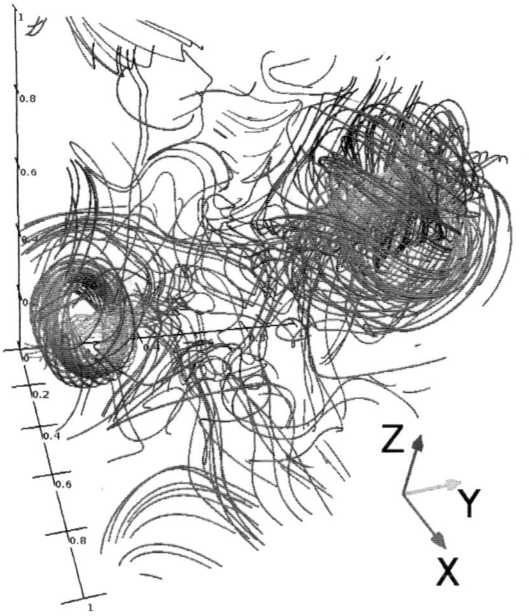

Figure 7. Magnetic streamlines of the Borromean rings configuration after a decay over 78 Alfvenic times. The colors represent the strength of the magnetic field. Two helical structures appear which are spatially separated. Their helicity content impedes any fast decay of magnetic energy. Taken from (Candelaresi & Brandenburg, 2011).

knots and links would enhance our understanding of topological constraints and magnetic reconnection.

Acknowledgments
We thank Dr. Anthony van Eysden for suggestions in improving the text's quality. We also thank the Swedish National Allocations Committee for providing computing resources at the National Supercomputer Centre in Linkoping and the Center for Parallel Computers at the Royal Institute of Technology in Sweden. This work was supported in part by the Swedish Research Council, Grant 621-2007-4064, the European Research Council under the AstroDyn Research Project 227952, and the National Science Foundation under Grant No. NSF PHY05-51164.

References

ARNOLD, V. I. 1974 The asymptotical hopf invariant and its applications. *Sel. Math. Sov.* **5**.

BROWN, R. F. 1971 **The Lefschetz Fixed Point Theorem**. Scott Foresman, London.

CANDELARESI, S. & BRANDENBURG, A. 2011 Decay of helical and nonhelical magnetic knots. *Phys. Rev. E* **84**, 016406.

CANFIELD, R. C., HUDSON, H. S. & MCKENZIE, D. E. 1999 Sigmoidal morphology and eruptive solar activity. *Geophys. Res. Lett.* **26**, 627–630.

DEL SORDO, F., CANDELARESI, S. & BRANDENBURG, A. 2010 Magnetic-field decay of three interlocked flux rings with zero linking number. *Phys. Rev. E* **81**, 036401.

MOFFATT, H. K. 1969 The degree of knottedness of tangled vortex lines. *J. Fluid Mech.* **35**, 117–129.

Moffatt, H. K. 1985 Magnetostatic equilibria and analogous Euler flows of arbitrarily complex topology. I - Fundamentals. *Journal of Fluid Mechanics* **159**, 359–378.

Motojima O. et al. 2005 Overview of confinement and mhd stability in the large helical device. *Nuclear Fusion* **45** (10), S255.

Ruzmaikin, A. & Akhmetiev, P. 1994 Topological invariants of magnetic fields, and the effect of reconnections. *Phys. Plasmas* **1**, 331–336.

Taylor, J. B. 1974 Relaxation of toroidal plasma and generation of reverse magnetic fields. *Phys. Rev. Lett.* **33**, 1139–1141.

Taylor, J. B. 1986 Relaxation and magnetic reconnection in plasmas. *Rev. Mod. Phys.* **58**, 741–763.

Yeates, A. R., Hornig, G. & Wilmot-Smith, A. L. 2010 Topological constraints on magnetic relaxation. *Phys. Rev. Lett.* **105**, 085002.

Journal of Physics: Conference Series **544** (2014) 012007 doi:10.1088/1742-6596/544/1/012007

Tightening elastic $(n, 2)$-torus knots

E. L. Starostin, G. H. M. van der Heijden

Department of Civil, Environmental & Geomatic Engineering,
University College London,
Gower Street, London WC1E 6BT, UK

E-mail: ucesest@ucl.ac.uk, g.heijden@ucl.ac.uk

Abstract. We present a theory for equilibria of elastic torus knots made of a single thin, uniform, homogeneous, isotropic, inextensible, unshearable rod of circular cross-section. The theory is formulated as a special case of an elastic theory of geometrically exact braids consisting of two rods winding around each other while remaining at constant distance. We introduce braid strains in terms of which we formulate a second-order variational problem for an action functional that is the sum of the rod elastic energies and constraint terms related to the inextensibility of the rods. The Euler–Lagrange equations for this problem, partly in Euler–Poincaré form, yield a compact system of ODEs suitable for numerical solution. By solving an appropriate boundary-value problem for these equations we study knot equilibria as the dimensionless ropelength parameter is varied. We are particularly interested in the approach of the purely geometrical ideal (tightest) limit. For the trefoil knot the tightest shape we could get has a ropelength of 32.85560666, which is remarkably close to the best current estimate. For the pentafoil we find a symmetry-breaking bifurcation.

Contact problems in the theory of elasticity have seen a surge of interest recently. They are challenging because of the one-sided constraints they introduce. Determining the topology and geometry of the contact set given the boundary conditions is probably the hardest part of any contact problem. Although great progress has been made in some of the 1D continua problems (Schuricht & von der Mosel, 2003; Gonzalez *et al.*, 2002; van der Heijden *et al.*, 2006) we are still far from a full understanding of the solutions of the Euler–Lagrange equations for general contact problems. Ideal knots (links) (Stasiak *et al.*, 1998) deliver a particular example of a geometrical contact problem where ropelength (the length-to-thickness ratio of a tube embedded into space) is minimised instead of elastic energy. Here we consider an elastic version of the knot tightening problem by applying what seem to be the first equilibrium equations for elastic knots made of an incompressible tube. The theory of elastic braids (Starostin & van der Heijden, 2014) deals with a special type of (self-)contact problems of elastic rods. Our special case of contact involves only equality constraints and is in this sense simpler allowing progress to be made. We assume the topology of the contact set to be a circle, but we make no a priori assumption on the geometry. Rather, we obtain the shape of the contact curve as part of the solution.

We apply the theory of equilibria of braids made of two elastic rods (2-braids) (Starostin & van der Heijden, 2014) to compute tight elastic $(n, 2)$-torus knots. We recall the essential equations, then formulate boundary conditions that smoothly seal the ends of the rods so that a knot is formed (Fig. 1) and after that discuss numerical solutions obtained when we inflate the incompressible tubes modelling the rods.

1. Theory of elastic braids made of two strands at constant distance

1.1. Jointly parametrised curves, reference frames and strains

Consider a pair of smooth curves $\boldsymbol{r}_1(s) \in \mathbb{R}^3$, $s \in [0, L_1]$, $\boldsymbol{r}_2(\sigma) \in \mathbb{R}^3$, $\sigma \in [0, L_2]$, both parametrised by arclength, that serve as centrelines of two inextensible elastic rods of length $L := L_1$ and L_2, respectively. We denote by $\boldsymbol{t}_1(s) = \frac{\mathrm{d}\boldsymbol{r}_1(s)}{\mathrm{d}s}$ and $\boldsymbol{t}_2(\sigma) = \frac{\mathrm{d}\boldsymbol{r}_2(\sigma)}{\mathrm{d}\sigma}$ the unit tangents to each of the curves. We define the point-to-point squared distance function $D_2(s,\sigma) = \boldsymbol{\rho}^2(s,\sigma)$, where $\boldsymbol{\rho}(s,\sigma) = \boldsymbol{r}_2(\sigma) - \boldsymbol{r}_1(s)$ is a chord vector connecting the two curves. We assume that there is a one-to-one mapping $[0, L_1] \leftrightarrow [0, L_2] : s \leftrightarrow \sigma(s)$ between the two curves such that D_2 is bicritical at corresponding points, i.e., $\frac{\partial D_2}{\partial s}(s,\sigma)\Big|_{\sigma=\sigma(s)} = -2\boldsymbol{\rho}(s,\sigma(s)) \cdot \boldsymbol{t}_1(s) = 0$ and $\frac{\partial D_2}{\partial \sigma}(s,\sigma)\Big|_{\sigma=\sigma(s)} = 2\boldsymbol{\rho}(s,\sigma(s)) \cdot \boldsymbol{t}_2(\sigma(s)) = 0$. Then $D_2(s,\sigma(s))$ is constant, say equal to Δ^2, and the two curves are said to be at constant distance Δ. Moreover the chords $\boldsymbol{\rho}(s,\sigma(s))$ are orthogonal to the curve tangents at both ends. Because of the one-to-one mapping there exists a common parametrisation for both curves. Taking arclength along the first curve as the common parameter, we can introduce the unit chord vector $\boldsymbol{d}_1(s) = \frac{1}{\Delta}\boldsymbol{\rho}(s,\sigma(s)) = \frac{1}{\Delta}[\boldsymbol{r}_2(\sigma(s)) - \boldsymbol{r}_1(s)]$. We shall henceforth write expressions like $\boldsymbol{r}_2(s)$ instead of $\boldsymbol{r}_2(\sigma(s))$.

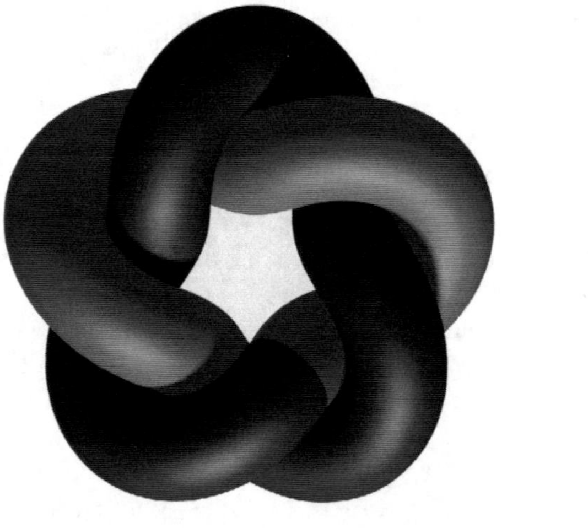

Figure 1. Elastic pentafoil knot as a closed 2-braid with two 'strands' (of unequal length) painted in different colours. The ropelength is 49.01152520.

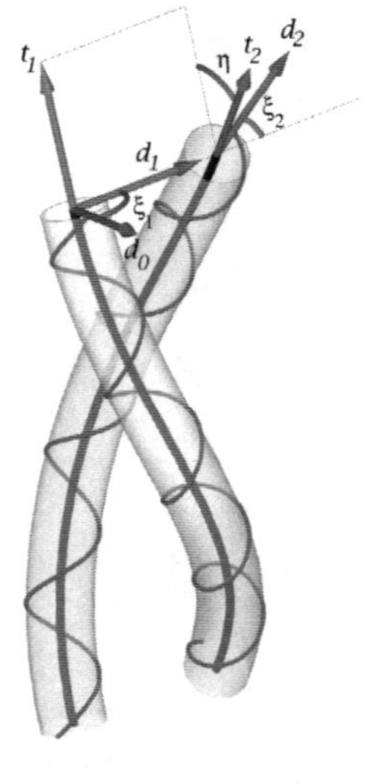

Figure 2. A braid made of two rods. Centrelines are drawn as thick blue curves while red helices show how the material twists.

Along each curve we define two moving frames as follows (Fig. 2). We define $\boldsymbol{u}_1 := \boldsymbol{t}_1 \times \boldsymbol{d}_1$ so that the vectors $\boldsymbol{t}_1$, $\boldsymbol{d}_1$ and $\boldsymbol{u}_1$ form an orthonormal frame that we call the first braid frame. We also define an orthonormal material frame on the first rod: $\{\boldsymbol{t}_1, \boldsymbol{d}_0, \boldsymbol{v}_1\}$ ($\boldsymbol{d}_0$ along arbitrary normal direction). The two frames differ by a rotation about $\boldsymbol{t}_1$ through an angle ξ_1 measured from $\boldsymbol{d}_0$ to $\boldsymbol{d}_1$. Correspondingly, we define two moving orthonormal frames on the second curve $\boldsymbol{r}_2$. The first one (called the second braid frame) is made up of the vectors $\boldsymbol{t}_2$, $\boldsymbol{d}_1$ and $\boldsymbol{u}_2 := \boldsymbol{t}_2 \times \boldsymbol{d}_1$, while the second one (called the second material frame) is $\{\boldsymbol{t}_2, \boldsymbol{d}_2, \boldsymbol{v}_2\}$. They differ by a rotation about the tangent vector $\boldsymbol{t}_2$ through an angle ξ_2, measured from $\boldsymbol{d}_1$ to $\boldsymbol{d}_2$.

After choosing a coordinate system we may identify the orientations of the above four reference frames with elements of the group of orthogonal 3×3 matrices:

$$R_{ij}(s) := (\boldsymbol{t}_j(s), \boldsymbol{d}_i(s), \boldsymbol{\beta}_{ij}(s)) \in SO(3), \quad \boldsymbol{\beta}_{ij}(s) = \boldsymbol{t}_j(s) \times \boldsymbol{d}_i(s) = i(2-i)\boldsymbol{u}_j(s) + |i-1|\boldsymbol{v}_j(s),$$

where $i \in \{0,1,2\},\ j \in \{1,2\},\ ij \neq 02, 21$. These define four skew-symmetric 3×3 matrices in the Lie algebra $\mathfrak{so}(3)$ as follows:

$$\widehat{\omega} = R_{11}^{\mathsf{T}} R_{11}', \qquad \widehat{\Omega} = R_{12}^{\mathsf{T}} R_{12}', \qquad \widehat{\widetilde{\omega}} = R_{01}^{\mathsf{T}} R_{01}', \qquad \widehat{\widetilde{\Omega}} = R_{22}^{\mathsf{T}} R_{22}', \tag{1}$$

where we have introduced the 'hat' isomorphism between skew-symmetric matrices $\widehat{\mathsf{w}} = \begin{pmatrix} 0 & -w_3 & w_2 \\ w_3 & 0 & -w_1 \\ -w_2 & w_1 & 0 \end{pmatrix}$ in $\mathfrak{so}(3)$ and axial (rotation) vectors $\mathsf{w} = (w_1, w_2, w_3)^{\mathsf{T}}$ in $\mathbb{R}^3$. [1] Here and in the following a prime denotes differentiation with respect to s. Thus we have defined four axial vectors $\omega, \Omega, \widetilde{\omega}, \widetilde{\Omega}$, where $\omega = (\omega_1, \omega_2, \omega_3)^{\mathsf{T}}$ is the rotation vector of the braid frame $\{\boldsymbol{t}_1, \boldsymbol{d}_1, \boldsymbol{u}_1\}$, $\Omega = (\Omega_1, \Omega_2, \Omega_3)^{\mathsf{T}}$ is the rotation vector of the braid frame $\{\boldsymbol{t}_2, \boldsymbol{d}_1, \boldsymbol{u}_2\}$, $\widetilde{\omega} = (\widetilde{\omega}_1, \widetilde{\omega}_2, \widetilde{\omega}_3)^{\mathsf{T}}$ is the rotation vector of the material frame $\{\boldsymbol{t}_1, \boldsymbol{d}_0, \boldsymbol{v}_1\}$ and $\widetilde{\Omega} = (\widetilde{\Omega}_1, \widetilde{\Omega}_2, \widetilde{\Omega}_3)^{\mathsf{T}}$ is the rotation vector of the material frame $\{\boldsymbol{t}_2, \boldsymbol{d}_2, \boldsymbol{v}_2\}$.

The orthonormal frames form a sequence under consecutive rotations about $\boldsymbol{t}_1$, $\boldsymbol{d}_1$ and $\boldsymbol{t}_2$. Thus

$$R_{11} = R_{01} R_1(\xi_1), \qquad R_{12} = R_{11} R_2(\eta), \qquad R_{22} = R_{12} R_1(\xi_2), \tag{2}$$

where

$$R_1(\xi_i) = \exp(\xi_i \widehat{\mathsf{e}}_1) = \begin{pmatrix} 1 & 0 & 0 \\ 0 & \cos\xi_i & -\sin\xi_i \\ 0 & \sin\xi_i & \cos\xi_i \end{pmatrix} \in SO(3),\ i = 1,2, \qquad \mathsf{e}_1 = (1,0,0)^{\mathsf{T}},$$

$$R_2(\eta) = \exp(\eta \widehat{\mathsf{e}}_2) = \begin{pmatrix} \cos\eta & 0 & \sin\eta \\ 0 & 1 & 0 \\ -\sin\eta & 0 & \cos\eta \end{pmatrix} \in SO(3), \qquad \mathsf{e}_2 = (0,1,0)^{\mathsf{T}},$$

and η is the angle, about $\boldsymbol{d}_1$, from the first tangent, $\boldsymbol{t}_1$, to the second, $\boldsymbol{t}_2$ (see Fig. 2). From Eqs. (1), (2) it follows that the rotation vectors of the material and braid frames are related as

$$\widehat{\omega} = R_1^{\mathsf{T}}(\xi_1) \widehat{\widetilde{\omega}} R_1(\xi_1) + R_1^{\mathsf{T}}(\xi_1) R_1'(\xi_1), \tag{3}$$

$$\widehat{\Omega} = R_2^{\mathsf{T}}(\eta) \widehat{\omega} R_2(\eta) + R_2^{\mathsf{T}}(\eta) R_2'(\eta), \tag{4}$$

$$\widehat{\widetilde{\Omega}} = R_1^{\mathsf{T}}(\xi_2) \widehat{\Omega} R_1(\xi_2) + R_1^{\mathsf{T}}(\xi_2) R_1'(\xi_2). \tag{5}$$

Owing to the inextensibility condition $|\frac{\mathrm{d}\boldsymbol{r}_2}{\mathrm{d}\sigma}| = 1$, the arclength parameter σ along the second centreline satisfies

$$\sigma' = \sqrt{(\Delta\omega_1)^2 + (\Delta\omega_3 - 1)^2}. \tag{6}$$

We express the angle η as a function of the components of ω as

$$\sin\eta = -\Delta\omega_1/\sigma', \qquad \cos\eta = (1 - \Delta\omega_3/)\sigma', \tag{7}$$

and hence

$$\omega_1 = \left(\omega_3 - \frac{1}{\Delta}\right) \tan\eta. \tag{8}$$

[1] Throughout we adopt the notation that for any vector $\boldsymbol{v} \in \mathbb{R}^3$ the sans-serif symbol v denotes the triple of components $(v_1, v_2, v_3)^{\mathsf{T}} = (\boldsymbol{v} \cdot \boldsymbol{t}_1, \boldsymbol{v} \cdot \boldsymbol{d}_1, \boldsymbol{v} \cdot \boldsymbol{u}_1)^{\mathsf{T}}$ in the first braid frame. An exception is however made for sans-serif rotation vectors, which are always triples of components in their corresponding frames.

Eqs. (3)–(5) allow us to express the strand strains $(\widetilde{\omega}_1, \widetilde{\omega}_2, \widetilde{\omega}_3, \widetilde{\Omega}_1, \widetilde{\Omega}_2, \widetilde{\Omega}_3)$ in terms of the variables $(\omega_1, \omega_2, \omega_3, \xi_1, \xi_2)$, which we shall call the braid strains. In much of the following it will be convenient to use Eq. (8) to eliminate ω_1 in favour of the angle η, which also has an intuitive physical meaning as braid angle. With either choice of variables we have an unconstrained description in which, by construction, the constant-distance constraint is automatically satisfied.

We note that if $\omega_1 = 0$ and $\omega_3 \neq 1/\Delta$, then $\eta = 0$ and the two strands are parallel (this condition does not restrict in any way the shape of the first centreline). When $\omega_3 = 1/\Delta$, and $\omega_1 \neq 0$, the second strand becomes orthogonal to the first one. If both $\omega_1 = 0$ and $\omega_3 = 1/\Delta$ then $\sigma' = 0$, i.e., the induced parametrisation of the second centreline is singular. The angle η is then not defined. We need to rule out this case.

The tubes of radius $\Delta/2$ built around each of the centrelines should not overlap. Locally it means that the points connected by the bicritical chords must be the closest points of the centrelines. It may be shown that this requires $\eta \in (-\pi/2, \pi/2)$ (Starostin & van der Heijden, 2014).

1.2. Equations for the standard 2-braid

For the elastic energy we make the usual assumption of frame indifference, i.e., the energy is invariant under Euclidean motions. It will then depend only on the strains (and possibly arclength) and not, for instance, on the centreline $\boldsymbol{r}_1$. We have already assumed both rods in the braid to be inextensible and unshearable, so there is no elastic energy on account of stretches. Apart from this restriction we allow for arbitrary hyperelastic rods (Antman, 2005) and write

$$U_1 = \int_0^{L_1} f_1(\widetilde{\omega}) \, \mathrm{d}s \qquad \text{and} \qquad U_2 = \int_0^{L_2} f_2(\widetilde{\Omega}(\sigma)) \, \mathrm{d}\sigma = \int_0^{L_1} f_2(\widetilde{\Omega}/\sigma') \, \sigma' \, \mathrm{d}s \tag{9}$$

for the elastic strain energy of the first and second rod, respectively, where f_1 and f_2 are the strand strain energy densities, and the explicit argument of $\widetilde{\Omega}$ in the σ integral emphasises that $\widetilde{\Omega}$ is to be regarded as a function of arclength σ of the second rod in this integral.

Assuming fixed ends of the first strand, we add $-\boldsymbol{F} \cdot [\boldsymbol{r}_1(L) - \boldsymbol{r}_1(0)] = -\int_0^L \boldsymbol{F} \cdot \boldsymbol{t}_1 \, \mathrm{d}s$ as an isoperimetric constraint expression with $\boldsymbol{F}$ a (constant) Lagrange multiplier (to become the internal force in the braid). Also, by inextensibility of the second rod we require the length of the second strand

$$L_2 = \int_0^{L_2} \mathrm{d}\sigma = \int_0^{L} \sigma' \, \mathrm{d}s =: \int_0^{L} f_\sigma(\omega_1, \omega_3) \, \mathrm{d}s \tag{10}$$

to be constant, giving a second isoperimetric constraint.

We can express the arguments of the integrands in Eqs. (9) and (10) in terms of the braid strains (ω, ξ_1, ξ_2) and their derivatives by using Eqs. (3)–(5) and (6) and thus formulate a (second-order) variational problem for the reduced functional $l : 2\mathfrak{so}(3) \times \mathbb{R}^3 \times TS^1 \times TS^1 \to \mathbb{R}$, $l(\omega, \omega', \mathsf{F}, \xi_1, \xi_1', \xi_2, \xi_2') = f_1(\widetilde{\omega}) + f_2(\widetilde{\Omega}(\sigma)) + h\sigma' - \mathsf{F} \cdot \mathsf{t}_1$,

$$\delta \int_0^L l(\omega, \omega', \mathsf{F}, \xi_1, \xi_1', \xi_2, \xi_2') \, \mathrm{d}s = 0, \tag{11}$$

for variations that keep the end positions and orientations of the rods fixed; h is a (constant) Lagrange multiplier.

The Euler–Lagrange equations are derived partly in Euler–Poincaré form and comprise (a) balance equations for the components of the overall braid force $\mathsf{F} = (F_1, F_2, F_3)^{\mathsf{T}}$ and braid moment $\mathsf{M} = (M_1, M_2, M_3)^{\mathsf{T}}$ expressed in the first braid frame (Starostin & van der Heijden, 2014)

$$\mathsf{F}' + \omega \times \mathsf{F} = 0, \tag{12}$$

$$\mathsf{M}' + \omega \times \mathsf{M} + \mathsf{t}_1 \times \mathsf{F} = 0, \tag{13}$$

(b) the 'constitutive' equations

$$M_j = \frac{\partial l}{\partial \omega_j} - \frac{\mathrm{d}}{\mathrm{d}s}\frac{\partial l}{\partial \omega_j'}, \quad j = 1, 2, 3, \tag{14}$$

and (c) the phase equations for the twist angles of the rods

$$\frac{\partial l}{\partial \xi_i} - \frac{\mathrm{d}}{\mathrm{d}s}\frac{\partial l}{\partial \xi_i'} = 0, \quad i = 1, 2. \tag{15}$$

Equations (12) and (13) are the familiar $\boldsymbol{F}' = \mathbf{0}$, $\boldsymbol{M}' + \boldsymbol{r}_1' \times \boldsymbol{F} = \mathbf{0}$ in an inertial frame. It follows that $|\mathsf{F}|$ and $\mathsf{F} \cdot \mathsf{M}$ are first integrals.

Equations (15) can be written as a set of four first-order equations by introducing new variables T_i, $i = 1, 2$:

$$T_i = \frac{\partial l}{\partial \xi_i'}, \qquad T_i' = \frac{\partial l}{\partial \xi_i}. \tag{16}$$

If we eliminate the variable ω_1 in favour of η by using $\eta = \eta(\omega_1, \omega_3)$ given in Eq. (7), the reduced density l takes the form

$$l = g(\omega_2, \omega_3, \eta, \eta', F_1, \xi_1, \xi_1', \xi_2, \xi_2'), \tag{17}$$

while Eq. (14) transforms into (Starostin & van der Heijden, 2009)

$$M_1 = \frac{\partial \eta}{\partial \omega_1}\mathcal{E}_\eta(g) = \frac{\Delta \cos^2 \eta}{\Delta\omega_3 - 1}\left[\frac{\partial g}{\partial \eta} - \left(\frac{\partial g}{\partial \eta'}\right)'\right], \tag{18}$$

$$M_2 = \frac{\partial g}{\partial \omega_2}, \tag{19}$$

$$M_3 = \frac{\partial g}{\partial \omega_3} + \frac{\partial \eta}{\partial \omega_3}\mathcal{E}_\eta(g) = \frac{\partial g}{\partial \omega_3} - \frac{\Delta \sin\eta\cos\eta}{\Delta\omega_3 - 1}\left[\frac{\partial g}{\partial \eta} - \left(\frac{\partial g}{\partial \eta'}\right)'\right] \tag{20}$$

(cf. (Starostin & van der Heijden, 2007)).

We now specialise our theory to the case where the two strands are linearly elastic, uniform, isotropic, initially straight rods with bending stiffness B_1 and B_2 and torsional stiffness C_1 and C_2, respectively. We allow the strands to be intrinsically twisted with twist rates ω_{01} and ω_{02}. We call this special case the standard 2-braid. The density g in Eq. (17) then takes the form

$$g = g_e(\omega_2, \omega_3, \eta, \eta', \xi_1', \xi_2') + h(1 - \Delta\omega_3)/\cos\eta - F_1, \tag{21}$$

with the elastic energy density g_e $(= f_1 + f_2)$, the sum of bending and torsional energy densities of the two strands, given by (using Eqs. (3)–(5))

$$\begin{aligned} g_e = \frac{1}{2}\left[B_1(\widetilde{\omega}_2^2 + \widetilde{\omega}_3^2) + \frac{B_2}{\sigma'}(\widetilde{\Omega}_2^2 + \widetilde{\Omega}_3^2) + C_1(\widetilde{\omega}_1 - \omega_{01})^2 + \frac{C_2}{\sigma'}\left(\widetilde{\Omega}_1 - \sigma'\omega_{02}\right)^2\right] = \\ = \frac{1}{2}\left\{B_1(\omega_2^2 + \omega_3^2) + B_2\left[\frac{\cos\eta}{1 - \Delta\omega_3}(\omega_2 + \eta')^2 + \frac{(\omega_3 - \sin^2\eta/\Delta)^2}{(1 - \Delta\omega_3)\cos\eta}\right] + \right. \\ +C_1\left[\left(\omega_3 - \frac{1}{\Delta}\right)\tan\eta - \xi_1' - \omega_{01}\right]^2 + \\ \left. +C_2\left[\frac{\cos\eta}{1 - \Delta\omega_3}\left(\frac{\sin\eta}{\Delta} - \xi_2'\right)^2 + 2\omega_{02}\left(\frac{\sin\eta}{\Delta} - \xi_2'\right) + \omega_{02}^2\,\frac{1 - \Delta\omega_3}{\cos\eta}\right]\right\}. \end{aligned} \tag{22}$$

This gives a system of 13 ODEs in $(F_1, F_2, F_3, M_1, M_2, M_3, \eta, \omega_2, \omega_3, \xi_1, T_1, \xi_2, T_2)$: Eqs. (12), (13), (16), (18), (19) and (20). The last three in fact contain two algebraic equations because g does not depend on ω_2' and ω_3', but these can be turned into ODEs by differentiation (and the algebraic equations used to solve for η' and h). Also note that since g does not depend on the phase angles ξ_1 and ξ_2, the T_i (the strand torques) are constants.

1.3. Kinematics equations

Reconstruction of the centreline of the first strand requires solving for the tangent $\boldsymbol{t}_1$ and integrating this to get $\boldsymbol{r}_1$. We choose a parametrisation of the first braid frame $\{\boldsymbol{t}_1, \boldsymbol{d}_1, \boldsymbol{u}_1\}$ in terms of four Euler parameters (or quaternions) $\boldsymbol{q} = (q_0, q_1, q_2, q_3)$ subject to the normalisation condition $q_0^2 + q_1^2 + q_2^2 + q_3^2 = 1$ (Hanson, 1994) and write

$$\boldsymbol{t}_1 = \begin{pmatrix} q_0^2 + q_1^2 - q_2^2 - q_3^2 \\ 2q_1q_2 + 2q_0q_3 \\ 2q_1q_3 - 2q_0q_2 \end{pmatrix}, \quad \boldsymbol{d}_1 = \begin{pmatrix} 2q_1q_2 - 2q_0q_3 \\ q_0^2 - q_1^2 + q_2^2 - q_3^2 \\ 2q_2q_3 + 2q_0q_1 \end{pmatrix}, \quad \boldsymbol{u}_1 = \begin{pmatrix} 2q_1q_3 + 2q_0q_2 \\ 2q_2q_3 - 2q_0q_1 \\ q_0^2 - q_1^2 - q_2^2 + q_3^2 \end{pmatrix}.$$

These Euler parameters, unlike Euler angles, give a singularity-free description of arbitrary rotations in space and are therefore convenient for numerical computations. The kinematics equations are obtained by differentiating the above vectors and using Eq. (1):

$$\begin{pmatrix} q_0 \\ q_1 \\ q_2 \\ q_3 \end{pmatrix}' = \frac{1}{2} \begin{pmatrix} 0 & -\omega_1 & -\omega_2 & -\omega_3 \\ \omega_1 & 0 & \omega_3 & -\omega_2 \\ \omega_2 & -\omega_3 & 0 & \omega_1 \\ \omega_3 & \omega_2 & -\omega_1 & 0 \end{pmatrix} \begin{pmatrix} q_0 \\ q_1 \\ q_2 \\ q_3 \end{pmatrix}, \tag{23}$$

where we recall from Eq. (8) that $\omega_1 = \left(\omega_3 - \frac{1}{\Delta}\right) \tan\eta$. It is straightforward to verify that $\boldsymbol{q} \cdot \boldsymbol{q}$ is a first integral of this system of equations. To find the centreline $\boldsymbol{r}_1$ we solve Eq. (23) in conjunction with the equation $\boldsymbol{r}_1' = \boldsymbol{t}_1$. The second centreline is then given by $\boldsymbol{r}_2 = \boldsymbol{r}_1 + \Delta \boldsymbol{d}_1$.

2. Numerical solution

In this section we present results obtained by numerically solving boundary-value problems for the standard 2-braid equations using the continuation and bifurcation code AUTO (Doedel, E. et al., 2007). AUTO solves ODEs using collocation and uses pseudo-arclength continuation to advance solutions as a control (or bifurcation) parameter is incremented. The code is also able to detect pitchfork or other bifurcation points along a solution branch where this branch intersects another solution branch and to switch to and compute this second branch. The thickness Δ will be chosen as a bifurcation parameter.

Here we consider only the closed braid problem where the two rods form a $(n, 2)$-torus knot. We start from an analytical solution (two linked rings) and perform continuations mimicking the cutting, twisting and resealing of the rings to obtain different torus knots (and links) (Starostin & van der Heijden, 2014). We compute solutions with different linking numbers for which we derive boundary conditions that ensure smooth closure of the braid into a knot (for semi-integer linking number of the braid).

For a 2-braid knot we define a linking number by following Fuller's extension of the concept of a linking number to a (not necessarily closed) ribbon (Fuller, 1978). By a ribbon is here meant a curve with a field of normal vectors of constant length. First note that the contact curve of our knot $\frac{1}{2}(\boldsymbol{r}_1(s) + \boldsymbol{r}_2(s)) = \boldsymbol{r}_1(s) + \frac{\Delta}{2}\boldsymbol{d}_1(s)$, $s \in [0, L]$, is a (smoothly) closed curve and the director $\frac{\Delta}{2}\boldsymbol{d}_1(s)$ is an orthogonal vector defined for all $s \in [0, L]$. It therefore forms a non-closed ribbon. We define the linking number of this ribbon as $\mathcal{L}k = \mathcal{T}w + \mathcal{W}r$, where $\mathcal{W}r$ is the writhing number of the contact curve and $\mathcal{T}w = \frac{1}{2\pi}\int_0^L tw(s)\,ds$ is the total twist (the twisting number) of the braid with twist rate tw. To have a closed single centreline, $\mathcal{L}k$ has to be semi-integer, $\mathcal{L}k = n/2, n \in \mathbb{Z}$. Alternatively, if we think of the ribbon as extended in both $\pm\frac{\Delta}{2}\boldsymbol{d}_1$ directions, we may say that, for our knots, this ribbon is one-sided (like a Möbius strip (Starostin & van der Heijden, 2007)) and its edge (given by the centreline of the rod) is a single knotted curve.

In addition to the 13-dimensional ODE derived in Section 1.2 we solve the 7 kinematics equations for $\boldsymbol{q}$ and $\boldsymbol{r}_1$ of Section 1.3, so as to be able to apply displacement boundary conditions

Journal of Physics: Conference Series **544** (2014) 012007 doi:10.1088/1742-6596/544/1/012007

and plot solution shapes, and equation (6) to compute the second arclength. Integration of this last equation allows us to impose the total-length constraint $L_{\text{tot}} := L + L_2 = \text{const.}$, the natural constraint to consider for knot solutions. Since $\boldsymbol{q} \cdot \boldsymbol{q}$ is an integral of the kinematics equations, the quaternion normalisation condition $|\boldsymbol{q}| = 1$ can simply be imposed by choosing boundary conditions for the q_i that satisfy it.

Given a solution of this 21D system of ODEs we can compute other properties such as the normal contact pressure p or the curvature κ_c of the contact curve. Frictionless hardcore contact is assumed. First integrals are monitored throughout to keep track of the numerical accuracy of solutions.

2.1. Boundary conditions

To compute knot solutions of total length L_{tot} we apply the following boundary conditions

$$x_1(0) = 0, \quad (24)$$
$$y_1(0) = 0, \quad (25)$$
$$z_1(0) = 0, \quad (26)$$
$$q_0(0) = 0, \quad (27)$$
$$q_1(0) = -1/\sqrt{2}, \quad (28)$$
$$q_2(0) = -1/\sqrt{2}, \quad (29)$$
$$q_3(0) = 0, \quad (30)$$
$$\xi_1(0) = 0, \quad (31)$$
$$\xi_2(0) = 0, \quad (32)$$
$$T_1(0) = t_{10}, \quad (33)$$
$$T_2(0) = t_{20}, \quad (34)$$
$$\sigma(0) = 0, \quad (35)$$

$$x_1(L) - x_1(0) = \Delta, \quad (36)$$
$$y_1(L) = 0, \quad (37)$$
$$z_1(L) = 0, \quad (38)$$
$$q_0(L) = q_3(L), \quad (39)$$
$$q_1(L) = -q_2(L), \quad (40)$$
$$4q_0(L)q_1(L) = \sin\eta(L), \quad (41)$$
$$\eta(L) = \eta(0), \quad (42)$$
$$[M_2(0) - B_1\omega_2(0)]/B_2 = -\omega_2(L), \quad (43)$$
$$(1 - \Delta\omega_3(0))/\cos\eta(0) = \cos\eta(L)/(1 - \Delta\omega_3(L)), \quad (44)$$
$$\sigma(L) = L_2, \quad (45)$$
$$L + L_2 = L_{\text{tot}}, \quad (46)$$

where t_{10} and t_{20} are constants.

The first nine of these conditions ensure smooth closure of the braid into a doubly-covered ring, smooth here meaning continuity up to second derivatives of $\boldsymbol{r}_i$, i.e., curvatures. Conditions Eqs. (36)–(38) place the end of the first centreline $\boldsymbol{r}_1$ at the beginning of the second centreline $\boldsymbol{r}_2$, both strands thus forming a single closed rod. Conditions Eqs. (39) and (40) imply $\boldsymbol{d}_1(L) = -\boldsymbol{d}_1(0)$. Condition Eq. (41) guarantees that $\boldsymbol{t}_1(L)$ makes an angle $\eta(L)$ with $\boldsymbol{t}_1(0)$ of the same strand. Together with condition Eq. (42) this ensures matching of the tangents: the tangent at $s = L$ of the first strand aligns with the tangent at $s = 0$ of the second strand, and vice versa. Conditions Eqs. (42)–(44) are equivalent to $\Omega_1(0)/\sigma'(0) = \omega_1(L)$, $\Omega_2(0)/\sigma'(0) = -\omega_2(L)$ and $\Omega_3(0)/\sigma'(0) = -\omega_3(L)$, which ensure that the curvatures at the end of the first strand match the curvatures at the beginning of the second strand.

The total number of boundary conditions is 23 and the three free parameters required for solution branches are L, L_2 and Δ. Since for a knot the two strands are part of a single closed rod, it only makes good sense to take $B_2 = B_1$, $C_2 = C_1$ and $t_{20} = -t_{10}$. Note that the material frames of these knot solutions will in general not be closed. Material closure requires $\xi_1(0) + \xi_2(L) = \pi \pmod{2\pi}$ and can be achieved by inserting the right amount of twist by varying t_{10} and t_{20}.

3. Tightening the trefoil and pentafoil knots

There is considerable recent interest in the configuration and contact set of the tightest possible (tubular) knot within a given topological class, i.e., the closed tube embedded in 3D space of shortest ropelength $2L_{\text{tot}}/\Delta$ (Stasiak *et al.*, 1998). Note that local non-self-intersection requires that the radius of curvature of the centreline must be bounded from below by the radius $\Delta/2$ of the tube. These tightest knots, called ideal knots, appear naturally in various physical problems (Calvo *et al.*, 2005). They allow, for instance, the prediction of electrophoretic mobility of DNA knots through gel (Stasiak *et al.*, 1996). Large numerical efforts have in particular focused on the trefoil knot for which the best current upper bound for the shortest ropelength is 32.742934547 (Przybyl & Pieranski, 2014). It is known from numerical simulations that in the ideal configuration each point along the knot's centreline has two (distant) closest-approach points. The best numerical results available also indicate that in the ideal shape the centreline curvature κ attains its maximum allowed value, $\kappa = 2/\Delta$, over six arclength intervals. Our theory, assuming one-to-one arclength mapping, is of course not designed for situations with multiple points of contact. Nevertheless, it is instructive to see how close our elastic knots, not merely geometrical solutions but equilibria, can be inflated to the ideal shape.

Figure 3 shows an example of a trefoil knot solution. It is used as starting solution of a continuation run in which the thickness Δ is increased while applying the boundary conditions of Section 2.1. Two trefoil knots computed in this run are shown in Fig. 4. All computed

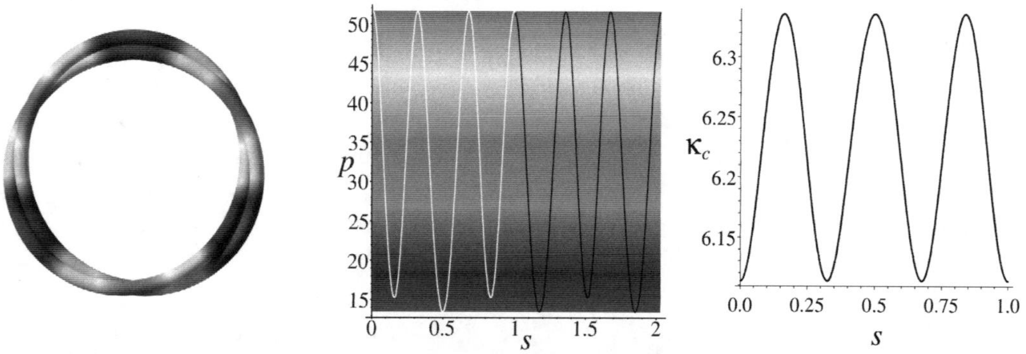

Figure 3. Trefoil 3_1 knot with $\mathcal{L}k = -\frac{3}{2}$ (*left*), the contact pressure p (*middle*) and the curvature κ_c of the contact curve (*right*). Ropelength 202.7096000, $L = 1$, $\Delta = 0.02$, $B_1 = B_2 = 1$, $t_{10} = t_{20} = 0$. The force vector is perpendicular to the plane of the figure.

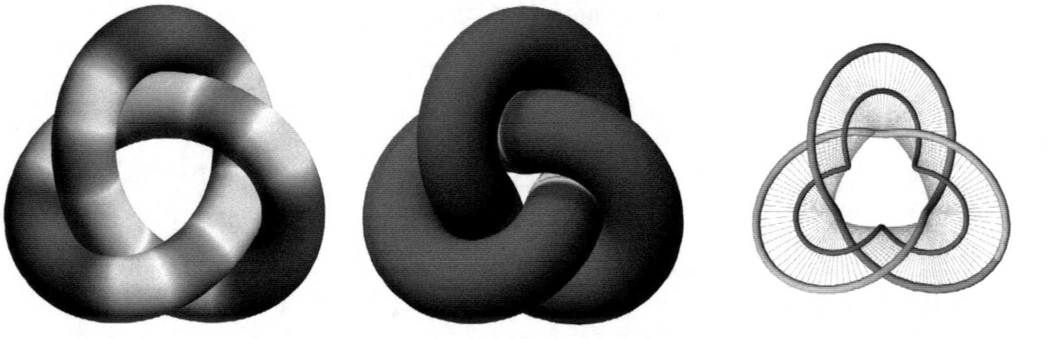

Figure 4. Elastic trefoil knot. Individually adjusted colouring of the tubes shows the contact pressure (everywhere positive). Ropelength 39.50609870 (*left*) 32.85560666 (*middle* and *right*). *Right:* Contact set (pink) and trace of the closest-approach chord.

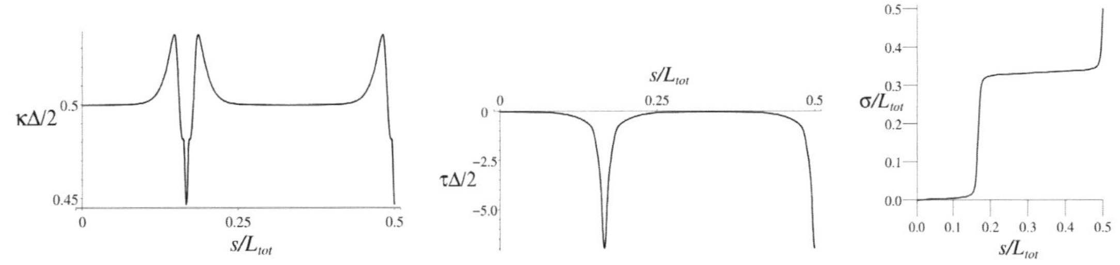

Figure 5. Normalised plots for the tight trefoil knot of Fig. 4, middle. Curvature (*left*) and torsion (*middle*) of the first strand. *Right:* Arclength σ of the second strand against s.

knots are untwisted ($t_{10} = t_{20} = 0$), so we are effectively only minimising the bending energy of the centrelines (i.e., we may set $C_1 = C_2 = 0$). The solution in Fig. 4 with the ropelength of $2(L + L_2)/\Delta = 32.85560666$ is the tightest we could get. This value is remarkably close to the best current estimate given above. Careful examination reveals that the solution is free from self-intersections. Figure 5 (*left*) shows that the centreline curvature stays well below the maximum allowed value $2/\Delta = 16.2170$, Fig. 5 (*middle*) gives a plot of the torsion of the centreline, while Fig. 5 (*right*) shows the alternating steep-flat nature of the arclength mapping $s \to \sigma$ close to the limiting ideal state. For our elastic trefoil the contact set always remains a topological circle (see Fig. 4 (*right*), which also illustrates the arclength mapping by showing the closest-approach chords), while for the ideal shape the best available numerics suggests it is a trefoil knot (Przybyl & Pieranski, 2014).

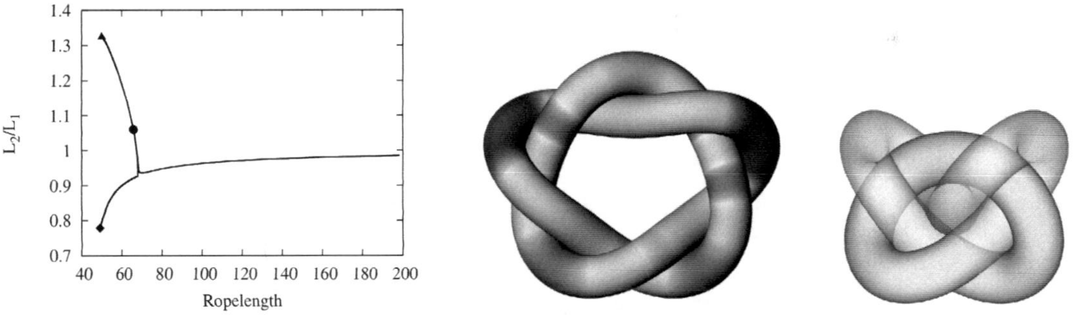

Figure 6. Bifurcation diagram (*left*) for elastic pentafoil knots and two examples on the C_2-symmetric branch: ● (*middle*) ropelength 65.90515002, no overlapping; ▲ (*right*) ropelength 49.73900556, the second strand tube (transparent) overlaps itself. ♦ marks the D_5-symmetric knot shown in Fig. 1.

We carried out similar numerical tightening of the pentafoil knot 5_1. It is known that, contrary to the trefoil, its ideal shape lacks D_5 symmetry (Cantarella *et al.*, 2014). However, with increase of Δ the solution path bifurcates and we may follow either a D_5-symmetric or a C_2-symmetric branch. The bifurcation diagram is presented in Fig. 6. (The symmetry is checked up to numerical accuracy.) So, both D_5-symmetric and C_2-symmetric pentafoil knots are in equilibrium for some interval of thickness. We could keep tightening D_5-symmetric shapes up to ropelength 49.01152520 (Fig. 1) without self-overlapping of the tube. The tightest D_5-symmetric pentafoil computed in (Cantarella *et al.*, 2014) has ropelength 48.23. We see that we do not approach this value as closely as the limiting value for the trefoil. The code stops working for thicker shapes because the angle η between strands approaches the singular value $\pi/2$. Nevertheless, for the pentafoil η gets substantially closer to $\pi/2$ than for the trefoil. This suggests that for the ideal D_5-symmetric pentafoil the two centrelines get closer to orthogonality than for the trefoil. This is not surprising as the braid axis of the D_n-symmetric knot becomes less curved as n increases, in the limit reaching the tightest double helix with exactly orthogonal strands (Neukirch & van der Heijden, 2002). As to the non-symmetric equilibria (Fig. 6), they begin self-intersecting earlier, i.e., at larger ropelength.

Acknowledgments
This work was supported by the UK's Engineering and Physical Sciences Research Council (grant no. EP/H009736/1) and the Human Frontier Science Program (grant no. RGP0049/2010).

References

Antman, S. S. 2005 **Nonlinear Problems of Elasticity**, 2nd edn. Springer.

Calvo, J. A., Millett, K. C., Rawdon, E. J. & Stasiak, A. (ed.) 2005 **Physical and Numerical Models in Knot Theory Including Applications to the Life Sciences**. *Series on Knots and Everything*, vol. 36. World Scientific.

Cantarella, J., Fu, J. H. G., Mastin, M. & Royal, E. J. 2014 Symmetric criticality for tight knots. *Journal of Knot Theory and Its Ramifications* **23** (2), 1450008.

Doedel, E. et al. 2007 AUTO: Software for continuation and bifurcation problems in ordinary differential equations. User's manual. Concordia University, Montreal, Canada.

Fuller, F. B. 1978 Decomposition of the linking number of a closed ribbon: A problem from molecular biology. *Proc. Natl. Acad. Sci. USA* **75** (8), 3557–3561.

Gonzalez, O., Maddocks, J. H., Schuricht, F. & von der Mosel, H. 2002 Global curvature and self-contact of nonlinearly elastic curves and rods. *Calculus of Variations and Partial Differential Equations* **14**, 29–68.

Hanson, A. J. 1994 Quaternion Frenet frames: making optimal tubes and ribbons from curves. Technical report TR407. Indiana Univ., Dept. of Computer Science.

van der Heijden, G. H. M., Peletier, M. A. & Planqué, R. 2006 Self-contact for rods on cylinders. *Archive for Rational Mechanics and Analysis* **182**, 471–511.

Neukirch, S. & van der Heijden, G. H. M. 2002 Geometry and mechanics of uniform n-plies: From engineering ropes to biological filaments. *Journal of Elasticity* **69** (1-3), 41–72.

Przybyl, S. & Pieranski, P. 2014 High resolution portrait of the ideal trefoil knot. *Journal of Physics A: Mathematical and Theoretical* **47** (28), 285201.

Schuricht, F. & von der Mosel, H. 2003 Euler-Lagrange equations for nonlinearly elastic rods with self-contact. *Archive for Rational Mechanics and Analysis* **168**, 35–82.

Starostin, E. L. & van der Heijden, G. H. M. 2007 The shape of a Möbius strip. *Nature Materials* **6** (8), 563–567.

Starostin, E. L. & van der Heijden, G. H. M. 2009 Force and moment balance equations for geometric variational problems on curves. *Physical Review E* **79** (6), 066602.

Starostin, E. L. & van der Heijden, G. H. M. 2014 Theory of equilibria of elastic 2-braids with interstrand interaction. *Journal of the Mechanics and Physics of Solids* **64**, 83–132.

Stasiak, A., Katritch, V., Bednar, J., Michoud, D. & Dubochet, J. 1996 Electrophoretic mobility of DNA knots. *Nature* **384**, 122.

Stasiak, A., Katritch, V. & Kauffman, L. H. (ed.) 1998 **Ideal Knots**. *Series on Knots and Everything*, vol. 19. World Scientific.

Journal of Physics: Conference Series **544** (2014) 012008 doi:10.1088/1742-6596/544/1/012008

Generalized helicity and Beltrami fields

Roman V. Buniy[1], **Thomas W. Kephart**[2]

[1] Schmid College of Science, Chapman University, Orange, CA 92866, USA

[2] Department of Physics and Astronomy, Vanderbilt University, Nashville, TN 37235, USA

E-mail: roman.buniy@gmail.com,tom.kephart@gmail.com

Abstract. We propose Lorentz-covariant generalizations of the magnetic helicity and Beltrami equation. The gauge invariance, variational principle, conserved current, energy-momentum tensor and choice of boundary conditions elucidate the subject. In particular, we prove that any extremal of the Maxwell action functional $\frac{1}{4}\int_\Omega F_{\mu\nu}F^{\mu\nu}\,d^4x$ subject to the local constraint $\varepsilon^{\mu\nu\alpha\beta}F_{\mu\nu}F_{\alpha\beta}=0$ satisfies the covariant Beltrami equation.

1. Introduction

The introduction of the concept of magnetic helicity (Woltier, 1958; Moffatt, 1969, 1985) revolutionized our understanding of plasma physics phenomena, from dynamos (Moffatt, 1978) to the solar wind, to the operation of controlled fusion devices (Marsh, 1996; Bellan, 2000, 2006), and it plays a central role when applied to a variety of concepts such as the Beltrami equation and force-free fields in the form of Taylor states (Taylor, 1974, 1986). Helicity was introduced in a three-dimensional context, but we find it useful to express it covariantly (Carter, 1978; Bekenstein, 1987). We propose and explore a covariantization of the Beltrami equation, and in particular study its relation to helicity conservation. For instance, new terms can arise that vanish in the non-relativistic case, but which can contribute when we have a multi-component helicity system where the components are moving with relativistic velocities with respect to each other. This could be the case for helicity in relativistic plasmas (Lichnerowicz, 1967) ejected from astrophysical objects colliding with another plasma clouds (Goedbloed *et al.*, 2010), but it can also apply in particle physics to the hadronization process where relativistic flux tubes interact (Casher *et al.*, 1979; Neuberger, 1979; Casher *et al.*, 1980; Buniy & Kephart, 2003, 2005; Buniy *et al.*, 2014) or to the early universe as it cools through various epochs.

2. Non-covariant case

We first briefly review the three-dimensional Beltrami equation and helicity.

Consider a region $\Omega \subset \mathbb{R}^3$ and let (x, ∇) be the Cartesian coordinates and the derivative operator in Ω. A vector field B in Ω is called a Beltrami vector field if it satisfies

$$B \times (\nabla \times B) = 0, \tag{1}$$

where $\times$ is the vector product in $\mathbb{R}^3$. Eq. (1) implies that the vectors $\nabla \times B$ and B are parallel, which leads to the Beltrami equation

$$\nabla \times B = \lambda B, \tag{2}$$

Journal of Physics: Conference Series **544** (2014) 012008 doi:10.1088/1742-6596/544/1/012008

where λ is a scalar function in Ω.

Eq. (2) shows that Beltrami fields are eigenfields of the curl operator. Eigenfields of the curl operator can be related to more familiar functions by using the identity

$$\nabla \times \nabla \times B = \nabla(\nabla \cdot B) - \Delta B, \tag{3}$$

where $\cdot$ is the scalar product and Δ is the Laplace operator in $\mathbb{R}^3$. It follows that the square of the curl operator, when restricted to the space of divergence-free vector fields, is the negative of the Laplace operator $-\Delta$. Thus, in some sense, the curl operator is the square root of the operator $-\Delta$ (which itself is a positive operator).

The restriction to the space of divergence-free vector fields is not accidental, but is required by physical considerations of B being a magnetic field. In such a case, the divergence-free condition $\nabla \cdot B = 0$ for B in (2) implies

$$B \cdot \nabla\lambda = 0, \tag{4}$$

so that λ is constant along any field line of B. Eq. (4) is the consistency condition for (2). According to the Beltrami equation (2), the Maxwell current $J = \nabla \times B$ is parallel to the magnetic field, $J = \lambda B$. Note that the current conservation $\nabla \cdot J = 0$ also implies (4).

To learn more about a Beltrami field B, it is instructive to consider a vector potential A such that $B = \nabla \times A$. Since a vector potential is defined only up to the gradient of an arbitrary function, it will be important to ensure gauge invariance of various physical quantities under a gauge transformation

$$A \mapsto A + \nabla g, \tag{5}$$

where g is an arbitrary real-valued function in Ω. The two simplest such gauge invariant quantities are the energy W and helicity H of the field B in the region Ω,

$$W = \int_\Omega \tfrac{1}{2}\|B\|^2\, d^3x, \tag{6}$$

$$H = \int_\Omega A \cdot B\, d^3x, \tag{7}$$

where $\|\ \|$ is the scalar norm in $\mathbb{R}^3$.

Convergence of the integrals in (6) and (7) imposes certain restrictions on A and B. We are concerned here with the case of a non-compact Ω and restrictions derived from the required behavior of A and B for $\|x\| \to \infty$. In such a case, the integral in (7) converges if

$$A = O(\|x\|^p),\ \|x\| \to \infty,\ p < -1. \tag{8}$$

This leads to $B = O(\|x\|^{p-1})$, $\|x\| \to \infty$, which means that allowed field configurations do not include magnetic monopoles. It follows that the integral in (6) also converges for such fields.

The gauge invariance of the energy is obvious, while the corresponding gauge transformation of the helicity is

$$H \mapsto H + \int_{\partial\Omega} g(n \cdot B) d^2\sigma, \tag{9}$$

where $\partial\Omega$ is the boundary of Ω, n is the unit normal vector to $\partial\Omega$, and $d^2\sigma$ is the area differential on $\partial\Omega$. Since we require gauge invariance of H, we set the boundary condition

$$(n \cdot B)|_{\partial\Omega} = 0, \tag{10}$$

Journal of Physics: Conference Series **544** (2014) 012008 doi:10.1088/1742-6596/544/1/012008

which means that the field lines do not cross the boundary. For a non-compact Ω, the asymptotic behavior $B = O(\|x\|^{p-1})$, $\|x\| \to \infty$ ensures the gauge invariance of H as well since the boundary integral in (9) vanishes.

The helicity is often conserved in physical systems involving magnetic fields, and this restricts their dynamics. For example, suppose that B is a field in Ω satisfying the boundary condition (10) which minimizes its energy W and conserves its helicity H. The resulting variational problem is equivalent to finding B minimizing the functional

$$W - \tfrac{1}{2}\lambda H = \int_\Omega L\, d^3x \tag{11}$$

with the Lagrangian

$$L = \tfrac{1}{2}\|B\|^2 - \tfrac{1}{2}\lambda A \cdot B. \tag{12}$$

(We have chosen the form of the Lagrange multiplier λ which leads to the conventional form of the Beltrami equation.) For an infinitesimal variation of the vector potential δA, we find

$$\delta L = (\nabla \times B - \lambda B) \cdot \delta A + \nabla \cdot \big[(-B + \tfrac{1}{2}\lambda A) \times \delta A\big], \tag{13}$$

which leads to

$$\delta(W - \tfrac{1}{2}\lambda H) = \int_\Omega (\nabla \times B - \lambda B) \cdot \delta A\, d^3x + \int_{\partial\Omega} n \cdot \big[(-B + \tfrac{1}{2}\lambda A) \times \delta A\big]\, d^2\sigma. \tag{14}$$

We eliminate the boundary term in (14) by setting the boundary condition

$$\delta A|_{\partial\Omega} = 0. \tag{15}$$

The variation (14) vanishes for any δA satisfying (15) if and only if $\nabla \times B = \lambda B$. Thus, a Beltrami field with $\lambda =$ const is a stationary point of the energy functional W subject to the condition $H =$ const. It can be further proved that such a field is a local minimum of W with constant H.

Another aspect of the helicity relates to the conserved Noether current. Gauge transformations are the symmetry operations of the theory defined by the Lagrangian L. The proof of the invariance of the theory requires showing (without using the equation of motion) that L is changed only by the divergence term. (For the following derivation we assume λ is constant.) Indeed, for a gauge transformation (5) with $\delta A = \nabla g$, (13) becomes

$$\delta L = \nabla \cdot \Gamma, \tag{16}$$

$$\Gamma = -\tfrac{1}{2}\lambda g B. \tag{17}$$

On the other hand, using the equation of motion we find

$$\delta L = \nabla \cdot \left(\frac{\partial L}{\partial \nabla A_k} \delta A_k \right). \tag{18}$$

Equating (16) and (18), we arrive at the conserved Noether current ($\nabla \cdot j = 0$),

$$j = \frac{\partial L}{\partial \nabla A_k} \delta A_k - \Gamma, \tag{19}$$

$$j = (-B + \tfrac{1}{2}\lambda A) \times \nabla g + \tfrac{1}{2}\lambda g B. \tag{20}$$

Since the gauge function g is arbitrary, we can define another conserved Noether current k by

$$gk = j + \nabla \times \left[(-B + \tfrac{1}{2}\lambda A)g\right]. \tag{21}$$

and find

$$k = -\nabla \times B + \lambda B. \tag{22}$$

Now the Beltrami equation (2) gives $k = 0$, so that the Noether current associated with the helicity, λB, equals the Maxwell current $J = \nabla \times B$. As expected, there is only one independent conserved current in the problem.

Computing the Noether energy-momentum tensor

$$\begin{aligned} \theta^i{}_j &= \frac{\partial L}{\partial(\nabla_i A_k)} \nabla_j A_k - \delta^i{}_j L \\ &= (B_l - \tfrac{1}{2}\lambda A_l)\varepsilon^{lik}\nabla_j A_k - \delta^i{}_j\left(\tfrac{1}{2}\|B\|^2 - \tfrac{1}{2}\lambda A \cdot B\right), \end{aligned} \tag{23}$$

we find that its divergence

$$\nabla_i \theta^i{}_j = (-\nabla \times B + \lambda B)^k \nabla_j A_k \tag{24}$$

vanishes for any solution of the Beltrami equation (2), which implies the conservation equation $\nabla_i \theta^i{}_j = 0$.

We now derive the lower bound for the energy in terms of helicity and constant λ (Arnold & Khesin, 1978). We first integrate the Beltrami equation (2) once to obtain

$$\nabla \times A = \lambda A + \nabla \varphi, \tag{25}$$

where φ is an arbitrary scalar function in Ω. We can now use the gauge transformation (5) with $g = -\lambda^{-1}\varphi$ to replace (25) with

$$\nabla \times A = \lambda A, \tag{26}$$

which is of the same form as (2). Hence in this gauge $B = \lambda A$, which gives

$$W = \tfrac{1}{2}\lambda H. \tag{27}$$

Although (26) is not gauge invariant, its consequence, (27), is gauge invariant. We conclude that the minimal value of the variational functional $W - \frac{1}{2}\lambda H$ equals zero for any solution of the Beltrami equation with constant λ.

The field satisfying $B = \lambda A$ saturates the lower bound for the energy in terms of helicity (Arnold & Khesin, 1978). To derive this, we consider a non-local operator curl^{-1} acting on the space of divergence-free vector fields. We use the Schwarz inequality

$$\left| \int_\Omega B \cdot \mathrm{curl}^{-1} B \, d^3x \right| \leq \left[\int_\Omega \|B\|^2 \, d^3x \right]^{1/2} \left[\int_\Omega \|\mathrm{curl}^{-1} B\|^2 \, d^3x \right]^{1/2} \tag{28}$$

and the Poincaré inequality

$$\int_\Omega \|\mathrm{curl}^{-1} B\|^2 \, d^3x \leq C^{-2} \int_\Omega \|B\|^2 \, d^3x, \tag{29}$$

where $C > 0$ is a certain constant depending on Ω. Combination of the two inequalities gives $W \geq \frac{1}{2}C\,|H|$. Finally, using the Rayleigh min-max theorem

$$B \cdot \mathrm{curl}^{-1} B \leq |\mu|_{\max} \|B\|^2, \tag{30}$$

where

$$|\mu|_{\max} = \max_a |\mu_a|, \tag{31}$$

$$\mathrm{curl}^{-1} B_a = \mu_a B_a, \tag{32}$$

we see that we can use $C = 2\,|\mu|_{\max}^{-1}$ and find

$$W \geq \tfrac{1}{2}\,|\mu|_{\max}^{-1}\,|H|\,. \tag{33}$$

It is clear that the field satisfying $B = \lambda A$ saturates the bound (33) since in this case we have $|\mu|_{\max}^{-1} = |\lambda|$ and $W = \frac{1}{2}\,|\lambda|\,|H|$.

3. Covariant case

The proceeding non-covariant analysis is sufficient for the description of magnetic fields in nonrelativistic plasmas. Generalizations to the electric case have been carried out (Ranada, 1989) and applied (Irvine & Bouwmeester, 2008; Arrayas & Trueba, 2012), but relativistic plasmas require a full covariant analysis. In particular, this applies to Beltrami fields and helicity.

To derive the covariant forms of equations obtained in the preceding section (Buniy & Kephart, 2014), we consider Lorentzian $(\mathbb{R}^{1,3}, \Omega, x, \nabla)$, where $\mathbb{R}^{1,3}$ has a constant pseudo-Riemannian metric with signature $(1,3)$ and $\Omega = [t_1, t_2] \times \Omega'$, $\Omega' \subset \mathbb{R}^3$. The magnetic field B is now a part of the gauge field strength tensor $F_{\mu\nu}$. Using

$$B_i = \tfrac{1}{2}\varepsilon_{ijk}F^{jk}, \tag{34}$$

we first write (1), (2), (4) in the form

$$F_i{}^j \nabla^k F_{jk} = 0, \tag{35}$$

$$\nabla^j F_{ij} = \tfrac{1}{2}\lambda\varepsilon_{ijk}F^{jk}, \tag{36}$$

$$\varepsilon^{ijk}F_{jk}\nabla_i\lambda = 0, \tag{37}$$

then setting

$$E_i = F_{i0}, \tag{38}$$

$$\varepsilon_{0ijk} = \varepsilon_{ijk}, \tag{39}$$

$$\lambda_0 = \lambda, \tag{40}$$

we arrive at the covariant form of (35), (36), (37),

$$F_\alpha{}^\mu \nabla^\nu F_{\mu\nu} = 0, \tag{41}$$

$$\nabla^\nu F_{\mu\nu} = \tfrac{1}{2}\varepsilon_{\mu\nu\alpha\beta}\lambda^\nu F^{\alpha\beta}, \tag{42}$$

$$\varepsilon^{\mu\nu\alpha\beta}F_{\alpha\beta}\nabla_\mu\lambda_\nu = 0. \tag{43}$$

These covariant expressions require that we identify non-covariant λ with the time component of a 4-vector λ_μ. Note that the left-hand side of (43) vanishes identically if we set

$$\nabla_\mu\lambda_\nu - \nabla_\nu\lambda_\mu = 0, \tag{44}$$

Journal of Physics: Conference Series **544** (2014) 012008 doi:10.1088/1742-6596/544/1/012008

The requirement (44) will appear later in the variational formulation of the problem.

Similarly to (2) implying (4) and (4) not implying (2) for $\nabla \cdot B = 0$, we have (42) implying (43) and (43) not implying (42). However, although (1) and (2) are equivalent, their covariant counterparts (41) and (42) are not equivalent; in fact, none of the two implies the other.

In terms of the E and B fields, the time and space components of (41) become

$$E \cdot \nabla_0 E - E \cdot (\nabla \times B) = 0, \tag{45}$$

$$E(\nabla \cdot E) + B \times (\nabla_0 E) - B \times (\nabla \times B) = 0, \tag{46}$$

the time and space components of (42) become

$$-\nabla \cdot E = \lambda \cdot B, \tag{47}$$

$$-\nabla_0 E + \nabla \times B = \lambda_0 B + \lambda \times E, \tag{48}$$

and (43) becomes

$$-B \cdot \nabla\lambda_0 + B \cdot \nabla_0 \lambda + E \times (\nabla \times \lambda) = 0. \tag{49}$$

Note that the left-hand side of (49) vanishes identically if we set $\nabla\lambda_0 - \nabla_0\lambda = 0$ and $\nabla \times \lambda = 0$, which combine to give (44).

A consistency condition is required for compatibility of (41) and (42) for arbitrary λ. Indeed, combining these equations, we find

$$F^{\gamma\mu}\varepsilon_{\mu\nu\alpha\beta}\lambda^{\nu}F^{\alpha\beta} = 0. \tag{50}$$

Considering the values $\gamma = 0$ and $\gamma = i$ in (50), yields $\lambda^0 E \cdot B = 0$ and $\lambda^i E \cdot B = 0$, respectively. This requires the same consistency condition $E \cdot B = 0$ for each values of γ, which we write in the covariant form

$$\varepsilon^{\mu\nu\alpha\beta}F_{\mu\nu}F_{\alpha\beta} = 0. \tag{51}$$

We could have arrived at the consistency condition $E \cdot B = 0$ also by noting that it is an appropriate covariant form of the three-dimensional constraint $E = 0$.

The covariant analogue of the energy W is the negative of the Maxwell action

$$\begin{aligned} W &= \int_\Omega \tfrac{1}{4}F_{\mu\nu}F^{\mu\nu}d^4x, \\ &= \int_\Omega \left(-\tfrac{1}{2}\|E\|^2 + \tfrac{1}{2}\|B\|^2\right) d^4x, \end{aligned} \tag{52}$$

(We have introduced the sign difference in the definition of W so that the non-covariant W is a limiting case of the covariant W.) As a covariant form of the helicity H, we propose

$$\begin{aligned} H(f) &= -\int_\Omega \tfrac{1}{2}\varepsilon^{\mu\nu\alpha\beta}(\nabla_\mu f)A_\nu F_{\alpha\beta}\, d^4x \\ &= \int_\Omega \tfrac{1}{2}\big[(\nabla_0 f)(A \cdot B) - A_0(B \cdot \nabla f) - \nabla f \cdot (A \times E)\big] d^4x, \end{aligned} \tag{53}$$

where f is an arbitrary scalar function in Ω. (It will become clear in what follows why in (53) we use $\lambda_\mu = \nabla_\mu f$ instead of a general λ_μ.)

For a non-compact Ω, the integral in (53) converges if

$$A = O(\|x\|^p), \ \|x\| \to \infty, \ p < -1 - \tfrac{1}{2}q, \tag{54}$$

Journal of Physics: Conference Series **544** (2014) 012008 doi:10.1088/1742-6596/544/1/012008

where we assumed

$$f = O(\|x\|^q), \ \|x\| \to \infty \tag{55}$$

for a certain q. Since $F = O(\|x\|^{p-1})$, $\|x\| \to \infty$, the integral in (52) now converges for $p < -1$.

Our definition (53) is motivated by the following limiting case of covariant helicity $H(f)$. Suppose $\Omega = [t_1, t_2] \times \Omega'$, where $\Omega' \subset \mathbb{R}^3$, and f is a function of $x^0 = t$ only. It follows that

$$H(f) = \int_{t_1}^{t_2} H'(t)(\partial f/\partial t)dt, \tag{56}$$

where $H'(t)$ is the non-covariant helicity of the vector potential $A_i(t, x)$. In particular, for the conserved non-covariant helicity H', we find

$$H(f) = [f(t_2) - f(t_1)]\, H', \tag{57}$$

More generally, for an arbitrary f, (53) implies

$$H(f) = \tilde{H}(f) - \int_{\partial\Omega} \tfrac{1}{2}\varepsilon^{\mu\nu\alpha\beta} f A_\nu F_{\alpha\beta}\, d^3\sigma_\mu, \tag{58}$$

$$\tilde{H}(f) = \int_\Omega \tfrac{1}{4}\varepsilon^{\mu\nu\alpha\beta} f F_{\mu\nu} F_{\alpha\beta}\, d^4x, \tag{59}$$

which means that $H(f)$ is a boundary term when consistency condition (51) is satisfied.

Under a gauge transformation

$$A_\mu \mapsto A_\mu + \nabla_\mu g, \tag{60}$$

where g is an arbitrary real-valued function in Ω, the gauge invariance of W is obvious, while the corresponding gauge transformation of the helicity is

$$H(f) \mapsto H(f) + \int_{\partial\Omega} \tfrac{1}{2}\varepsilon^{\mu\nu\alpha\beta}(\nabla_\nu f) g F_{\alpha\beta}\, d^3\sigma_\mu. \tag{61}$$

Since we require gauge invariance of $H(f)$, we set

$$\left[\varepsilon^{\mu\nu\alpha\beta} n_\mu (\nabla_\nu f) F_{\alpha\beta}\right]_{\partial\Omega} = 0, \tag{62}$$

where n is the 4-vector normal to $\partial\Omega$. Using now (42) with $\lambda_\mu = \nabla_\mu f$, we find

$$(n^\mu \nabla^\nu F_{\mu\nu})_{\partial\Omega} = 0, \tag{63}$$

which is a covariant version of the boundary condition (10). For a non-compact Ω, the asymptotic behavior $F = O(\|x\|^{p-1})$, $\|x\| \to \infty$ ensures the gauge invariance of $H(f)$ as well, since the boundary integral in (61) vanishes.

In terms of the E and B fields, the boundary condition (63) becomes

$$[-n_0(\nabla f \cdot B) + (\nabla_0 f)(n \cdot B) + n \cdot (\nabla f \times E)]_{\partial\Omega} = 0. \tag{64}$$

In particular, for time-like and space-like hypersurface $\partial\Omega$ we have

$$(\nabla f \cdot B)_{\partial\Omega} = 0 \text{ for time-like } \partial\Omega, \tag{65}$$

$$[(\nabla_0 f)(n \cdot B) + n \cdot (\nabla f \times E)]_{\partial\Omega} = 0 \text{ for space-like } \partial\Omega. \tag{66}$$

We further emphasize the choice of the definition (53) by the following theorem.

Theorem 1. *Any extremal of the action functional* $W = \int_\Omega \frac{1}{4} F_{\mu\nu} F^{\mu\nu} d^4x$ *subject to the constraint* $\varepsilon^{\mu\nu\alpha\beta} F_{\mu\nu} F_{\alpha\beta} = 0$ *and the boundary condition* $\delta A|_{\partial\Omega} = 0$ *satisfies the covariant Beltrami equation* $\nabla^\nu F_{\mu\nu} = \frac{1}{2}\varepsilon_{\mu\nu\alpha\beta}\lambda^\nu F^{\alpha\beta}$ *for* $\lambda_\mu = \nabla_\mu f$*, where* $f = f(x)$ *is an arbitrary function.*

Proof. Any extremal of (52) subject to the local constraint (51) must be an extremal of the functional $W - \frac{1}{2}\tilde{H}(f)$, where $f(x)$ is a space-time dependent Lagrange multiplier (Elsgolc, 1961; Akhiezer, 1962). For an arbitrary variation of the gauge potential δA, we find

$$\delta(W - \tfrac{1}{2}\tilde{H}(f)) = \int_\Omega \left[-(\nabla_\mu F^{\mu\nu}) + \tfrac{1}{2}\varepsilon^{\mu\nu\alpha\beta}(\nabla_\mu f)F_{\alpha\beta}\right]\delta A_\nu \, d^4x + \int_{\partial\Omega} (F^{\mu\nu} - \tfrac{1}{2}\varepsilon^{\mu\nu\alpha\beta} f F_{\alpha\beta})\delta A_\nu \, d^3\sigma_\mu. \tag{67}$$

Using the boundary condition

$$\delta A|_{\partial\Omega} = 0, \tag{68}$$

we arrive at (42) with $\lambda_\mu = \nabla_\mu f$, which proves the theorem. □

Note that $\lambda_\mu = \nabla_\mu f$ derived in the proof implies (44), which we have already seen as a sufficient condition for (43) to be satisfied identically.

We now consider the covariant version of the conserved Noether current (Jackiw, 1985). Equations (11), (12), (16), (19), (20), (22) become

$$W - \tfrac{1}{2}\tilde{H}(f) = \int_\Omega L(f)\, d^4x, \tag{69}$$

$$L(f) = \tfrac{1}{4}F_{\mu\nu}F^{\mu\nu} - \tfrac{1}{8}\varepsilon^{\mu\nu\alpha\beta} f F_{\mu\nu}F_{\alpha\beta}, \tag{70}$$

$$\delta L(f) = 0, \tag{71}$$

$$j^\mu(f) = \frac{\partial L(f)}{\partial \nabla_\mu A_\nu}\delta A_\nu, \tag{72}$$

$$j^\mu(f) = \left(F^{\mu\nu} - \tfrac{1}{2}\varepsilon^{\mu\nu\alpha\beta} f F_{\alpha\beta}\right)\nabla_\nu g, \tag{73}$$

$$k^\mu(f) = -\nabla_\nu F^{\mu\nu} + \tfrac{1}{2}\varepsilon^{\mu\nu\alpha\beta}(\nabla_\nu f)F_{\alpha\beta}. \tag{74}$$

Now using the Beltrami equation (42), we find $k^\mu(f) = 0$, so that the Noether current associated with the helicity, $\frac{1}{2}\varepsilon^{\mu\nu\alpha\beta}(\nabla_\nu f)F_{\alpha\beta}$, equals the negative of the Maxwell current $J^\mu = -\nabla_\nu F^{\mu\nu}$. As expected, there is only one independent conserved current in the problem.

The Noether energy-momentum tensor is

$$\begin{aligned}\theta^\mu{}_\nu(f) &= \frac{\partial L(f)}{\partial(\nabla_\mu A_\sigma)}\nabla_\nu A_\sigma - \delta^\mu{}_\nu L(f), \\ &= (F^{\mu\sigma} - \tfrac{1}{2} f\varepsilon^{\mu\sigma\alpha\beta}F_{\alpha\beta})\nabla_\nu A_\sigma - \delta^\mu{}_\nu(\tfrac{1}{4}F_{\alpha\beta}F^{\alpha\beta} - \tfrac{1}{8}\varepsilon^{\alpha\beta\gamma\delta} f F_{\alpha\beta}F_{\gamma\delta})\end{aligned} \tag{75}$$

and the corresponding energy-momentum 4-vector is

$$P_\nu(f) = \int_{\Omega'} \theta^0{}_\nu(f)\, d^3x, \tag{76}$$

$$P_0(f) = -\int_{\Omega'} (\tfrac{1}{2}\|E\|^2 + \tfrac{1}{2}\|B\|^2)\, d^3x + \int_{\partial\Omega'} n\cdot(E + fB)A_0\, d^2\sigma, \tag{77}$$

$$P_i(f) = -\int_{\Omega'} (E \times B)_i \, d^3x + \int_{\partial\Omega'} n \cdot (E + fB) A_i \, d^2\sigma, \tag{78}$$

where we assumed $\Omega = [t_1, t_2] \times \Omega'$, $\Omega' \subset \mathbb{R}^3$. We set the boundary condition

$$n \cdot (E + fB)|_{\partial\Omega'} = 0 \tag{79}$$

and obtain the relation $P_\nu(f) = P_\nu(0)$ which is consitent with $L(f) - L(0)$ being a topological term. Also note that although $\theta^\mu{}_\nu(f)$ is not gauge invariant, the resulting $P_\nu(f)$ is.

To prove conservation of $\theta^\mu{}_\nu(f)$, we need to use the Beltrami equation. Indeed, in the expression

$$\nabla_\mu \theta^\mu{}_\nu(f) = \big((\nabla_\mu F^{\mu\sigma}) - \tfrac{1}{2}\varepsilon^{\mu\sigma\alpha\beta}(\nabla_\mu f) F_{\alpha\beta}\big)\nabla_\nu A_\sigma + \tfrac{1}{8}\varepsilon^{\alpha\beta\gamma\delta}(\nabla_\nu f) F_{\alpha\beta} F_{\gamma\delta}, \tag{80}$$

the first term on the right-hand side vanishes for any solution of (42) and the second term vanishes due to the constraint (51). Since (51) follows from (42), we conclude that the conservation equation $\nabla_\mu \theta^\mu{}_\nu(f) = 0$ holds for any solution of the Beltrami equation.

It is straightforward to carry out this analysis for non-abelian fields. See (Buniy & Kephart, 2014), where our results are also conveniently expressed in terms of differential forms.

4. Conclusions

We have generalized the magnetic helicity and Beltrami equation to the relativistic form. In the process, we discussed various interconnected features associated with this generalization. In particular, we found that the helicity is related to the Chern-Simons action and can also be viewed as a constraint requiring the vanishing of a generalized instanton term.

Besides its theoretical appeal, the covariant formulation of the magnetic helicity and Beltrami equation has an experimental advantage as well. It turns out that, for an ideal nonrelativistic plasma, charges flow until the electric field is completely shorted out. In the relativistic case, even for an ideal plasma, however, the current flow may not be able to keep up, and so the electric fields do not necessarily always vanish. Some possible applications of our results for the relativistic generalization of the Beltrami equation may be found for dynamos inside millisecond pulsars, pulsar and quasar atmospheres, collisions of plasma shock waves with other shocks or gas clouds and nuclear fusion via laser confinement.

Explicit solutions of the covariant Beltrami equation are of particular interest for applications, and we will address these elsewhere.

Acknowledgments

We thank Keith Moffatt for pointing out the early work by Carter on generalizing helicity. We are grateful to the Isaac Newton institute for hospitality while this work was being carried out. RVB thanks Chapman University for support. The work of TWK was supported by U.S. DoE grant number DE-FG05-85ER40226 and Vanderbilt University College of Arts and Sciences.

References

Akhiezer, N. I. 1962 **The Calculus of Variations**. Blaisdell, New York.

Arnold, V. I. & Khesin, B. A. 1978 **Topological Methods in Hydrodynamics**. Springer, New York.

Arrayas, M. & Trueba, J. L. 2012 Exchange of helicity in a knotted electromagnetic field. *Annalen Phys.* **524**, 71–75.

Bekenstein, J. D. 1987 Helicity conservation laws for fluids and plasmas. *Astroph. J.* **319**, 207.

Bellan, P. M. 2000 **Spheromaks: a Practical Application of Magnetohydrodynamic Dynamos and Plasma Self-Organization**. Imperial College Press, London, UK.

Bellan, P. M. 2006 **Fundamentals of Plasma Physics**. Cambridge University Press, Cambridge, UK.

Buniy, R. V., Cantarella, J., Kephart, T. W. & Rawdon, E. 2014 The tight knot spectrum in QCD. *Phys.Rev.* **D89**, 054513.

Buniy, R. V. & Kephart, T. W. 2003 A model of glueballs. *Phys. Lett.* **B576**, 127–134.

Buniy, R. V. & Kephart, T. W. 2005 Glueballs and the universal energy spectrum of tight knots and links. *Int. J. Mod. Phys.* **A20**, 1252–1259.

Buniy, R. V. & Kephart, T. W. 2014 Generalized helicity and Beltrami fields. *Annals Phys.* **344**, 179–193.

Carter, B. 1978 Perfect fluid and magnetic field conservation laws in the theory of black hole accretion rings. In *Active Galactic Nuclei* (ed. C. Hazard & S. Mitton), p. 273. Cambridge University Press, Cambridge, UK.

Casher, A., Neuberger, H. & Nussinov, S. 1979 Chromoelectric Flux Tube Model of Particle Production. *Phys. Rev.* **D20**, 179–188.

Casher, A., Neuberger, H. & Nussinov, S. 1980 Multiparticle Production by Bubbling Flux Tubes. *Phys. Rev.* **D21**, 1966.

Elsgolc, L. E. 1961 **Calculus of Variations**. Pergamon Press, Oxford, UK.

Goedbloed, J. P., Keppens, R. & Poedts, S. 2010 **Advanced magnetohydrodynamics: with applications to laboratory and astrophysical plasmas**. Cambridge University Press, Cambridge, UK.

Irvine, W. T. M. & Bouwmeester, D. 2008 Linked and knotted beams of light. *Nature Phys.* **4**, 716.

Jackiw, R. 1985 Topological investigations of quantized gauge theories. In *Current Algebra and Anomalies* (ed. S. B. Treiman, R. Jackiw, B. Zumino & E. Witten). World Scientific, Singapore.

Lichnerowicz, A. 1967 **Relativistic hydrodynamics and magnetohydrodynamics: lectures on the existence of solutions**. W.A. Benjamin, New York, NY.

Marsh, G. E. 1996 **Force-Free Magnetic Fields: Solutions, Topology and Applications**. World Scientific, Singapore, River Edge, NJ.

Moffatt, H. K. 1969 The degree of knottedness of tangled vortex lines. *J. Fluid Mech.* **35**, 117.

Moffatt, H. K. 1978 **Magnetic Field Generation in Electrically Conducting Fluids**. Cambridge University Press, Cambridge, UK.

Moffatt, H. K. 1985 Magnetostatic equilibria and analogous Euler flows of arbitrary complex topology. Part 1. Fundamentals. *J. Fluid Mech.* **159**, 359.

Neuberger, H. 1979 Finite Time Corrections to the Chromoelectric Flux Tube Model. *Phys. Rev.* **D20**, 2936.

Ranada, A. F. 1989 A Topological Theory of the Electromagnetic Field. *Lett. Math. Phys.* **18**, 97–106.

Taylor, J. B. 1974 Relaxation of toroidal plasma and generation of reverse magnetic fields. *Phys. Rev. Lett.* **33**, 1139.

Taylor, J. B. 1986 Relaxation and magnetic reconnection in plasmas. *Rev. Mod. Phys.* **58**, 741.

Woltier, L. 1958 A theorem on force-free magnetic fields. *Proc. Nat. Acad. Sci.* **44**, 489.

Journal of Physics: Conference Series **544** (2014) 012009 doi:10.1088/1742-6596/544/1/012009

Variational analysis of topological stationary barotropic MHD in the case of single-valued magnetic surfaces

A. Yahalom[1,2]

[1] Ariel University, Ariel 40700, Israel

[2] Isaac Newton Institute for Mathematical Sciences, 20 Clarkson Road, Cambridge CB3 0EH, United Kingdom

E-mail: asya@ariel.ac.il

Abstract. Variational principles for magnetohydrodynamics have been introduced by previous authors both in Lagrangian and Eulerian form. Yahalom & Lynden-Bell (2008) have previously introduced simpler Eulerian variational principles from which all the relevant equations of barotropic magnetohydrodynamics can be derived. These variational principles were given in terms of six independent functions for non-stationary barotropic flows with given topologies and three independent functions for stationary barotropic flows. This is less then the seven variables which appear in the standard equations of barotropic magnetohydrodynamics which are the magnetic field $\vec{B}$ the velocity field $\vec{v}$ and the density ρ. Later, Yahalom (2010) introduced a simpler variational principle in terms of four functions for non-stationary barotropic magnetohydrodynamics.

It was shown that the above variational principles are also relevant for flows of non-trivial topologies and in fact using those variational variables one arrives at additional topological conservation laws in terms of cuts of variables which have close resemblance to the Aharonov-Bohm phase (Yahalom (2013)).

In previous examples (Yahalom & Lynden-Bell (2008); Yahalom (2013)) the magnetic field lines with non-trivial topology were at the intersection of two surface one of which was always multivalued; in this paper an example is introduced in which the magnetic helicity is not zero yet both surfaces are single-valued.

1. Introduction

Variational principles for magnetohydrodynamics have been introduced by previous authors both in Lagrangian and Eulerian form. Sturrock (1994) has discussed in his book a Lagrangian variational formalism for magnetohydrodynamics. Vladimirov & Moffatt (1994) in a series of papers have discussed an Eulerian variational principle for incompressible magnetohydrodynamics. However, their variational principle contained three more functions in addition to the seven variables which appear in the standard equations of magnetohydrodynamics which are the magnetic field $\vec{B}$ the velocity field $\vec{v}$ and the density ρ. Kats (2003) has generalized Moffatt's work for compressible non-barotropic flows but without reducing the number of functions and the computational load. Sakurai (1979) has introduced a two-function Eulerian variational principle for force-free magnetohydrodynamics and used it as a basis of a numerical scheme; his method is discussed by Sturrock (1994) . Yahalom & Lynden-Bell (2008) have combined the Lagrangian of Sturrock (1994) with the Lagrangian of Sakurai (1979) to obtain an **Eulerian** variational principle which depends on only six functions for non-stationary flows. The variational derivative of this Lagrangian result in all the equations needed to describe barotropic magnetohydrodynamics without any additional constraints. The equations obtained resemble the Hamiltonian equations of Frenkel *et al.* (1982). Later it was shown by Yahalom (2010) that only four functions will suffice for barotropic magnetohydrodynamic non-stationary flows. Furthermore, it was shown that for stationary flows three functions will suffice in order to describe a Lagrangian principle for barotropic magnetohydrodynamics.

The non-singlevaluedness of the functions appearing in the reduced representation of barotropic magnetohydrodynamics was shown in Yahalom (2013) to be an essential feature of the case of magnetohydrodynamics with non trivial topology. In particular the discontinuities of ν and ζ were shown to be conserved and equal to the cross helicity per unit magnetic flux and the magnetic helicity per unit magnetic flux respectively.

Still the question remains what about the single valuedness of the rest of the variational functions. In the variational formalism the magnetic field is represented by Sakurai (1979) representation:

$$\vec{B} = \vec{\nabla}\chi \times \vec{\nabla}\eta \tag{1}$$

In previous examples of fields of non trivial topology which could be represented that way given in Yahalom & Lynden-Bell (2008) the function χ was single valued and the function η was multiple valued. Here we suggest an example in which both surfaces are single valued.

The plan of this paper is as follows: first we introduce the standard notations and equations of barotropic magnetohydrodynamics. Next we define the magnetic Load and Metage. This is followed by the derivation of a variational principle for stationary magnetohydrodynamics. Finally an example is given for a magnetic field of non-zero magnetic helicity which lies on the intersection of two single valued surfaces.

2. Basic equations

The standard set of equations solved for barotropic magnetohydrodynamics are given below:

$$\frac{\partial \vec{B}}{\partial t} = \vec{\nabla} \times (\vec{v} \times \vec{B}), \tag{2}$$

$$\vec{\nabla} \cdot \vec{B} = 0, \tag{3}$$

$$\frac{\partial \rho}{\partial t} + \vec{\nabla} \cdot (\rho \vec{v}) = 0, \tag{4}$$

$$\rho \frac{d\vec{v}}{dt} = \rho(\frac{\partial \vec{v}}{\partial t} + (\vec{v} \cdot \vec{\nabla})\vec{v}) = -\vec{\nabla} p(\rho) + \frac{(\vec{\nabla} \times \vec{B}) \times \vec{B}}{4\pi}. \tag{5}$$

The following notations are utilized: $\frac{\partial}{\partial t}$ is the temporal derivative, $\frac{d}{dt}$ is the temporal material derivative and $\vec{\nabla}$ has its standard meaning in vector calculus. $\vec{B}$ is the magnetic field vector, $\vec{v}$ is the velocity field vector and ρ is the fluid density. Finally $p(\rho)$ is the pressure which we assume depends on the density alone (barotropic case). The justification for those equations and the conditions under which they apply can be found in standard books on magnetohydrodynamics (see for example Sturrock (1994)). Equation (2) describes the fact that the magnetic field lines are moving with the fluid elements ("frozen" magnetic field lines), equation (3) describes the fact that the magnetic field is solenoidal, equation (4) describes the conservation of mass and equation (5) is the Euler equation for a fluid in which both pressure and Lorentz magnetic forces apply. The term:

$$\vec{J} = \frac{\vec{\nabla} \times \vec{B}}{4\pi}, \tag{6}$$

is the electric current density which is not connected to any mass flow. The number of independent variables for which one needs to solve is seven $(\vec{v}, \vec{B}, \rho)$ and the number of equations (2,4,5) is also seven. Notice that equation (3) is a condition on the initial $\vec{B}$ field and is satisfied automatically for any other time due to equation (2). Also notice that $p(\rho)$ is not a variable rather it is a given function of ρ.

3. The Load and Metage

In this section we define the Load and Metage which are quantities which are useful for defining variational variables. The treatment in this section will follow closely an analogous treatment for non-magnetic fluid dynamics given by Lynden-Bell & Katz (1981). We assume here a cylindrical or spherical topology, for more general topologies see Yahalom & Lynden-Bell (2008). Consider a thin tube surrounding a magnetic field line as described in figure 1:

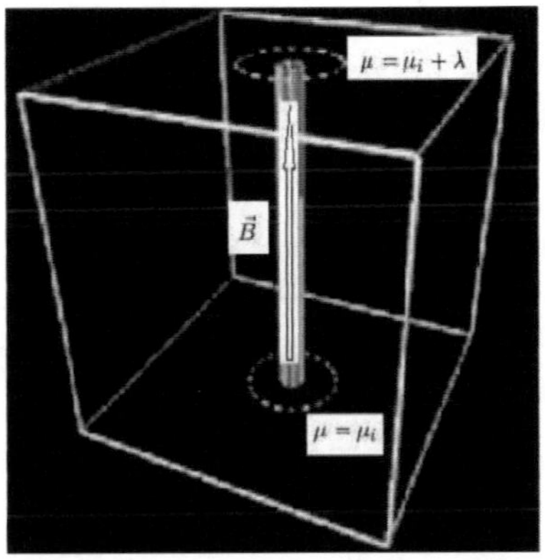

Figure 1. A thin tube surrounding a magnetic field line.

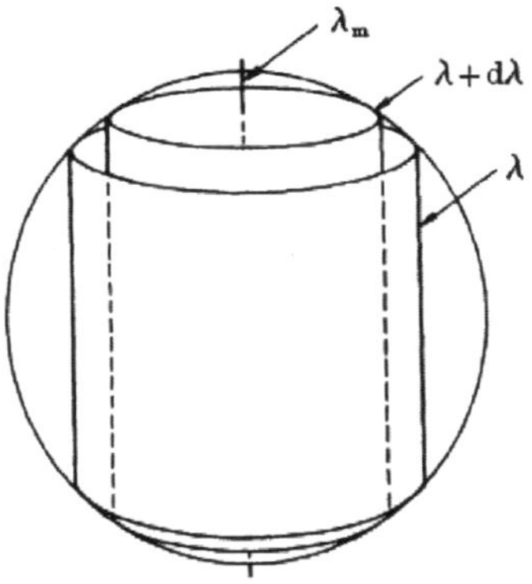

Figure 2. Surfaces of constant load.

The magnetic flux contained within the tube is:

$$\Delta\Phi = \int \vec{B} \cdot d\vec{S} \tag{7}$$

and the mass contained with the tube is:

$$\Delta M = \int \rho d\vec{l} \cdot d\vec{S}, \tag{8}$$

in which dl is a length element along the tube. Since the magnetic field lines move with the flow by virtue of equation (2) both the quantities $\Delta\Phi$ and ΔM are conserved and since the tube is thin we may define the conserved magnetic load:

$$\lambda = \frac{\Delta M}{\Delta\Phi} = \oint \frac{\rho}{B} dl, \tag{9}$$

in which the above integral is performed along the field line. Obviously the parts of the line which go out of the flow to regions in which $\rho = 0$ have a null contribution to the integral. Notice that λ is a **single valued** function that can be measured in principle. Since λ is conserved it satisfies the equation:

$$\frac{d\lambda}{dt} = 0. \tag{10}$$

By construction surfaces of constant magnetic load move with the flow and contain magnetic field lines. Hence the gradient to such surfaces must be orthogonal to the field line:

$$\vec{\nabla}\lambda \cdot \vec{B} = 0. \tag{11}$$

Now consider an arbitrary comoving point on the magnetic field line and denote it by i, and consider an additional comoving point on the magnetic field line and denote it by r. The integral:

$$\mu(r) = \int_i^r \frac{\rho}{B} dl + \mu(i), \tag{12}$$

is also a conserved quantity which we may denote following Lynden-Bell & Katz (1981) as the magnetic metage. $\mu(i)$ is an arbitrary number which can be chosen differently for each magnetic line. By construction:

$$\frac{d\mu}{dt} = 0. \tag{13}$$

Also it is easy to see that by differentiating along the magnetic field line we obtain:

$$\vec{\nabla}\mu \cdot \vec{B} = \rho. \tag{14}$$

Notice that μ will be generally a **non single valued** function. At this point we have two comoving coordinates of flow, namely λ, μ obviously in a three dimensional flow we also have a third coordinate. However, before defining the third coordinate we will find it useful to work not directly with λ but with a function of λ. Now consider the magnetic flux within a surface of constant load $\Phi(\lambda)$ as described in figure 2 (the figure was given by Lynden-Bell & Katz (1981)). The magnetic flux is a conserved quantity and depends only on the load λ of the surrounding surface. Now we define the quantity:

$$\chi = \frac{\Phi(\lambda)}{2\pi}. \tag{15}$$

Obviously χ satisfies the equations:

$$\frac{d\chi}{dt} = 0, \qquad \vec{B} \cdot \vec{\nabla}\chi = 0, \tag{16}$$

we will immediately show that this function is identical to Sakurai's function defined in equation (1). Let us now define an additional comoving coordinate η^* since $\vec{\nabla}\mu$ is not orthogonal to the $\vec{B}$ lines we can choose $\vec{\nabla}\eta^*$ to be orthogonal to the $\vec{B}$ lines and not be in the direction of the $\vec{\nabla}\chi$ lines, that is we choose η^* not to depend only on χ. Since both $\vec{\nabla}\eta^*$ and $\vec{\nabla}\chi$ are orthogonal to $\vec{B}$, $\vec{B}$ must take the form:

$$\vec{B} = A\vec{\nabla}\chi \times \vec{\nabla}\eta^*. \tag{17}$$

However, using equation (3) we have:

$$\vec{\nabla} \cdot \vec{B} = \vec{\nabla}A \cdot (\vec{\nabla}\chi \times \vec{\nabla}\eta^*) = 0. \tag{18}$$

Which implies that A is a function of χ, η^*. Now we can define a new comoving function η such that:

$$\eta = \int_0^{\eta^*} A(\chi, \eta^{'*}) d\eta^{'*}, \qquad \frac{d\eta}{dt} = 0. \tag{19}$$

In terms of this function we recover the Sakurai presentation defined in equation (1):

$$\vec{B} = \vec{\nabla}\chi \times \vec{\nabla}\eta. \tag{20}$$

Journal of Physics: Conference Series **544** (2014) 012009 doi:10.1088/1742-6596/544/1/012009

Hence we have shown how χ, η can be constructed for a known $\vec{B}, \rho$. Notice however, that η is defined in a non unique way since one can redefine η for example by performing the following transformation: $\eta \rightarrow \eta + f(\chi)$ in which $f(\chi)$ is an arbitrary function. The comoving coordinates χ, η serve as labels of the magnetic field lines. Moreover the magnetic flux can be calculated as:

$$\Phi = \int \vec{B} \cdot d\vec{S} = \int d\chi d\eta. \tag{21}$$

In the case that the surface integral is performed inside a load contour we obtain:

$$\Phi(\lambda) = \int_{\lambda} d\chi d\eta = \chi \int_{\lambda} d\eta = \left\{ \begin{array}{c} \chi[\eta] \\ \chi(\eta_{max} - \eta_{min}) \end{array} \right. \tag{22}$$

There are two cases involved; in one case the load surfaces are topological cylinders, in this case η is not single valued and hence we obtain the upper value for $\Phi(\lambda)$. In a second case the load surfaces are topological spheres, in this case η is single valued and has minimal η_{min} and maximal η_{max} values. Hence the lower value of $\Phi(\lambda)$ is obtained. For example in some cases η is identical to twice the latitude angle θ. In those cases $\eta_{min} = 0$ (value at the "north pole") and $\eta_{max} = 2\pi$ (value at the "south pole").

Comparing the above equation with equation (15) we derive that η can be either **single valued** or **not single valued** and that its discontinuity across its cut in the non single valued case is $[\eta] = 2\pi$.

4. Simplified variational principle of stationary barotropic magnetohydrodynamics

Consider equation (16), for a stationary flow it takes the form:

$$\vec{v} \cdot \vec{\nabla} \chi = 0. \tag{23}$$

Hence $\vec{v}$ can take the form:

$$\vec{v} = \frac{\vec{\nabla}\chi \times \vec{K}}{\rho}. \tag{24}$$

However, the velocity field must satisfy the stationary mass conservation equation (4):

$$\vec{\nabla} \cdot (\rho \vec{v}) = 0. \tag{25}$$

We see that a sufficient condition (although not necessary) for $\vec{v}$ to solve equation (25) is that $\vec{K}$ takes the form $\vec{K} = \vec{\nabla} N$, where N is an arbitrary function. Thus, $\vec{v}$ may take the form:

$$\vec{v} = \frac{\vec{\nabla}\chi \times \vec{\nabla} N}{\rho}. \tag{26}$$

Let us now calculate $\vec{v} \times \vec{B}$ in which $\vec{B}$ is given by Sakurai's presentation equation (1):

$$\begin{aligned} \vec{v} \times \vec{B} &= \left(\frac{\vec{\nabla}\chi \times \vec{\nabla} N}{\rho}\right) \times (\vec{\nabla}\chi \times \vec{\nabla}\eta) \\ &= \frac{1}{\rho} \vec{\nabla}\chi (\vec{\nabla}\chi \times \vec{\nabla} N) \cdot \vec{\nabla}\eta. \end{aligned} \tag{27}$$

Since the flow is stationary N can be at most a function of the three comoving coordinates $\chi, \mu, \bar{\eta} = \eta + t$[1], hence:

$$\vec{\nabla} N = \frac{\partial N}{\partial \chi} \vec{\nabla}\chi + \frac{\partial N}{\partial \mu} \vec{\nabla}\mu + \frac{\partial N}{\partial \bar{\eta}} \vec{\nabla}\bar{\eta}. \tag{28}$$

[1] While the physical variables are assumed to be stationary the potential need not be stationary, for more on this see Yahalom & Lynden-Bell (2008)

Inserting equation (28) into equation (27) will yield:

$$\vec{v} \times \vec{B} = \frac{1}{\rho} \vec{\nabla} \chi \frac{\partial N}{\partial \mu} (\vec{\nabla} \chi \times \vec{\nabla} \mu) \cdot \vec{\nabla} \bar{\eta}. \tag{29}$$

Rearranging terms and using Sakurai's presentation equation (1) we can simplify the above equation and obtain:

$$\vec{v} \times \vec{B} = -\frac{1}{\rho} \vec{\nabla} \chi \frac{\partial N}{\partial \mu} (\vec{\nabla} \mu \cdot \vec{B}). \tag{30}$$

However, using equation (14) this will simplify to the form:

$$\vec{v} \times \vec{B} = -\vec{\nabla} \chi \frac{\partial N}{\partial \mu}. \tag{31}$$

Now let us consider equation (2); for stationary flows this will take the form:

$$\vec{\nabla} \times (\vec{v} \times \vec{B}) = 0. \tag{32}$$

Inserting equation (31) into equation (32) will lead to the equation:

$$\vec{\nabla} (\frac{\partial N}{\partial \mu}) \times \vec{\nabla} \chi = 0. \tag{33}$$

However, since N is at most a function of $\chi, \mu, \bar{\eta}$ it follows that $\frac{\partial N}{\partial \mu}$ is some function of χ:

$$\frac{\partial N}{\partial \mu} = -F(\chi). \tag{34}$$

This can be easily integrated to yield:

$$N = -\mu F(\chi) + G(\chi, \bar{\eta}). \tag{35}$$

Inserting this back into equation (26) will yield:

$$\vec{v} = \frac{\vec{\nabla} \chi \times (-F(\chi) \vec{\nabla} \mu + \frac{\partial G}{\partial \bar{\eta}} \vec{\nabla} \bar{\eta})}{\rho}. \tag{36}$$

Let us now replace the set of variables $\chi, \bar{\eta}$ with a new set $\chi', \bar{\eta}'$ such that:

$$\chi' = \int F(\chi) d\chi, \qquad \bar{\eta}' = \frac{\bar{\eta}}{F(\chi)}. \tag{37}$$

This will not have any effect on the Sakurai representation given in equation (1) since:

$$\vec{B} = \vec{\nabla} \chi \times \vec{\nabla} \eta = \vec{\nabla} \chi \times \vec{\nabla} \bar{\eta} = \vec{\nabla} \chi' \times \vec{\nabla} \bar{\eta}'. \tag{38}$$

However, the velocity will have a simpler representation and will take the form:

$$\vec{v} = \frac{\vec{\nabla} \chi' \times \vec{\nabla}(-\mu + G'(\chi', \bar{\eta}'))}{\rho}, \tag{39}$$

in which $G' = \frac{G}{F}$. At this point one should remember that μ was defined in equation (12) up to an arbitrary constant which can vary between magnetic field lines. Since the lines are labelled by their $\chi', \bar{\eta}'$ values it follows that we can add an arbitrary function of $\chi', \bar{\eta}'$ to μ without effecting its properties. Hence we can define a new μ' such that:

$$\mu' = \mu - G'(\chi', \bar{\eta}'). \tag{40}$$

Notice that μ' can be multi-valued. Inserting equation (40) into equation (39) will lead to a simplified equation for $\vec{v}$:

$$\vec{v} = \frac{\vec{\nabla}\mu' \times \vec{\nabla}\chi'}{\rho}. \tag{41}$$

In the following the primes on $\chi, \mu, \bar{\eta}$ will be ignored. The above equation is analogous to Vladimirov & Moffatt (1994) equation 7.11 for incompressible flows, in which our μ and χ play the part of their A and Ψ. It is obvious that $\vec{v}$ satisfies the following set of equations:

$$\vec{v} \cdot \vec{\nabla}\mu = 0, \qquad \vec{v} \cdot \vec{\nabla}\chi = 0, \qquad \vec{v} \cdot \vec{\nabla}\bar{\eta} = 1, \tag{42}$$

to derive the right hand equation we have used both equation (13) and equation (1). Hence μ, χ are both comoving and stationary. It can be easily seen that if:

$$basis = (\vec{\nabla}\chi, \vec{\nabla}\bar{\eta}, \vec{\nabla}\mu), \tag{43}$$

is a local vector basis at any point in space than their exists a dual basis:

$$dual\ basis = \frac{1}{\rho}(\vec{\nabla}\bar{\eta} \times \vec{\nabla}\mu, \vec{\nabla}\mu \times \vec{\nabla}\chi, \vec{\nabla}\chi \times \vec{\nabla}\bar{\eta}) = (\frac{\vec{\nabla}\bar{\eta} \times \vec{\nabla}\mu}{\rho}, \vec{v}, \frac{\vec{B}}{\rho}). \tag{44}$$

Such that:

$$basis_i \cdot dual\ basis_j = \delta_{ij}, \qquad i, j \in [1, 2, 3], \tag{45}$$

in which δ_{ij} is Kronecker's delta. Hence while the surfaces $\chi, \mu, \bar{\eta}$ generate a local vector basis for space, the physical fields of interest $\vec{v}, \vec{B}$ are part of the dual basis. By vector multiplying $\vec{v}$ and $\vec{B}$ and using equations (41,1) we obtain:

$$\vec{v} \times \vec{B} = \vec{\nabla}\chi, \tag{46}$$

this means that both $\vec{v}$ and $\vec{B}$ lie on χ surfaces and provide a vector basis for this two dimensional surface. The above equation can be compared with Vladimirov & Moffatt (1994) equation 5.6 for incompressible flows in which their J is analogue to our χ.

5. The action principle

In the previous subsection we have shown that if the velocity field $\vec{v}$ is given by equation (41) and the magnetic field $\vec{B}$ is given by the Sakurai representation equation (1) than equations (2,3,4) are satisfied automatically for stationary flows. To complete the set of equations we will show how the Euler equations (5) can be derived from the action given in:

$$\begin{aligned} A &\equiv \int \mathcal{L} d^3x dt, \\ \mathcal{L} &\equiv \rho(\frac{1}{2}\vec{v}^2 - \varepsilon(\rho)) - \frac{\vec{B}^2}{8\pi}, \end{aligned} \tag{47}$$

in which both $\vec{v}$ and $\vec{B}$ are given by equation (41) and equation (1) respectively and the density ρ is given by equation (13):

$$\rho = \vec{\nabla}\mu \cdot \vec{B} = \vec{\nabla}\mu \cdot (\vec{\nabla}\chi \times \vec{\nabla}\eta) = \frac{\partial(\chi, \eta, \mu)}{\partial(x, y, z)}. \tag{48}$$

In this case the Lagrangian density of equation (47) will take the form:

$$\mathcal{L} = \rho(\frac{1}{2}(\frac{\vec{\nabla}\mu \times \vec{\nabla}\chi}{\rho})^2 - \varepsilon(\rho)) - \frac{(\vec{\nabla}\chi \times \vec{\nabla}\eta)^2}{8\pi} \tag{49}$$

and can be seen explicitly to depend on only three functions. Let us make arbitrary small variations $\delta\alpha_i = (\delta\chi, \delta\eta, \delta\mu)$ of the functions $\alpha_i = (\chi, \eta, \mu)$. Let us define a Δ variation that does not modify the α_i's, such that:

$$\Delta\alpha_i = \delta\alpha_i + (\vec{\xi} \cdot \vec{\nabla})\alpha_i = 0, \tag{50}$$

in which $\vec{\xi}$ is the Lagrangian displacement, thus:

$$\delta\alpha_i = -\vec{\nabla}\alpha_i \cdot \vec{\xi}. \tag{51}$$

Which will lead to the equation:

$$\vec{\xi} \equiv -\frac{\partial\vec{r}}{\partial\alpha_i}\delta\alpha_i. \tag{52}$$

Making a variation of ρ given in equation (48) with respect to α_i will yield:

$$\delta\rho = -\vec{\nabla} \cdot (\rho\vec{\xi}) \tag{53}$$

Furthermore, taking the variation of $\vec{B}$ given by Sakurai's representation (1) with respect to α_i will yield:

$$\delta\vec{B} = \vec{\nabla} \times (\vec{\xi} \times \vec{B}). \tag{54}$$

It remains to calculate $\delta\vec{v}$ by varying equation (41) this will yield:

$$\delta\vec{v} = -\frac{\delta\rho}{\rho}\vec{v} + \frac{1}{\rho}\vec{\nabla} \times (\rho\vec{\xi} \times \vec{v}). \tag{55}$$

Inserting equations (53,54,55) into:

$$\begin{aligned} \delta A &= \int \delta\mathcal{L} d^3x dt, \\ \delta\mathcal{L} &= \delta\rho(\frac{1}{2}\vec{v}^2 - w(\rho)) + \rho\vec{v} \cdot \delta\vec{v} - \frac{\vec{B} \cdot \delta\vec{B}}{4\pi}, \end{aligned} \tag{56}$$

will yield:

$$\begin{aligned} \delta\mathcal{L} &= \vec{v} \cdot \vec{\nabla} \times (\rho\vec{\xi} \times \vec{v}) - \frac{\vec{B} \cdot \vec{\nabla} \times (\vec{\xi} \times \vec{B})}{4\pi} - \delta\rho(\frac{1}{2}\vec{v}^2 + w) \\ &= \vec{v} \cdot \vec{\nabla} \times (\rho\vec{\xi} \times \vec{v}) - \frac{\vec{B} \cdot \vec{\nabla} \times (\vec{\xi} \times \vec{B})}{4\pi} + \vec{\nabla} \cdot (\rho\vec{\xi})(\frac{1}{2}\vec{v}^2 + w). \end{aligned} \tag{57}$$

Using the well known vector identity:

$$\vec{A} \cdot \vec{\nabla} \times (\vec{C} \times \vec{A}) = \vec{\nabla} \cdot ((\vec{C} \times \vec{A}) \times \vec{A}) + (\vec{C} \times \vec{A}) \cdot \vec{\nabla} \times \vec{A} \tag{58}$$

and the theorem of Gauss we can write now equation (56) in the form:

$$\begin{aligned} \delta A &= \int dt\{\oint d\vec{S} \cdot [\rho(\vec{\xi} \times \vec{v}) \times \vec{v} - \frac{(\vec{\xi} \times \vec{B}) \times \vec{B}}{4\pi} + (\frac{1}{2}\vec{v}^2 + w)\rho\vec{\xi}] \\ &+ \int d^3x\vec{\xi} \cdot [\rho\vec{v} \times \vec{\omega} + \vec{J} \times \vec{B} - \rho\vec{\nabla}(\frac{1}{2}\vec{v}^2 + w)]\}. \end{aligned} \tag{59}$$

The time integration is of course redundant in the above expression. Also notice that we have used the current definition equation (6) and the vorticity definition $\vec{\omega} = \vec{\nabla} \times \vec{v}$. Suppose now that $\delta A = 0$ for a $\vec{\xi}$ such that the boundary term (including both the boundary of the domain and relevant cuts) in the above equation is null but that $\vec{\xi}$ is otherwise arbitrary, then it entails the equation:

$$\rho \vec{v} \times \vec{\omega} + \vec{J} \times \vec{B} - \rho \vec{\nabla}(\frac{1}{2}\vec{v}^2 + w) = 0. \tag{60}$$

Using the well known vector identity:

$$\frac{1}{2}\vec{\nabla}(\vec{v}^2) = (\vec{v} \cdot \vec{\nabla})\vec{v} + \vec{v} \times (\vec{\nabla} \times \vec{v}) \tag{61}$$

and rearranging terms we recover the stationary Euler equation:

$$\rho(\vec{v} \cdot \vec{\nabla})\vec{v} = -\vec{\nabla} p + \vec{J} \times \vec{B}. \tag{62}$$

6. The Azimuthal Magnetic Star

At this penultimate section let us consider a spherical star of radius R and uniform density ρ, furthermore we assume that the star contains an azimuthal magnetic field:

$$\vec{B} = B_0 \sin(\theta)\hat{\phi} \tag{63}$$

(which need not be azimuthal outside the star) in which $\hat{\phi}$ is a unit vector along the azimuthal direction, θ is the zenith angle and B_0 is constant. The vector potential $\vec{A}$ can be taken as:

$$\vec{A} = \frac{1}{2}B_0 r \sin(\theta)\hat{\theta} + \frac{\alpha}{r\sin\theta}\hat{\phi}, \tag{64}$$

in which r is the radial coordinate, $\hat{\theta}$ is a unit vector along the zenith direction and α is constant inside the star and goes smoothly to zero on a infinitesimal shell outside the star. First let us calculate to load using equation (9) we obtain:

$$\lambda = 2\pi r \frac{\rho}{B_0}, \tag{65}$$

hence the load surfaces are spheres. The χ function can now be calculated according to equation (15) to yield inside the star the value:

$$\chi = \frac{B_0 r^2}{2\pi}. \tag{66}$$

Solving equation (20) for η we obtain the following non unique solution:

$$\eta = -\pi\cos(\theta). \tag{67}$$

Substituting equation (66) and equation (67) into:

$$\vec{A} = \chi\vec{\nabla}\eta + \vec{\nabla}\zeta, \tag{68}$$

will result in the following equation:

$$\vec{A} = \frac{1}{2}B_0 r \sin(\theta)\hat{\theta} + \vec{\nabla}\zeta. \tag{69}$$

Comparing equation (64) with equation (69) we can solve for ζ and obtain:

$$\zeta = \alpha\phi, \tag{70}$$

in which ϕ is the azimuthal angle. Hence ζ is non single valued and has a constant discontinuity value:

$$[\zeta] = 2\pi\alpha. \tag{71}$$

Thus we can easily integrate

$$\mathcal{H}_M = \int [\zeta]_{\chi,\eta} d\chi d\eta = \int [\zeta] d\Phi, \tag{72}$$

(Yahalom (2013)) in this simple case and obtain:

$$\mathcal{H}_M = [\zeta]\Phi = 2\pi\alpha\Phi. \tag{73}$$

In particular if are interested in calculating the magnetic helicity inside a surface of constant load (a sphere of any radius $r \leq R$) we can combine equation (15) with equation (73) to obtain:

$$\mathcal{H}_M = [\zeta]\Phi = (2\pi)^2\alpha\chi = 2\pi\alpha r^2 B_0, \tag{74}$$

in which equation (66) was taken into account. This result coincides with a direct calculation using the magnetic helicity definition given by:

$$\mathcal{H}_M \equiv \int \vec{B} \cdot \vec{A} d^3x, \tag{75}$$

The reader may wonder what is the reason that the second term in the vector potential that has a zero curl and hence produces a zero magnetic field contributes to the helicity. This has to do with the singularity of the term. Indeed this term produces a non zero "magnetic flux":

$$\oint \vec{A} \cdot d\vec{l} = 2\pi\alpha. \tag{76}$$

Thus introducing this term is equivalent to introducing an infinitesimal thin flux tube parallel to the z-axis in addition to the azimuthal magnetic field of equation (63), which is the reason for the non-zero helicity. Of course one may choose not to have such a thin flux tube in the model. If it does not exist than one should be careful in taking advantage of the vector potential gauge freedom since this freedom is only limited to regular functions. Singular gauges are not allowed as they have concrete physical effects as the above example clearly shows.

7. Conclusion

It was shown that unlike in previous examples (Yahalom & Lynden-Bell (2008)) a non-zero magnetic helicity can be obtained even if both χ and η are single valued, the price to pay is having a singular vector potential.

References

Frenkel, A., Levich, E. & Stilman, L. 1982 Hamiltonian description of ideal mhd revealing new invariants of motion. *Phys. Lett. A* **88**, 461–465.

Kats, A. V. 2003 Canonical description of ideal magnetohydrodynamics and integrals of motion. *JETP Lett.* **77**, 657.

LYNDEN-BELL, D. & KATZ, J. 1981 Isocirculational flows and their lagrangian and energy principles. *Proc. R. Soc. Lond. Series A, Mathematical and Physical Sciences* **378** (1773), 179–205.

SAKURAI, T. 1979 A new approach to the force-free field and its application to the magnetic field of solar active regions. *Pub. Ast. Soc. Japan* **31**, 209–230.

STURROCK, P. A. 1994 **Plasma Physics**. Cambridge University Press, Cambridge.

VLADIMIROV, V. A. & MOFFATT, H. K. 1994 On general transformations and variational principles for the magnetohydrodynamics of ideal fluids. part i. fundamental principles. *J. Fluid Mech.* **283**, 125–139.

YAHALOM, A. 2010 A four function variational principle for barotropic magnetohydrodynamics. *EPL* **89**, 34005.

YAHALOM, A. 2013 Aharonov-bohm effects in magnetohydrodynamics. *Phys. Lett. A* **377** (31-33), 1898–1904.

YAHALOM, A. & LYNDEN-BELL, D. 2008 Simplified variational principles for barotropic magnetohydrodynamics. *J. Fluid Mech.* **607**, 235–265.

The effect of tangled magnetic fields on instabilities in tokamak plasmas

A.J. Thornton[1], A. Kirk[1], J.R. Harrison[1], P. Cahyna[2], I.T. Chapman[1], E. Nardon[3] and the MAST team

[1] EURATOM/CCFE Fusion Association, Culham Science Centre, Abingdon, Oxon, 0X14 3DB, UK

[2] Institute of Plasma Physics AS CR v.v.i., Association EURATOM/IPP.CR, Prague, Czech Republic

[3] Association Euratom/CEA, CEA Cadarache, F-13108, St. Paul-lez-Durance, France

E-mail: andrew.thornton@ccfe.ac.uk

Abstract. The high pressure gradients in the edge of a tokamak plasma can lead to the formation of explosive plasma instabilities known as edge localised modes (ELMs). The control of ELMs is an important requirement for the next generation of fusion devices such as ITER. Experiments performed on the Mega Amp Spherical Tokamak (MAST) at Culham have shown that the application of non-axisymetric resonant magnetic perturbations (RMPs) can be used to mitigate ELMs. During the application of the RMPs, clear structures are observed in visible-light imaging of the X-point region. These lobes, or tangles, have been observed for the first time and their appearance is correlated with the mitigation of ELMs. Tangle formation is seen to be associated with the RMPs penetrating the plasma and may be important in explaining why the ELM frequency increases during ELM mitigation. Whilst the number and location of the tangles can be explained by vacuum magnetic field modelling, obtaining the correct radial extent of the tangles requires the plasma response to be taken into account.

1. Introduction

In tokamaks, when sufficient external heating power is applied, there is a transition between a low confinement mode (L mode) and a high confinement mode (H mode) (Wagner *et al.* (1982)). The H mode plasma is characterised by decreased cross field transport and steep gradients in both current and pressure at the plasma edge. The build up of pressure at the plasma edge continues until a stability limit is reached, at which point the pressure at the edge of the plasma collapses as shown in Figure 1 and a filament of plasma is ejected. The instability which results in the ejection of a filament of plasma is known as an edge localised mode (ELM) (Zohm (1996)). An example of an ELM from the Mega Amp Spherical Tokamak (MAST) is shown in Figure 2 and was captured using high speed visible imaging of the plasma. The typical duration of an ELM in MAST is 200 μs, with a single ELM carrying between 5% and 20% of the energy at the edge of the plasma (Kirk *et al.* (2004)). Following an ELM, the current and pressure profiles recover and the pressure begins to increase at the edge once more, the cycle then repeats itself thorough the duration of the discharge.

The energy carried by the ELM can give rise to significant heat fluxes to the exhaust system of the tokamak, known as the divertor. The high heat fluxes are a concern for the next generation

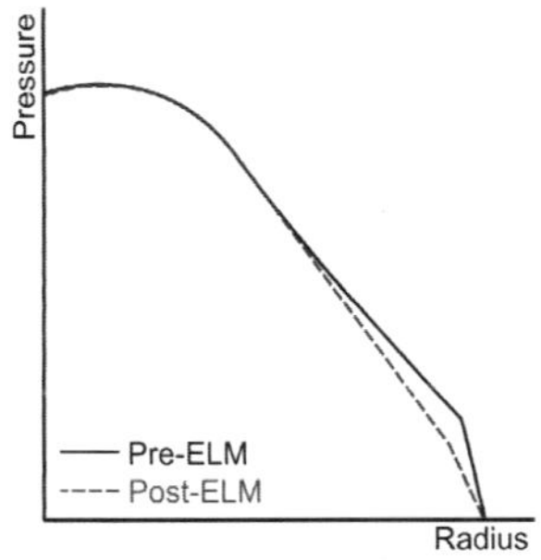

Figure 1. A schematic diagram showing the pressure before (black solid line) and the pressure after (red dashed line) an ELM event. The ELM carries away the energy at the edge of the plasma.

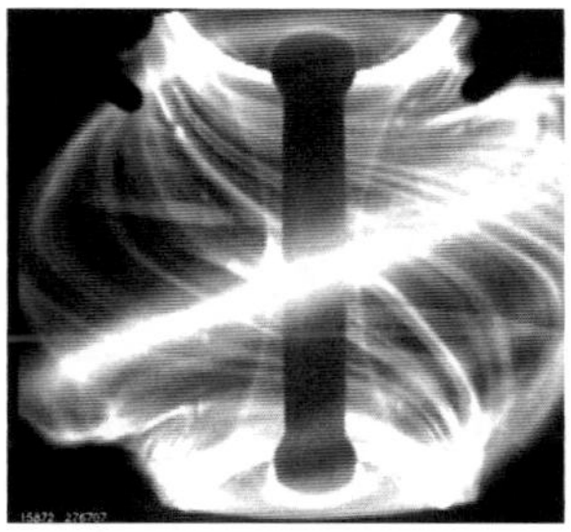

Figure 2. Visible emission from an ELM emitted during a MAST discharge. The filament is ejected out of the plasma and is aligned to the magnetic field.

of fusion devices, such as ITER where the energy released from an ELM is expected to be 20 MJ in 500 μs (Loarte *et al.* (2007)) producing a heat flux on the divertor surfaces of 20 MJ m^{-2}. The high heat loads generated by ELMs would limit the lifetime of the ITER divertor, resulting in the replacement of the divertor being necessary. It is clear that a means of ELM control is required to mitigate the heat loads to the divertor in future devices.

One scheme of ELM mitigation is the use of 3D magnetic perturbations to modify the nested 2D flux surfaces at the edge of the plasma. The modification of the flux surface is achieved using a non-axisymmetric perturbation which is applied using coils, known as a resonant magnetic perturbation (RMP) (Evans *et al.* (2004)). These RMP fields are a radial field at the plasma edge which is thought to generate magnetic islands. When the overlap of adjacent island chains forms a stochastic or ergodic region, this prevents the build up of pressure and leads to ELM mitigation.

In order to fully understand the effect of mitigation via RMP, the plasma response to the 3D field must be understood. The plasma response is the screening out of the external RMP field by the plasma which affects the penetration of the field into the plasma and the formation of the magnetic islands. The application of the RMP affect the frequency at which the ELMs occur and as such lead to a change in the stability of the plasma. The effect of the RMP on the stability of the plasma is another key area of research for RMP physics, along with the plasma response. In the case of ELM control, the reduction in the pressure profile could be a result of increased cross field transport, however, this would lead to reduction of the ELM frequency which is not always seen experimentally. This work aims to study the plasma response to the applied field and determine the implications the RMP have on plasma stability.

2. Mitigation of edge localised modes (ELMs)

ELM control via RMP can take two forms, firstly the technique can lead to the elimination of the ELMs from the discharge which is known as ELM suppression and has been achieved on DIII-D (Evans *et al.* (2004)). Alternatively, RMP field could lead to an increase in the ELM frequency with a corresponding decrease in the ELM size, which is known as ELM mitigation. The product of the ELM frequency and the ELM size has been found to be constant. The peak heat flux to the divertor surfaces, which is the source of the damage, scales with the ELM size, therefore a decrease in the ELM size decreases the effect on the divertor (Federici *et al.* (2003)).

MAST is fitted with 18 RMP coils for ELM mitigation studies, 12 coils are located below the midplane of the device, equally spaced around the circumference of the plasma, with the remaining 6 equally spaced above the midplane (Kirk *et al.* (2013)). Figure 3 shows the effect of applying the RMP coils on an ELMing MAST discharge. The D_α emission from the plasma in the coils off case is shown in the middle panel in Figure 3. The ELMs are the sudden increases in the D_α light and are generated by the ejected filament striking the neutral gas surrounding the plasma, causing it to emit. The H mode period starts at 0.24s and the first ELM occurs at 0.26s. The RMP coil current is shown in the top panel of Figure 3 and the effect on the ELMs in the bottom panel. It is clear that the ELM frequency increases during the coils on period and the amplitude of the D_α emission decreases showing that the ELM size also falls. When the coils are turned off the ELM size and frequency return to that seen in the coils off case.

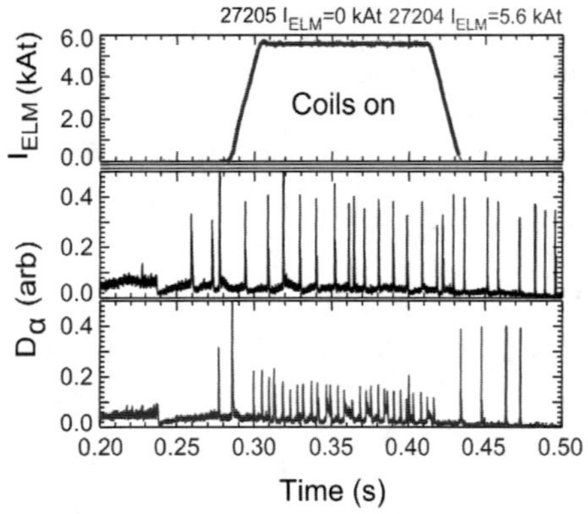

Figure 3. The effect of the application of the RMP coils is shown in these two discharges. The coil current in the coils is shown in the top panel. The centre and bottom panel shows the D_α light emitted from the plasma as a function of time for two shots, one without the coils on (centre panel) and one with the coils on (bottom panel).

3. Effect of 3D fields on tokamak equilibria

The magnetic field is toroidally symmetric prior to the application of the RMP coils, however, the RMP coils cause the toroidal symmetry to be lost and the field becomes three dimensional. The 3D field causes the field lines to behave as a trajectory on a Hamiltionian system when the coils are applied (Evans *et al.* (2005)). The effect of this can be seen in Figure 4 which shows the plasma boundary, known as the last closed flux surface (LCFS) or separatrix for a plasma with the coils off (panel (*a*)) and for the coils on (panel (*b*)). In the coils off case, if a field line on the LCFS is followed around the tokamak in one direction, and then in the other direction and a Poincare plot generated for each direction, the two surfaces overlap with the result shown in Figure 4 (*a*). The field lines all lie on the same surface, which is known as a manifold. When this procedure is repeated with the RMP coils applied, the two manifolds differ as a result of the 3D nature of the field, giving rise to structures at the X point as shown in Figure 4 (*b*) (Wingen *et al.* (2009)). The tangled field in the X point region leads to the formation of lobes.

The lobes generated by the RMP field can be modelled using the vacuum modelling code ERGOS (Nardon *et al.* (2007)). The vacuum modelling code takes the background magnetic field from the plasma and adds the field from the RMP coils, assuming that there is no plasma response. Figure 5 shows a Poincare plot of a poloidal cross section through the plasma with the RMP coils off with the unperterbed LCFS and X point shown. The colour indicates the minimum depth to which the field line penetrates into the plasma in terms of normalised flux

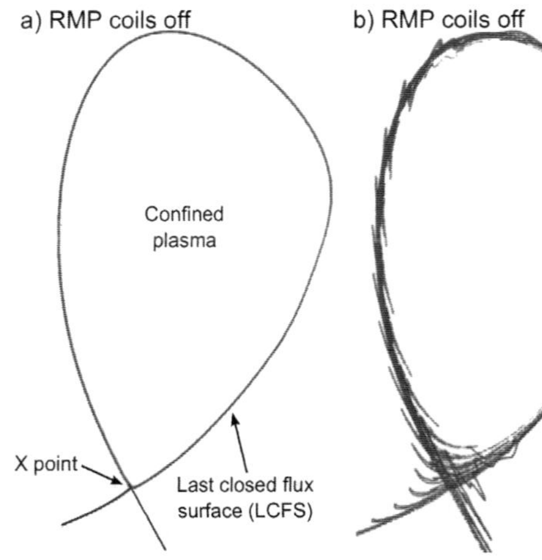

Figure 4. The stable and unstable manifolds in the plasma for the case of (a) no RMP field applied and (b) RMP field applied. The application of the RMP field gives rise to the formation of the lobes as a result of the intersection of the stable and unstable manifolds.

(Ψ_N), where $\Psi_N = 0$ is defined to be the axis of the plasma and $\Psi_N = 1$ at the LCFS. The nested flux surfaces expected in a confined plasma can be seen. In Figure 6 the Poincare plot is repeated with the coils on using a toroidal mode number of n=6 and 5.6 kAT in each coil. The modelled image clearly shows the formation of the lobes and the depth to which the field lines in the lobes penetrate into the plasma. The toroidal mode number of the applied perturbation determines the number of lobes present, with higher toroidal mode numbers of the applied RMP field giving rise to more lobes.

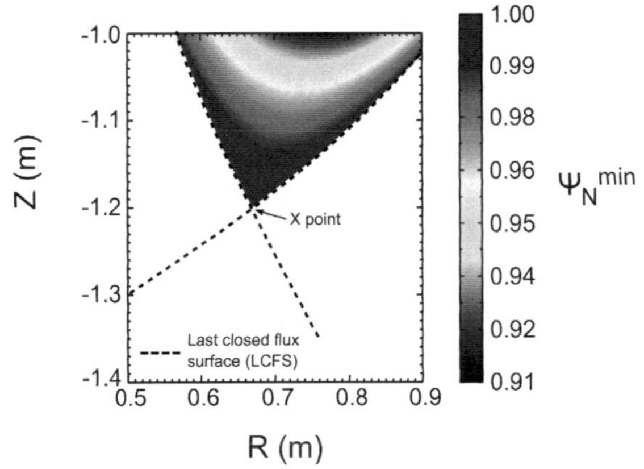

Figure 5. Output of field line tracing without the RMP coils applied. The colour indicates the depth into the plasma (in terms of normalised flux) to which the field line penetrates. The nested set of flux surface expected in a confined tokamak plasma can be seen.

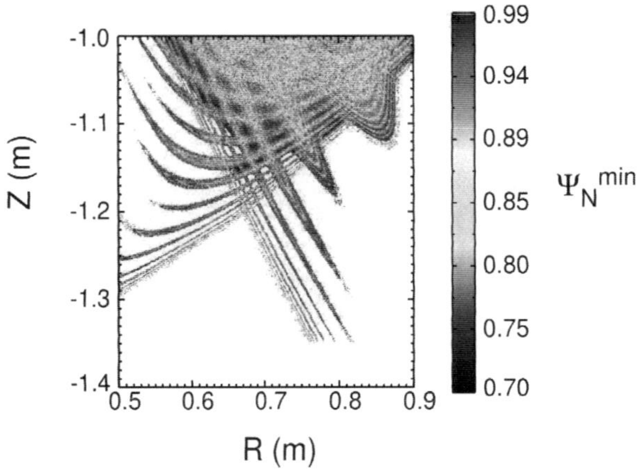

Figure 6. Field line tracing with the RMP coils applied and no screening of the RMP field from the plasma. The colour indicates the depth into the plasma to which the field line penetrates. The lobe structures can be clearly seen.

The lobe structures predicted by the modelling in Figure 6 have recently been observed on MAST via high speed imaging of the X point region (Kirk *et al.* (2012*a*)). The He II filtering has been used to image the plasma as emission at this wavelength is localised to the plasma edge. The imaging has a spatial resolution of 1.8 mm and a frame rate of 500 Hz during the period where the coils are applied. The imaging of the X point region is shown in Figure 7, with panel (*a*) showing the plasma prior to the coils being applied and panel (*b*) with the coils applied. The lobes can clearly be seen in the coils on image and the formation of the lobes is simulataneous with the increase in ELM frequency, confirming that the lobes and the effect of the RMP coils on the ELM frequency are associated with each other.

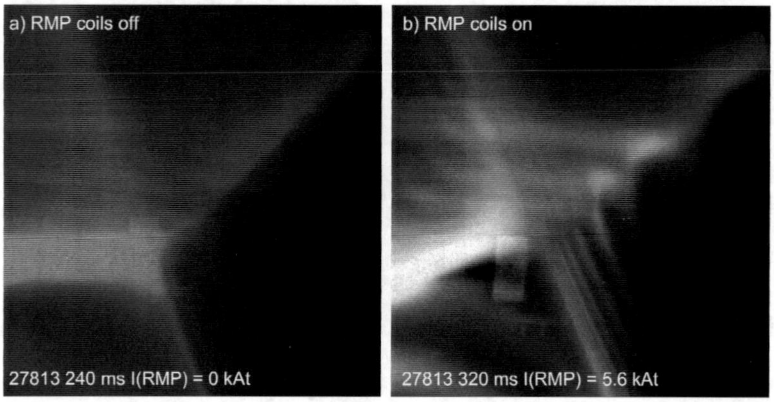

Figure 7. Filtered camera images (He II) showing the emission from the plasma edge in the case of (*a*) discharge with the RMP coils off and in (*b*) the coils on. The formation of the lobes between the coils on and the coils off case can clearly be seen. The applied RMP field has a toroidal mode, n=6, and a current of 5.6kAt.

The imaging of the lobes can be compared with modelling of the lobes (Kirk *et al.* (2012*b*)) using the vacuum field line tracing code. The intensity of the helium emission as a function of Ψ_N can be measured experimentally in the coils off case, this information can then be used to determine the intensity of the emission for an RMP applied case by using the result of the vacuum field line tracing shown in Figure 6 and forward modelling the emission expected for the lobes. The result of this modelling for the vacuum field line tracing result (Figure 6) is shown in Figure 8 (*a*) and can be compared with the image shown in Figure 7 (*b*). It is clear from comparing these figures that the modelling accurately reproduces the structure and location of the lobes expected when the coils are applied. However, the extent of the lobes as measured from the LCFS to the end of the lobe differs between the modelled and measured images. Simulation and experiment have shown that the lobe length is set by the size of the applied RMP field, with higher fields generating larger lobes (Cahyna *et al.* (2011)). As the modelled lobes are longer than those measured, this would suggest that the applied RMP field is being reduced by some mechanism. The vacuum modelling of the plasma neglects the effect of the plasma response to the applied external field, which acts to cancel out the applied field of the coils (Bécoulet *et al.* (2012)), thereby explaining the differences seen between the modelled images and the experimental measurements.

4. Plasma response to RMP fields

The screening of the applied field by the plasma is performed by currents induced inside the plasma. The induced currents on field lines which close on themselves after a finite number of poloidal turns (rational surfaces) oppose the formation of magnetic islands:. One method to

Journal of Physics: Conference Series **544** (2014) 012010 doi:10.1088/1742-6596/544/1/012010

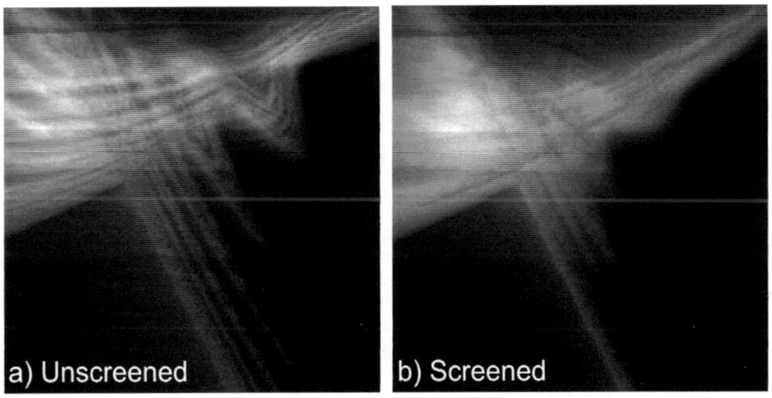

Figure 8. Modelling of the He II emission from the plasma based on the depth to which a field line penetrates into the plasma. The simulated image in panel (*a*) is for a coil current of 5.6kAt in an n=6 configuration without any screening of the RMP field from the plasma response taken into account. Panel (*b*) shows the same modelling repeated using an ad-hoc method to determine the screening of the RMP field by the plasma.

model the screening is to use an ad-hoc approach (Cahyna *et al.* (2011)). In this approach the screening response of the plasma is found by determining the RMP field at each of the rational surfaces and calculating the current required, on a given rational, surface to cancel out the field. The number of surfaces, and hence the radius of the plasma inside which the screening occurs, can be adjusted to best match the experimentally measured lobe size. Figure 9 is a Poincare plot showing the poloidal cross section of an unscreened plasma. The island chains are formed at the rational surfaces, with the largest islands present on the surface at $\Psi_N = 0.77$ which corresponds to a field line with a poloidal mode number of m=7. The screening effect is shown in Figure 10 where the current applied to the rational surface has cancelled out the islands, leading to the formation of a set of nested flux surfaces as seen in plasmas without the coils applied. The screening in Figure 10 acts over the surfaces $7 \leq m \leq 18$ with the shaded region in the figure showing the region of plasma which is unscreened.

The screened field can be used to generate a simulated image of the lobe structures as was performed for the unscreened field. The modelled image with screening applied is shown in Figure 8 (*b*). Comparison of Figure 8 (*b*) and the measured image (Figure 7 (*b*)) now show good agreement with the length of the lobes. The good match in the lobe length suggests that the RMP field is screened over 98% of the plasma radius, with penetration of the applied field across only 2-4% of the outer region of the plasma. Despite the low penetration of the applied field into the plasma, mitigation of the ELMs has been observed along with the formation of the lobe structures at the onset of the mitigation effect.

5. Conclusion

In a tokamak, the sudden ejection of a filament of plasma as a result of exceeding a critical pressure limit is known as an edge localised mode (ELM). These ELM events produce heat and particle fluxes to the divertor surfaces in the tokamak due to the energy contained within the filament of plasma. ELMs have the potential to limit the lifetime of the ITER divertor, and so a means of ELM control is required. One scheme of ELM control is the application of a non-axisymmetric field to perturb the plasma edge. This form of ELM control been seen to be manifest as ELM mitigation, (where the ELM frequency is increased which decreases the size of each ELM) or as ELM suppression, where the ELMs are removed completely. ELM mitigation

Journal of Physics: Conference Series **544** (2014) 012010 doi:10.1088/1742-6596/544/1/012010

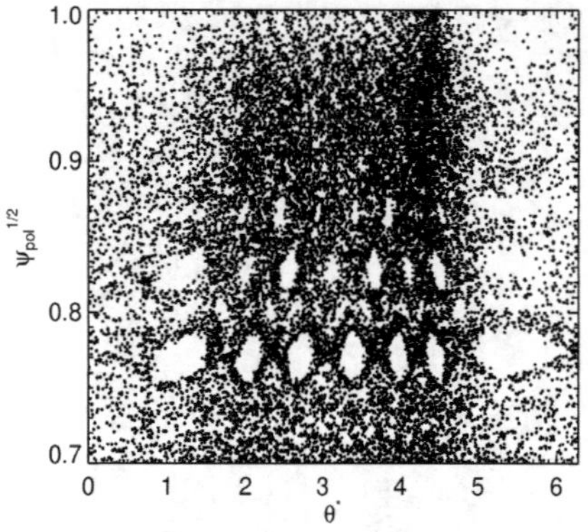

Figure 9. Poincare plot of the flux surface in a plasma with RMP coils applied at full current without any screening from the plasma response. The applied field leads to the formation of magnetic islands at the rational flux surfaces.

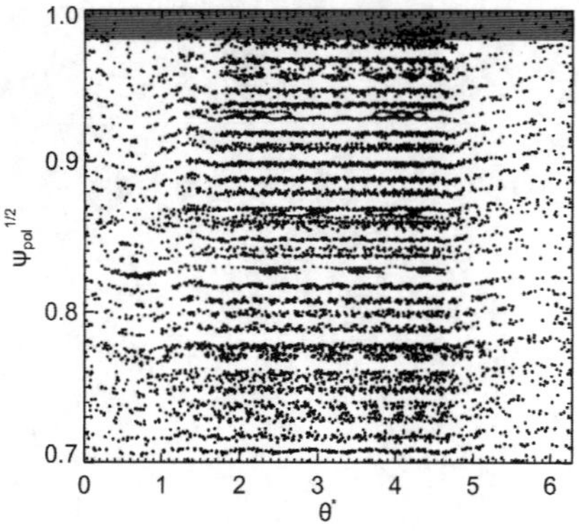

Figure 10. Poincare plot of the flux surfaces in the plasma with the RMP coils applied and a screening response from the plasma. The shaded region indicates where there is no screening and the RMP field penetrates the plasma.

has been achieved in MAST, with a five fold increase in the ELM frequency and corresponding decrease in the energy loss per ELM when a RMP field is applied. The applied RMP field produces a 3D magnetic field which gives rise to tangles in the magnetic field at the X point of the plasma. Imaging of the X point region of the plasma has confirmed that the tangled magnetic field produces lobes which form at the onset of the ELM mitigation. The plasma acts to screen the penetration of the RMP field from the plasma, thereby decreasing the amplitude of the RMP field experienced by the plasma. Modelling of the lobe structures including this screening has shown that the RMP field only needs to penetrate the outer 2-4% of the plasma radius to give rise to an effect on the ELMs. It is clear from the comparison of the vacuum modelling and the experimental data that there is significant screening of the applied perturbation as a result of the plasma response. The plasma response acts to reduce the region over which the magnetic islands are formed. The effect on the stability can also be understood by the formation of the lobes. The lobes act to perturb the plasma boundary and it has been suggested Chapman *et al.* (2012) that this acts to reduce the pressure gradient required for the onset of the ELMs. The reduction of the pressure gradient required by the lobes matches the experimental observations of ELM mitigation whereby the RMP coils act to increase the ELM frequency and decrease the ELM size. It should be noted that the modelling underlying the work on the effect of lobe stability is ideal, and a full treatment using resistive analysis is required which is the subject of future analysis.

Acknowledgments

This work was part funded by the RCUK Energy Programme under grant EP/I501045 and the European Communities under the Contract of Association between EURATOM and CCFE. The views and opinions expressed herein do not necessarily reflect those of the European Commission.

References

Bécoulet, M. *et al.* 2012 Screening of resonant magnetic perturbations by flows in tokamaks. *Nucl. Fusion* **52**, 054003.

CAHYNA, P. *et al.* 2011 Model for screening of resonant magnetic perturbations by plasma in a realistic tokamak geometry and its impact on divertor strike points. *J. Nucl. Mater.* **415**, S927–31.

CHAPMAN, I. *et al.* 2012 Towards understanding ELM mitigation: the effect of axisymmetric lobe structures near the X-point on ELM stability **52**, 123006.

EVANS, T. *et al.* 2004 Suppression of large edge-localized modes in high-confinement DIII-D plasmas with a stochastic magnetic boundary. *Phys. Rev. Lett.* **92**, 235003.

EVANS, T. *et al.* 2005 Experimental signatures of homoclinic tangles in poloidally diverted tokamaks. *J. Phys: Conf. Ser.* **7**, 174–90.

FEDERICI, G. *et al.* 2003 Assessment of erosion of the ITER divertor targets during type I ELMs. *Plasma Phys. Control. Fusion* **45**, 1523–47.

KIRK, A. *et al.* 2004 ELM characteristics in MAST. *Plasma Phys. Control. Fusion* **46**, 551–72.

KIRK, A. *et al.* 2012*a* Observation of lobes near the X point in resonant magnetic perturbation experiments on MAST. *Phys. Rev. Lett.* **108**, 255003.

KIRK, A. *et al.* 2012*b* Understanding ELM mitigation by resonant magnetic perturbations on MAST. Submitted to Nucl. Fusion.

KIRK, A. *et al.* 2013 Effect of resonant magnetic perturbations with toroidal mode numbers of 4 and 6 on edge-localized modes in single null h-mode plasmas in mast. *Plasma Phys. Control. Fusion* **55**, 015006.

LOARTE, A. *et al.* 2007 Chapter 4: Power and particle control. *Nucl. Fusion* **47**, S203–63.

NARDON, E. *et al.* 2007 Edge localized modes control by resonant magnetic perturbations. *J. Nucl. Mater.* **363-365**, 1071.

WAGNER, F. *et al.* 1982 Regime of improved confinement and high beta in neutral-beam-heated divertor discharges of the ASDEX tokamak. *Phys. Rev. Lett.* **49** (19), 1408–12.

WINGEN, A. *et al.* 2009 High resolution numerical studies of separatrix splitting due to non-axisymmetric perturbation in DIII-D. *Nucl. Fusion* **49**, 055027.

ZOHM, H. 1996 Edge localised modes (ELMs). *Plasma Phys. and Control. Fusion* **38**, 105–28.

The ideal versus the reality: topology and turbulence in the current density in tokamak

F. Spineanu, M. Vlad

National Institute of Laser, Plasma and Radiation Physics,
Magurele Bucharest, Romania

E-mail: `florin.spineanu@free.fr, madalina.vlad@euratom.ro`

Abstract. Models for the current density profile in magnetically confined plasma have been proposed starting from first principles, basically in analogy with the self-organization of the vorticity in the two-dimensional Euler fluid. The principle of maximum entropy has led to the derivation of the Liouville equation , under some criticism regarding the possibility of a statistical approach. We identify a class of asymptotic stationary states for the coupled fields of current-density and vorticity, using the Backlund transform. For this class of asymptotic states the maxima of the vorticity and current density must coincide. We also discuss a possible relevance of the field-theoretical solutions *sphaleron* for magnetic reconnection in low-collision regimes.

1. Introduction

The current and the magnetic configuration of the experimental thermonuclear fusion device "tokamak" offer an interesting field of application of methods derived from topological concepts. The high temperature plasma has low collisionality and the invariants are quasi-ideally preserved. On the other hand the large gradients of the plasma parameters (density, temperatures, electric current, etc.) are sources of free energy for instabilities that evolve to turbulence (Wesson, 2004). Therefore in the tokamak plasma both extremes are present: exact topological structures and turbulence, the same as, for instance, in astrophysical or solar plasmas.

Although the basic expectation is axisymmetry (invariance along the toroidal direction) the experiments reveal exceptions: a persistent, robust, helical structure of the magnetic field (island); generation of a stable filament of particle density and current density siting on a magnetic surface close to the axis ("snake"); a stable distribution of the current showing a hole centered on the magnetic axis, etc. These are nontrivial states not fully understood.

As many other systems, the physics of tokamak plasma is investigated using *conservation* equations (density, momentum, energy, etc.) which are hardly able to identify privileged states, if any. The signature that some states are special is the so-called "profile resiliency" , proved experimentally: even with high input of energy and high transport rate, the profiles of the main plasma parameters are almost unchanged. It can be explained by: (1) Self - Organized Criticality (SOC). This is an universal type of behavior of systems composed of a large number of quasi-independent sub-systems, each having a threshold-type dynamics (Bak, 1996). Under a slow drive the sub-systems evolve approaching the threshold for instability. When the limit is exceeded for one sub-system, its reaction of return to equilibrium (*i.e.* under the threshold) is very fast and is propagated to the neighbors that are, themselves, at marginal stability. They become active and the effect is a perturbation (an avalanche) which can extend on a wide range of scales, given by the space extensions of the clusters of marginally-stable sub-systems. On large time scale the SOC is a stationary state of statistical stability with minimum rate of entropy production. The Tokamak plasma there are instabilities with fast growth rate beyond a threshold given by the gradient of the temperature (Dimits *et al.*, 2000). Transport events of "avalanche" - type have been observed in experiments (Garbet *et al.*, 2004), suggesting that the SOC may be the adequate physical picture. However the large eddies and structures do not allow the simultaneous presence of all spatial scales of correlations, specific to criticality. (2) Natural or "priviledged" states. They are stable and stiff since they are the extrema of an action functional, defined on a space of functions representing the plasma states. The shape of the functional around the extremum may explain the resiliency. We are then led to examine the problem of natural profiles of the current density in tokamak. This problem is actually not solved at this time.

2. Natural current profiles in Tokamak

2.1. The relaxed state of the current in tokamak

For slow motion (relative to the Alfven speed), taking the rotational of the momentum equation at equilibrium, we have $\nabla\times(\mathbf{J}\times\mathbf{B})=0$ or

$$B_0\frac{dJ_z}{dz}+(\mathbf{B}_\perp\cdot\nabla_\perp)\,J_z=0 \tag{1}$$

where J_z is the current density along the toroidal direction, $\mathbf{B}_\perp$ is the magnetic field component laying in the meridional section of the torus, B_0 the magnitude of the main, toroidal, magnetic field. Introducing the scalar function ψ , the z component of the magnetic potential, $\mathbf{A}_z=\psi\widehat{\mathbf{e}}_z$, it results $\mathbf{B}_\perp=-\nabla\psi\times\widehat{\mathbf{e}}_z$ and $\Delta_\perp\psi=J_z$.

The magnetic surfaces are surfaces of constant ψ: $\mathbf{B}\cdot\nabla\psi=0$. The solution for the scalar function ψ is

$$\Delta_\perp\psi=J(\psi) \tag{2}$$

where J is an *arbitrary function* of ψ. Taylor (1993) assumed that the current density is in a slightly chaotic state. The current exists as *filaments* with unique magnitude j_0, and their intersection with the meridional plane of the torus is a set of points. Now consider that the z coordinate is like the *time* variable, $z\to t$. The interaction between these filaments is

$$j_0\frac{dx_i}{dt}=\frac{1}{B_0}\frac{\partial H}{\partial y_i}\quad,\quad j_0\frac{dy_i}{dt}=-\frac{1}{B_0}\frac{\partial H}{\partial x_i} \tag{3}$$

The Hamiltonian is $H=\sum\limits_{i>j}j_0^2U(\mathbf{r}_i,\mathbf{r}_j)$. The potential $U(\mathbf{r}_i,\mathbf{r}_j)$ is generated in the point $\mathbf{r}_i$ of the plane by the filament ("rod") located at $\mathbf{r}_j$. For a large plane box of linear dimension L, the potential is $U(\mathbf{r}_i,\mathbf{r}_j)\approx(2\pi)^{-1}\ln(|\mathbf{r}_i-\mathbf{r}_j|/L)$, the Green function of the Laplace operator.

The idea of Taylor was to use statistical methods to analyze the ensemble of discrete, point-like objects, the puncture in plane of the filaments of J_z . First it is discretized the plane box and it is "counted" the number of filaments in each elementary cell, n_k. The microcanonical ensemble is the statistical ensemble of realizations of the distributions of N current filaments, with two constants: energy $E=\sum\limits_k\sum\limits_i j_0^2n_kU_{ki}n_i$ and total number (*i.e.* total current) $N=\sum\limits_k n_k$. In analogy with the set of discrete point-like vortices of the Euler equation for the ideal fluid, it is calculated the distribution function of the microcanonical ensemble, $\rho(\{\mathbf{r}_i\})=\delta(E-H(\{\mathbf{r}_i\}))$.

Journal of Physics: Conference Series **544** (2014) 012011 doi:10.1088/1742-6596/544/1/012011

The average density of filaments is

$$\langle \rho (\mathbf{r}) \rangle = \int \delta \left[E - H \left(\{ \mathbf{r}, \mathbf{r}_i \} \right) \right] \prod_{i=2}^{N} d\mathbf{r}_i \tag{4}$$

This average can be calculated on the basis of the maximum entropy principle (for this system the statistical temperature is *negative*), under preservation of the total energy, E, and the total number of filaments, N.

$$\langle \rho (\mathbf{r}) \rangle = \sum_{\{n_i\}} \exp \left(- \sum_i n_i \ln n_i \right) \delta \left[E - \sum_k \sum_i j_0^2 n_k U_{ki} n_i \right] \delta \left(N - \sum_k n_k \right) \tag{5}$$

At the continuum limit one defines the scalar function $\psi (\mathbf{r}) = \int d\mathbf{r}' \; j_0 U (\mathbf{r}, \mathbf{r}') \rho (\mathbf{r}')$ and the distribution of the current density over the plane is (K is a constant)

$$J = j_0 \langle \rho (\mathbf{r}) \rangle = K \exp \left(-\mu j_0 \psi (\mathbf{r}) \right) \tag{6}$$

The equation for the current density is obtained from the equation for the magnetic potential

$$\Delta \psi = A \exp (-\lambda \psi) \tag{7}$$

which is the Liouville equation. Although criticized, this approach has a subtle difference relative to others: it formulates the problem in terms of matter (density of point-like objects in plane), field (long range, Coulombian potential) and interaction. This suggests a formalism of classical field theory.

2.2. A field theoretical formalism for the current in tokamak

Inspired by the field theoretical description of the coherent flows at relaxation of the $2D$ ideal and incompressible (Euler) fluid, we would try to build a similar formalism for the current density. It will be sufficient to do this for the two-dimensional plasma, *i.e.* the current is transversal on the plane representing the meridional cross-section of the tokamak. The current is non-zero within a circular region of radius a.

From the beginning we note that we do not attempt to tackle here the $2D$ MHD problem, where two fields $(\mathbf{v}, \mathbf{B})$ [and their rotationals (ω, j)] are evolving in interaction. The separation of the models for j and for ω requires the factorization of a unique model and may be possible at, and in close proximity of, the relaxation states.

The essential step in developing a field-theoretical framework for the Euler fluid in $2D$ was the equivalence of the physical model (the Euler equation) with the dynamics of the discrete set of point-like vortices interacting in plane by a self-determined, long-range, potential. For the latter model we have been able to write a Lagrangian density, constructed such as to reflect the Lorentz-type motion (relative gyration) of the elements of vorticity (Spineanu & Vlad, 2003, 2005). The nature of the elementary vortices is reflected in the fact that the scalar (matter) field ϕ must be a mixed spinor, an element of $sl(2, \mathbf{C})$ algebra. At self-duality this becomes the $su(2)$ algebra. As discussed before, there exists a discrete model for the the current density: it neglects the resistivity, the transport processes, the change in the electric field due to induction, the convection of the current by the plasma flow but retains the essential aspect: the Biot-Savart interaction of parallel current filaments leading to Lorentz-type relative motion. The model is then similar to the one for the Euler fluid and we expect to be able to write a Lagrangian density.

We immediately see that there is an essential difference between the elements of vorticity and the elements of current. The elementary vortex is a spin-like entity while the elementary current is not. Then the model for the latter should be Abelian. Assuming a certain universality which places the $2D$ point-like objects in the same class as interacting "charges" in plane, we take as reference Jackiew and Pi (1990) and the expression of the Lagrangian density is written:

$$\mathcal{L} = \frac{\kappa}{4}\varepsilon^{\alpha\beta\gamma}A_\alpha F_{\beta\gamma} + i\hbar\psi^*\left(\frac{\partial}{\partial t} + \frac{ie}{\hbar}A^0\right)\psi - \frac{\hbar^2}{2m}\left|\mathbf{D}\psi\right|^2 + \frac{g}{2}\left(\psi^*\psi\right)^2 \tag{8}$$

Here ψ is the matter field, A_μ is the gauge field that carries the interaction and the last term is the scalar - field nonlinearity. This model has actually been constructed to describe the quantum Hall model but its classical counterpart is adequate for the dynamics of point-like objects moving in plane according to the Lorentz-type interaction. It is found that, when the relationship $\frac{g}{2} \pm \frac{e^2\hbar}{2mc\kappa} = 0$ between the parameters exists one obtains the Self-Duality state, whose signature is the possibility of the Bogomolnyi procedure of writing the action as a sum of squares. The self-dual states of this model are solutions of $D_1\phi = \pm iD_2\phi$ or

$$D_-\phi = 0 \tag{9}$$

and this is formally solved by

$$\mathbf{A} = \nabla\chi \pm \frac{\hbar c}{2e}\nabla \times \ln\rho \tag{10}$$

where $\phi = \rho^{1/2}\exp(i\chi)$. For the scalar field one obtains

$$\nabla^2 \ln\rho = \pm 2\frac{e^2}{\hbar c\kappa}\rho \tag{11}$$

This is exactly the equation derived by Taylor (1993) for the natural profiles of the current density in $2D$ plasma: $\Delta\ln\rho = -\gamma\rho$, after the z component of the magnetic potential, ψ, is introduced by $\rho = \exp(\psi)$. It then results that the asymptotic stationary distribution of current density obeys the Liouville equation, confirming the Taylor's result derived from statistical physics. We recall that this result is for pure current, plasma is not taken into account.

Naturally, the simultaneous consideration of vorticity and current density makes the problem more difficult and returns to MHD relaxation.

3. MHD relaxation in $2D$: the simplest asymptotic states via Backlund transformations

At relaxation in the in $2D$ plasma there is a strong correlation between the spatial patterns of the vorticity and of the current density. The numerical simulations (Kinney *et al.*, 1994) show that in the stationary $2D$ MHD states exhibit coincidence of the local maxima of the current and of the vorticity. More generally, when ω and j are overlapped, they move together. When they are initially distinct they try to overlap (Yatsuyanagi *et al.*, 2002). In a plasma with a helical configuration where the current density $\mathbf{J}$ and the vorticity ω are parallel (*i.e.* aligned) the total rate of energy dissipation (ohmic and viscous) is minimal (Montgomery *et al.*, 1989). Then a cylindrical, axially periodic plasma which is initially axisymmetric will spontaneously bifurcate to a state where a first order helical configuration with $\mathbf{J}$ and ω aligned is present. This state is favorable since it has minimum rate of entropy production. Looking for states of minimum of the energy content under the constraint of constant cross helicity it is found the condition (Montgomery, 1992)

$$\bar{\mathbf{j}} = \left(\frac{\nu}{\eta}\right)^{1/2}\overline{\omega} \tag{12}$$

This suggests the existence of a certain similitude in the stationary asymptotic states of the fields ψ (fluid streamfunction) and a_z (z- component of the magnetic potential), with the vorticity $\omega\widehat{\mathbf{e}}_z \to \omega = \Delta\psi$ and the current density $J_z\widehat{\mathbf{e}}_z \to J_z = \Delta a_z$.

Consider the $2D$ MHD equations

$$\begin{aligned} \frac{d\omega}{dt} &= \nabla_{\|} J_{\|} = \left[\frac{\partial}{\partial z} + \left(-\nabla_{\perp} a_z \times \widehat{\mathbf{e}}_z\right) \cdot \nabla_{\perp}\right] J_{\|} \qquad (13) \\ \frac{\partial a_z}{\partial t} + \left(\mathbf{v} \cdot \nabla_{\perp}\right) a_z &= 0 \end{aligned}$$

The first equation can be written

$$\left[\frac{\partial}{\partial t} + \left(-\nabla_{\perp}\psi \times \widehat{\mathbf{e}}_z\right) \cdot \nabla_{\perp}\right] \Delta\psi = \left[\frac{\partial}{\partial z} + \left(-\nabla_{\perp} a_z \times \widehat{\mathbf{e}}_z\right) \cdot \nabla_{\perp}\right] \Delta a_z \qquad (14)$$

The structure of the two sides is identical, if we adopt the following approximation $J_{\|} \approx J_z$. Then the following limits of these two scalar fields are seen to solve the equations

$$\begin{aligned} \left[\frac{\partial}{\partial t} + \left(-\nabla_{\perp}\psi \times \widehat{\mathbf{e}}_z\right) \cdot \nabla_{\perp}\right] \Delta\psi &= 0 \qquad (15) \\ \left[\frac{\partial}{\partial z} + \left(-\nabla_{\perp} a_z \times \widehat{\mathbf{e}}_z\right) \cdot \nabla_{\perp}\right] \Delta a_z &= 0 \end{aligned}$$

in the following particular states

$$\begin{aligned} \psi &\equiv \text{stationary } \frac{\partial\psi}{\partial t} = 0 \qquad (16) \\ a_z &\equiv \text{invariant on the } z\text{-direction, } \frac{\partial a_z}{\partial z} = 0 \end{aligned}$$

In this case the first equation in Eq.(15) can be seen as the asymptotic form of the Euler equation and this is known to lead to states where the streamfunction $\psi(x, y)$ verifies the *sinh*-Poisson equation

$$\Delta\psi + \sinh\psi = 0 \qquad (17)$$

The second equation is formally identical and the solution a_z should be one of the solutions of the same equation

$$\Delta a_z + \sinh a_z = 0 \qquad (18)$$

Then from the first MHD equation it results that a possible class of stationary asymptotic MHD states consists of two functions $\psi(x, y)$ and $a_z(x, y)$ that are both solutions of the *sinh*-Poisson equation.

However there is the second MHD equation

$$\frac{\partial a_z}{\partial t} + \left[\left(-\nabla_{\perp}\psi \times \widehat{\mathbf{e}}_z\right) \cdot \nabla_{\perp}\right] a_z = 0 \qquad (19)$$

where we see that the scalar magnetic potential a_z is advected by the velocity of the *other* field, ψ.

If now we choose at random two different solutions of the *sinh*-Poisson equation, for ψ and respectively for a_z then introducing them in the equation above it will result in general a *time-variation* of the function a_z. We would like the system to be asymptotically stationary, $\partial/\partial t \equiv 0$.

Then one needs that the two solutions of the *sinh*-Poisson equation that have been chosen to be such that

$$[(-\nabla_\perp \psi \times \widehat{\mathbf{e}}_z) \cdot \nabla_\perp] \, a_z = 0 \tag{20}$$

One possibility consists of simply taking a solution of *sinh*-Poisson for a_z with the gradient ∇a_z perpendicular in plane on the streamlines of the other field ψ

$$-\nabla_\perp \psi \times \widehat{\mathbf{e}}_z \perp \nabla a_z \tag{21}$$

This can be made by either choosing for ψ and for a_z the *same* solution of the *sinh*-Poisson Eq., or by choosing a solution for ψ and then choosing another solution for a_z but with equilines that are everywhere *parallel* to the streamlines of ψ. Then, whatever is the amplitude of a_z in the point (x, y) (different of the amplitude of $\psi(x, y)$) the geometrical form ensures the fulfillment of the condition.

The question is: is it possible to find two different solutions of the *sinh*-Poisson equation that have the same geometrical form of the equilines? We now recall that a fundamental property of the *sinh*-Poisson equation is the exact integrability. For integrable equations in general there exists the property that solutions are connected between them by Backlund transforms. Since both a_z and ψ are solutions of the *sinh*-Poisson equation we must first find a solution (say ψ) then find a_z by a Backlund transform and look for the degenerate case Eq.(20).

3.1. Backlund transform for the sinh *- Poisson equation*

Indeed it is possible to find solutions of Eq.(17) and Eq.(18) that also verifies Eq.(20). The first suggestion for a positive answer comes form the properties of an equation which is similar to *sinh*-Poisson, the *sine* - Gordon equation. One starts from a pseudospherical surface Σ which is a surface with a negative total curvature (Rogers & Schief, 2002)

$$\mathcal{K} = -\frac{1}{\rho^2} \tag{22}$$

Consider a point P on Σ and a segment of fixed length PP' tangent to Σ in P. If the way this segment is moved on Σ is given by the equations of a Backlund transformation between solutions of *sine* - Gordon equation, then the end point P' will trace another surface Σ' which is also pseudospherical and has the *same curvature* as the initial Σ. We take this as a suggestion to look for geometrical structures where the *sinh* - Poisson equation may be of certain relevance and try to extract from the properties of the geometrical structures results for the this equation. Fortunately, such connection exists.

The *sinh* - Poisson equation (also known as "elliptic *sinh* - Gordon") is not only the equation describing the asymptotic states of the $2D$ Euler fluid but also the equation of the conformal metric of Constant Mean Curvature (CMC) surfaces in the $3D$ Euclidean space (Bobenko, 1991): the first differential form can be written $ds^2 = 4\exp(\psi)\left(dx^2 + dy^2\right)$, with ψ verifying the *sinh* - Poisson equation. For these surfaces there is a property of periodicity: a surface that is parallel to a CMC surface and is at a distance $1/H$ has the same property, *i.e.* has constant mean curvature. This chain of periodicity defines a family of CMC surfaces and implicitly a family of solutions of the *sinh* - Poisson equation. The two surfaces are *dual* to each other and the associated streamfunctions are in the relation

$$\psi_2 = -\psi_1 + \ln\left(4/H^2\right) \tag{23}$$

Our conjecture is that the stationary asymptotic states of the $2D$ MHD can be realized by streamfunction ψ and magnetic potential a_z belonging to this family of solutions. If this is unique or if there are other Backlund transforms, remains to be checked.

Journal of Physics: Conference Series **544** (2014) 012011 doi:10.1088/1742-6596/544/1/012011

An alternative way is to start directly from the exact analytical solution of the *sinh*-Poisson equation (Ting *et al.*, 1987)

$$u(x,y) = 2\ln\frac{\Theta\left(\mathbf{l}+\frac{1}{2}\mathbf{1}\right)}{\Theta\left(\mathbf{l}\right)} + K \tag{24}$$

where K is a constant, Θ is the Riemann's hyperelliptic *theta* function and $\mathbf{l} = \mathbf{k}x + \omega y + \mathbf{l}_0$ is the vector of linear combinations arising from the linearization on the Jacobi torus. Finding (ψ, a_z) that verify the constraint Eq.(20) now consists of looking for adequate constants in Eq.(24).

4. The possible ideal route to magnetic reconnection: the *Sphaleron* solution

The fundamental content of a magnetic reconnection event is the change of the topology. If this is to be realized in the absence of any dissipation the field must traverse a singularity. An elementary representation of the singularity can be obtained as follows. Consider two closed, double periodic lines winding around a torus, given by $\xi_1 = m_1\theta - n_1\varphi$ and $\xi_2 = m_2\theta - n_2\varphi$, with different pairs of integers $(m_1, n_1) \neq (m_2, n_2)$. The lines belong to distinct topological classes and no smooth (homotopic) deformation can connect one to the other. They can only be connected if one of them traverses the singular line which is the axis of the torus. There θ is not defined so that state is singular. The process consists of the continuous reduction of the radius on which we imagine that the line (m_1, n_1) is sitting, with preservation of the periodicity, until the radius becomes zero and the line (m_1, n_1) is superposed with the axis of the torus. We actually may call this process straightening of the line (m_1, n_1). From the axis of the torus, which is a singular state, the line emerges with a different periodicity (m_2, n_2). This is the topological transition.

The basic idea is that the process that takes place at the reconnection is the *alignment* of the magnetic field lines that arrive from both sides to the X point. Only if the two lines are *aligned* they are indistinguishable and so can be considered reconnected.

Plasma at high temperature (like in tokamak reactor) or at very low density (like in astrophysics) has weak collisionality which only produces a slow rate of magnetic reconnection. This is frequently in contradiction with the experiments where fast rates of reconnection are observed. We are led to look for mechanisms of reconnection, which are fast even at very low collisionality. The *sphaleron* transition may be one of them.

The *sphaleron* (from the greek "ready to fall") has been found in field theory and represents an unstable solution which connects in the space of functions solutions that are separated by a barrier. The initial and final states can have different topologies. In magnetic reconnection, we expect the sphaleron transition to consist of: (1) starting from a topology (m, n); (2) first straightening the line, *i.e.* the line goes through the singularity where (m) is not defined; (3) emerge from the straight (singular) state with a different (m, n). This time the "line" is moving in $3D$ during this process *i.e.* the point of intersection of the line with the transversal plane comes from the up region, goes through the X point and moves away along the neutral plane, expelled from the X point.

In order to study such possibility we need a formulation in which the magnetic field lines are described in terms of field theory. For only an illustration we consider the Abelian-Higgs with the spatial coordinate compactified to a circle (Park *et al.*, 2001). The space is (x, t) with $x \in S^1$ and the Euclidean action is

$$S = \int dxdt \left[\frac{1}{4}F_{\mu\nu}F_{\mu\nu} + (D_\mu\phi)^*(D_\mu\phi) + \frac{\lambda}{4}\left(|\phi|^2 - \frac{v^2}{2}\right)^2\right] \tag{25}$$

where $D_\mu = \frac{\partial}{\partial x^\mu} - igA_\mu$.The gauge field has two components in this two-dimensional space

$A_\mu \equiv (A_0, A_1)$ and the gauge is fixed by $A_0 = 0$. It is found the sphaleron

$$\begin{aligned} A_{1sph} &= A \equiv \text{const} \\ \phi_{sph} &= \frac{kb(k)}{\sqrt{\lambda}} \exp(igAx)\,\text{sn}\,[b(k)\,x] \end{aligned} \quad (26)$$

where sn is the Jacobi function and $b(k) = \sqrt{\frac{\lambda}{2}} v \left(\frac{2}{1+k^2}\right)^{1/2}$.

5. Conclusions

The problems raised by the tokamak plasma are common to other plasma systems, in particular astrophysical plasma. We have shown that stationary asymptotic states of the current density, which have been suggested to be derived from extremum entropy principle can also be obtained from a field theoretical formulation. For $2D$ MHD, (which is a good approximation of the tokamak plasma), we found that the asymptotic states of vorticity and current density can be connected by Backlund transforms that correspond to a geometrical properties of Constant Mean Curvature surfaces. For magnetic reconnection at very low collisionality we have proposed a description based on the *sphaleron* solutions of the field theoretical formulation.

Acknowledgments

This work is partly supported by the grant 4 ERC - Like of UEFISCDI Romania.

References

BAK, P. 1996 **How nature works: the science of self-organized criticality**. Copernicus New York.

BOBENKO, A. 1991 All constant mean curvature tori in R^3, S^3 and H^3 in terms of theta functions. *Math. Ann.* **290**, 209–245.

DIMITS, A., BATEMAN, G., BEER, M., COHEN, B., DORLAND, W., HAMMET, G., KIM, C., KINSLEY, J., KOTSCHENREUTHER, M., KRITZ, A., LAO, L., MADREKAS, J., NEVINS, W., PARKER, S., REDD, A., SHUMAKER, D., SYDORA, R. & WEILAND, J. 2000 Comparisons and physics basis of tokamak transport models and turbulence simulations. *Phys. Plasmas* **7**, 969–983.

GARBET, X., MANTICA, P., RYTER, F., G. CORDEY, F. IMBEAUX, C. S., ASP, E., PARAIL, V., WOLF, R. & JET EFDA CONTRIBUTORS 2004 Profile stiffness and global confinment. *Plasma Phys. Cont. Fus.* **46**, 1351–1374.

JACKIEW, R. & PI, S.-Y. 1990 Classical and quantal nonrelativistic Chern-Simons theory. *Phys. Rev. D* **42** (10), 2500–3513.

KINNEY, R., TAJIMA, T., MCWILLIAMS, J. & PETVIASHVILI, N. 1994 Filamentary magnetohydrodynamic plasmas. *Physics of Plasmas* **1**, 260–280.

MONTGOMERY, D. 1992 "Reduced" manetohydrodynamics and minimum dissipation rates. *Phys. Fluids* **B4**, 292–293.

MONTGOMERY, D., PHILLIPS, L. & THEOBALD, M. 1989 Helical, dissipative magnetohydrodynamic states with flow. *Phys. Rev. A* **40**, 1515–1523.

PARK, D., MULLER-KIRSTEN, H., LIANG, J. & SHURGAIA, A. 2001 Instanton - sphaleron transitions in the d=2 Abelian-Higgs model on a circle. *http://arXiv.org/0005207* .

ROGERS, C. & SCHIEF, W. 2002 **Backlund and Darboux transformations**. Cambridge University Press.

Journal of Physics: Conference Series **544** (2014) 012011 doi:10.1088/1742-6596/544/1/012011

Spineanu, F. & Vlad, M. 2003 Self-duality of the asymptotic relaxation states of fluids and plasmas. *Phys. Rev. E* **67**, 046309.

Spineanu, F. & Vlad, M. 2005 Stationary vortical flows in two-dimensional plasma and in planetary atmospheres. *Phys. Rev. Lett.* **94**, 235003.

Taylor, J. 1993 Filamentation, current profile and transport in a tokamak. *Phys. Fluids* **B5**, 4378–4383.

Ting, A., Chen, H. & Lee, Y. 1987 Exact solutions of a nonlinear boundary value problem: the vortices of the two-dimensional sinh-poisson equation. *Physica* **26D**, 37–66.

Wesson, J. 2004 **Tokamaks**. Oxford University Press.

Yatsuyanagi, Y., Ebisuzaki, T., Hatori, T. & Kato, T. 2002 Formation of current-vortex filaments. *Physics of Plasmas* **9** (2), 446–456.

Journal of Physics: Conference Series **544** (2014) 012012 doi:10.1088/1742-6596/544/1/012012

Large scale reconstruction of the solar coronal magnetic field

T. Amari[1], J.-J. Aly[2], P. Chopin [1], A. Canou[1], & Z. Mikic[3]
[1]CNRS, Centre de Physique Théorique de l'Ecole Polytechnique,
F-91128 Palaiseau Cedex, France
[2]AIM - Unité Mixte de Recherche CEA - CNRS - Université Paris VII - UMR n^0 7158,
Centre d'Études de Saclay, F-91191 Gif sur Yvette Cedex, France
[3]Predictive Science Inc., San Diego, CA 92121, USA

E-mail: amari@cpht.polytechnique.fr

Abstract. It is now becoming necessary to access the global magnetic structure of the solar low corona at a large scale in order to understand its physics and more particularly the conditions of energization of the magnetic fields and the multiple connections between distant active regions (ARs) which may trigger eruptive events in an almost coordinated way. Various vector magnetographs, either on board spacecraft or ground-based, currently allow to obtain vector synoptic maps, composite magnetograms made of multiple interactive ARs, and full disk magnetograms. We present a method recently developed for reconstructing the global solar coronal magnetic field as a nonlinear force-free magnetic field in spherical geometry, generalizing our previous results in Cartesian geometry. This method is implemented in the new code XTRAPOLS, which thus appears as an extension of our active region scale code XTRAPOL. We apply our method by performing a reconstruction at a specific time for which we dispose of a set of composite data constituted of a vector magnetogram provided by SDO/HMI, embedded in a larger full disk vector magnetogram provided by the same instrument, finally embedded in a synoptic map provided by SOLIS. It turns out to be possible to access the large scale structure of the corona and its energetic contents, and also the AR scale, at which we recover the presence of a twisted flux rope in equilibrium.

1. Introduction

The problem of the reconstruction of the solar coronal magnetic field from photospheric boundary data has been a very active topic of research in recent years due to the arrival of high resolution and low noise vector magnetographs either on the ground, such as THEMIS and SOLIS, or on board present solar missions, such as HINODE and SDO/HMI, and future ones, such as EST and SOLAR-ORBITER. We refer the reader to Aly & Amari (2007) and Wiegelmann & Sakurai (2012) for a review of all existing reconstruction methods. We just recall here for the purpose of this Paper that the main methods proposed so far are the optimisation methods (Wiegelmann, 2004), which use all photospheric data and try to minimize a cost function measuring the difference between the computed transverse field and the measured one, the magnetohydrodynamic relaxation methods (Valori *et al.*, 2005), and the Grad-Rubin methods (Sakurai, 1981; Amari *et al.*, 1999; Inhester & Wiegelmann, 2006; Wheatland, 2007). Recently Amari *et al.* (2006) presented two methods that attempt to solve the Grad-Rubin Boundary

Journal of Physics: Conference Series **544** (2014) 012012 doi:10.1088/1742-6596/544/1/012012

Value Problem (BVP), namely XTRAPOL and FEMQ, the former based on a finite difference approximation and the latter on a finite element approximation.

Most of these methods have addressed the active region (AR) scale and have thus used the Cartesian geometry framework in which Sun's curvature is neglected. Such an approximation, however, becomes invalid when one attempts to understand the large scale coronal structures and their dynamics (Mackay & Yeates, 2012). With the arrival of SOLIS and SDO/HMI, it is possible to obtain composite high resolution vector magnetograms made of several active regions, or embedded in a full disk vector magnetogram and/or synoptic vector magnetograms. Furthermore vector magnetic fields are also measured on the whole surface of many stars with different techniques, although at a lower resolution and with less accuracy (Donati *et al.*, 2007).

This raises the issue of building up reconstruction models at global scale. Extension of the Cartesian models to spherical ones has thus recently become an important question, and actually such an extension has already been done in the optimization methods settings (Wiegelmann, 2007). It must be noted that, either for full disk or for synoptic maps, strong constraints on spatial resolution are imposed due to the fact that ARs represent only a small fraction of the Sun's surface. Therefore neither uniform nor non-uniform structured meshes can fit such structures in an optimal way.

In the present Paper, we present a new code, XTRAPOLS, that allows the reconstruction of the nonlinear force-free field in the exterior of a spherical region. This code is based on the iterative Grad-Rubin scheme that we have already implemented in Cartesian geometry (Amari *et al.*, 2006) and is now implemented in spherical coordinates. We favor this approach because it corresponds to a well-posed BVP. Moreover it has already proved to be quite efficient in the previously considered Cartesian context, and has allowed us to obtain significant quantitative results when applied to observational data (Bleybel *et al.*, 2002; Régnier *et al.*, 2002; Régnier & Amari, 2004; Régnier *et al.*, 2005; DeRosa *et al.*, 2009; Canou *et al.*, 2009; Canou & Amari, 2010; Petrie *et al.*, 2011).

XTRAPOLS is a massively parallel code, and as such it is able to handle the large amounts of data currently provided by vector magnetographs, with fields being measured on a very large scale and with a very high resolution. We apply this code to the reconstruction of the global coronal field at some particular date, using as boundary conditions a composite data set obtained by using vector magnetograms provided by SDO/HMI at both the AR scale and the full disk scale, and a SOLIS synoptic map on the rest of the Sun. We show that the model allows to determine both the large scale structure of the corona and the small scale structure of an AR. In particular, we find the presence of a twisted flux rope in equilibrium, which can possibly be responsible for the energization of the AR and the subsequent occurrence of a large scale eruptive event. More details on the computational method and the testing of the code can be found in Amari *et al.* (2013)., which was submitted for publication after the ICMS Workshop.

2. Method

2.1. The continuous BVP

In the theoretical model we consider, the corona and the photosphere are represented by a domain Ω that is the exterior of a sphere rather than the half-space $D = \{z > 0\}$ considered previously. This model thus differs from the previously developed Ω is assumed to be filled up with a low beta plasma embedded in a magnetic field $\mathbf{B}$ that is taken to be force-free and to decrease sufficiently fast to zero at infinity. Therefore, it does obey the equations

$$\nabla \times \mathbf{B} = \alpha \mathbf{B}, \tag{1}$$

$$\nabla \cdot \mathbf{B} = 0. \tag{2}$$

It results at once from Eqs. (1)-(2) that the function α satisfies the constraint

$$\mathbf{B} \cdot \nabla \alpha = 0, \tag{3}$$

which merely states that α keeps a constant value along any field line.

The set of equations (1)-(3) has a mixed elliptic-hyperbolic structure. Basically, it can be decomposed into an elliptic part for $\mathbf{B}$ (Eqs. (1)-(2) with α given), and an hyperbolic part for α (Eq. (3) with $\mathbf{B}$ given). To solve for the elliptic part, one should give the normal component B_n of the magnetic field on $\partial\Omega$, while to solve for the hyperbolic one, Eq. (3) indicates that the value of α should be given on the part $\partial\Omega^+$ of $\partial\Omega$ where $B_n > 0$, say. This leads us to consider the BVP first introduced by Grad & Rubin (1958). It consists of Eqs. (1)-(3) along with the boundary conditions

$$B_n|_{\partial\Omega} = b_0\,, \tag{4}$$
$$\alpha|_{\partial\Omega^+} = \alpha_0\,, \tag{5}$$

where b_0 and α_0 are two given regular functions, and the asymptotic condition

$$\lim_{|\mathbf{r}|\to\infty} |\mathbf{B}| = 0\,. \tag{6}$$

In the numerical practice we consider the bounded domain $\Omega_b = \{r_0 < r < r_1\}$, limited by the spherical surfaces $S_0 = \{r = r_0\}$ (the photosphere) and $S_1 = \{r = r_1\}$, instead of the unbounded domain Ω. Then the asymptotic condition (6) is no longer relevant and the boundary conditions (4)-(5) are imposed on the whole $\partial\Omega_b$.

2.2. Principle of the Iterative Grad-Rubin method

In the Grad-Rubin method, the BVP above is solved iteratively, with its elliptic and hyperbolic parts being considered in turn at each step. More precisely, we look for a sequence $(\mathbf{B}^{(n)}, \alpha^{(n)})$ solution of

$$\mathbf{B}^{(n)}.\nabla\alpha^{(n)} = 0 \quad \text{in} \quad \Omega_b\,, \tag{7}$$
$$\alpha^{(n)}|_{\partial\Omega_b^+} = \alpha_0\,, \tag{8}$$

and

$$\nabla\times\mathbf{B}^{(n+1)} = \alpha^{(n)}\mathbf{B}^{(n)} \quad \text{in} \quad \Omega_b\,, \tag{9}$$
$$\nabla\cdot\mathbf{B}^{(n+1)} = 0 \quad \text{in} \quad \Omega_b\,, \tag{10}$$
$$B_r^{(n+1)}|_{\partial\Omega_b} = b_0\,. \tag{11}$$

The iteration process is initialized by choosing for $\mathbf{B}^{(0)}$ the potential field with the normal component b_0, i.e., the unique solution of

$$\nabla\times\mathbf{B}^{(0)} = 0 \quad \text{in} \quad \Omega_b, \tag{12}$$
$$B_r^{(0)}|_{\partial\Omega_b} = b_0\,. \tag{13}$$

A fundamental property of our method is that it is well-posed – in particular it is stable with respect to changes in the boundary conditions. This makes a major difference with the methods, recalled above, that use the three components of the photospheric field as boundary conditions: as there is most generally no exact force-free solutions matching such conditions, these methods have to deal with ill-posedness. Another point is worth noticing. In the Grad-Rubin BVP one fixes the values of α in only one polarity. The values of α in the other polarity are obtained by computing the solution, and do not match the observed values α_0. Then reversing the role of the two polarities leads to a different solution. A unique solution for which the computed

Journal of Physics: Conference Series **544** (2014) 012012 doi:10.1088/1742-6596/544/1/012012

α is as close as possible to the observed α_0 on the whole S can however be obtained by using the recently proposed Grad-Rubin-Optimization-Method (Amari & Aly, 2010). The idea is to impose, at each iteration step, the value of α on a line to be a weighted average of the values of α_0 at its two footprints. This new approach has been shown to give good and robust results, but we shall not use it here.

The previous scheme is implemented in spherical coordinates, using a numerical method which is able to run on massively parallel computers. This is required if we want to deal in an efficient way with the large amount of data provided by the observations.

To address the $\nabla \cdot \mathbf{B} = 0$ constraint, XTRAPOLS introduces a vector potential $\mathbf{A}^{(n)}$ and ensures that the iterated numerical field $\mathbf{B}^{(n)} = \nabla \times \mathbf{A}^{(n)}$. To determine $\mathbf{A}^{(n)}$ in a unique way, we impose the gauge conditions

$$\nabla \cdot \mathbf{A}^{(n)} = 0 \quad \text{in } \Omega_b, \tag{14}$$

$$\mathbf{A}_t^{(n)} = \mathbf{A}_{\pi t} \quad \text{on } \partial\Omega_b, \tag{15}$$

where $\mathbf{X}_t$ denotes the component of $\mathbf{X}$ tangent to the boundary and $\mathbf{A}_\pi$ is a specific vector potential of the current-free field $\mathbf{B}_\pi$, i.e.,

$$\mathbf{B}_\pi = \nabla \times \mathbf{A}_\pi. \tag{16}$$

$\mathbf{A}_\pi$ is required to satisfy the gauge conditions

$$A_{\pi r}(r, \theta, \varphi) = 0, \tag{17}$$

$$\nabla \cdot \mathbf{A}_\pi(r_0, \theta, \varphi) = 0, \tag{18}$$

which implies that $\nabla \cdot \mathbf{A}_\pi = 0$ in Ω_b. To compute $\mathbf{A}_\pi$ on the boundary: (i) We introduce a scalar potential $\chi_\pi(r, \theta, \varphi)$ such that $\mathbf{A}_\pi = \nabla \chi_\pi \times \hat{\mathbf{r}}$ in Ω_b. (ii) We impose $\chi(r_0, \theta, \varphi)$ to be the unique solution with zero average value of the equation $-\nabla_\perp^2 \chi = B_r$ on S_0. Solving this equation determines $\mathbf{A}_\pi$ on S_0. (iii) We introduce the standard representation $\mathbf{B}_\pi = \nabla \psi_\pi$, and compute the scalar potential ψ_π by solving Laplace equation $-\nabla^2 \psi_\pi = 0$ in Ω_b along with the boundary condition $\partial_r \psi_\pi = B_r$ on $S_0 \cup S_1$. (iv) χ_π is thus given by

$$\chi_\pi(r, \theta, \varphi) = \chi_\pi(r_0, \theta, \varphi) + \int_{r_0}^{r} \psi_\pi(r', \theta, \varphi)\, \mathrm{d}r', \tag{19}$$

from which we get $\mathbf{A}_\pi$ on S_1.

In our vector potential formulation, α is determined by solving

$$\nabla \times \mathbf{A}^{(n)} . \nabla \alpha^{(n)} = 0 \quad \text{in} \quad \Omega_b\,, \tag{20}$$

$$\alpha^{(n)}|_{\partial\Omega_b^+} = \alpha_0\,, \tag{21}$$

which is just a rewriting of Eqs. (7)-(8). The boundary value of α is thus directly imposed at each iterative step, rather than being progressively increased to its nominal value through a second iterative loop as was done in Amari *et al.* (1999). In other words, we keep only the inner Grad-Rubin iteration loop in the two-level iteration procedure used in the latter paper.

Eqs. (20)-(21) are solved by the method of characteristics. The characteristic $\mathbf{X}(s, \mathbf{r})$ passing through the point $\mathbf{r} \in \Omega_b$ is a solution of (Amari *et al.*, 1999, 2006)

$$\mathbf{X}' = \mathbf{B}(\mathbf{X})\,, \tag{22}$$

$$\mathbf{X}(0) = \mathbf{r}\,, \tag{23}$$

Journal of Physics: Conference Series **544** (2014) 012012 doi:10.1088/1742-6596/544/1/012012

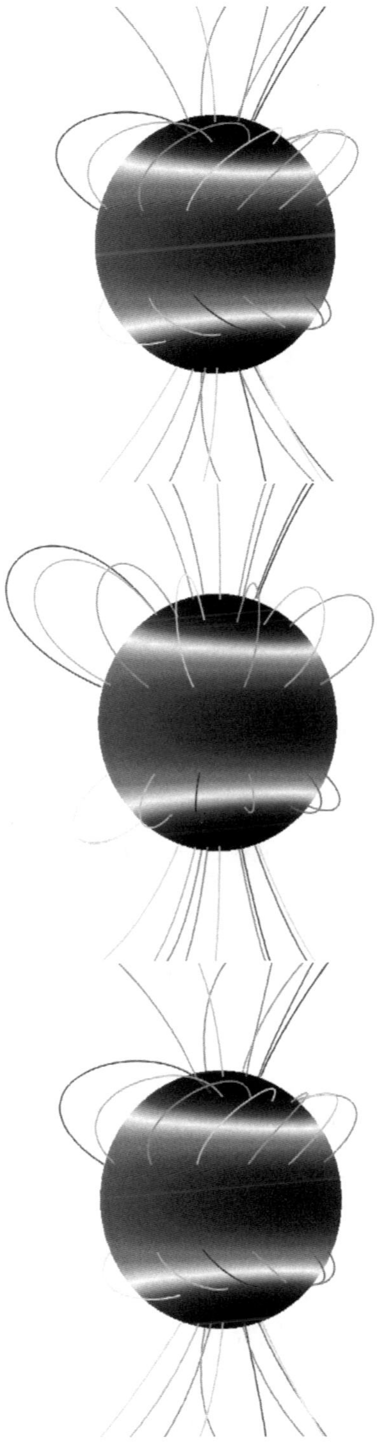

Figure 1. Reconstruction of the test field. We have drawn a set of field lines of the exact Low-Lou solution (top), one of the potential (current-free) field (middle), and one of the solution computed with our algorithm with a resolution of $80 \times 160 \times 320$ (bottom).

Figure 2. Set of field lines of the reconstructed global solar force-free magnetic field. The calculation is done with resolution $171 \times 251 \times 363$ and use AR scale and full disk magnetograms from SDO/HMI, and synoptic map from SOLIS.

where the prime symbol stands for differentiation with respect to the parameter s running along that curve. Then for any node $\mathbf{r}_h \in \Omega_b$ at which α is defined, one gets α_h as

$$\alpha(\mathbf{r}_h) = \alpha_0(\mathbf{X}^+(\mathbf{r}_h)) \,, \tag{24}$$

where $\mathbf{X}^+(\mathbf{r}_h)$ is the intersection of $\{\mathbf{X}(\mathbf{r}; s) : s < 0\}$ with $\partial\Omega_b^+$. Since α_0 is known at nodes that do not in general coincide with $\mathbf{X}^+(\mathbf{r}_h)$, we use an interpolation from their four nearest neighbors.

We define the location of α on the cell vertices. We use either a high order Adams-Bashforth integration scheme with adaptive step size or a 4th order Runge Kutta scheme, which also allows us to capture accurately the ending point of the characteristics defining the limits of the computational box. As already discussed in Cartesian geometry (Amari & Aly, 2010), it is possible to have at the k-th iteration of the scheme some field lines that do not intersect the part of the photospheric boundary where $B_r > 0$ but emerge from the upper boundary. In that case we set

$$\alpha^{(k)}(\mathbf{r}) = \alpha_0(\mathbf{X}^-(\mathbf{r})). \tag{25}$$

Our convergence criterion for the sequence of fields $\mathbf{B}^{(n)}$ is chosen to be

$$\frac{|| \, \mathbf{B}^{(n+1)} - \mathbf{B}^{(n)} \, ||_{L^2(\Omega_b)}}{|| \, \mathbf{B}^{(n)} \, ||_{L^2(\Omega_b)}} \;<\; \epsilon \,, \tag{26}$$

with $\epsilon = 10^{-6}$. We found that a total number of iterations of about $30-50$ is sufficient to achieve convergence in most cases, and the stopping criteria is robust with respect to changes in the data put as boundary conditions. Compared to the Cartesian version, to solve the hyperbolic problem we need to use at a given point in the domain a mapping from spherical to Cartesian coordinates while going back along a characteristic. The algorithm presented is parallel for both the Elliptic and the Hyperbolic problems (only with one global field for the latter). We use domain decomposition with the MPI interface on our staggered mesh.

3. Application to a semi analytic case

We have applied successfully our solver to the reconstruction of the nonlinear force-free Low-Lou solution (Low & Lou, 1990). In an auxiliary spherical coordinate system (r', θ', φ') centered at O', this axisymmetric field has the representation

$$\mathbf{B} = \nabla A \times \frac{\hat{\varphi}'}{r' \sin\theta'} + F \frac{\hat{\varphi}'}{r' \sin\theta'}, \tag{27}$$

with the poloidal flux function $A(r', \theta')$ and current function $F(r', \theta')$ being given by, respectively,

$$A(r', \theta') = \frac{P(\mu')}{(r')^n}, \tag{28}$$

$$F(r', \theta') = aA^{1+1/n} = a\frac{P^{1+1/n}(\mu')}{(r')^{n+1}}. \tag{29}$$

Here $\mu' = \cos\theta'$, and P is a solution of a nonlinear ordinary differential equation satisfying $P(\pm 1) = 0$.

For testing our numerical procedure with an asymmetric field depending on the three coordinates r, θ, φ, we introduce the solution having the quadrupolar structure and the "stretching parameter" $n = 1$ into our model by imposing its axis of symmetry $O'z'$ to coincide with the axis Oz of our main frame, with the center O' being located at the position $\overrightarrow{OO'} = d\hat{\mathbf{z}}$. We take for d the value selected by Wiegelmann (2007), who has already used this shifting procedure, thus making it possible to compare his results with ours.

Table 1 shows some of the diagnostics, often called figures of merit, already used for Cartesian coordinates calculations in Amari *et al.* (1999, 2006) and in Schrijver *et al.* (2006), to which we refer for precise definitions. The various quantities are evaluated for $\mathbf{u}$ equal to the exact solution and $\mathbf{v}$ equal to the numerical solution obtained with XTRAPOLS. They have to be compared with the reference values $VC(\mathbf{u}, \mathbf{u}) = CS(\mathbf{u}, \mathbf{u}) = 1$, $NVE(\mathbf{u}, \mathbf{u}) = MVE(\mathbf{u}, \mathbf{u}) = 0$, $\epsilon_M = 1$. We also found important to report estimates of $\nabla \cdot \mathbf{B}$, as well as the CPU time, both figures providing additional measures of the quality of the solution.

Table 1. Various error diagnostics allowing to evaluate the quality of the test field reconstructions performed with XTRAPOLS for three numerical resolutions:

RESOLUTION	VC	CS	NVE	MVE	ϵ_M	$\|\| \text{div}\mathbf{B} \|\|_{L^\infty}$	Time
$20 \times 40 \times 80$	0.9972	0.9980	0.0696	0.0783	1.008	1.53510^{-13}	61s
$40 \times 80 \times 160$	0.9983	0.9985	0.0650	0.0610	1.002	$1.278\ 10^{-12}$	800s
$80 \times 160 \times 320$	0.9992	0.9993	0.0612	0.0587	1.001	1.18410^{-12}	10077s

As shown on Fig. 1 the field lines of the numerically computed nonlinear force-free configuration are very similar to those of the exact solution. Note that they are quite far from those of the potential (current-free) field.

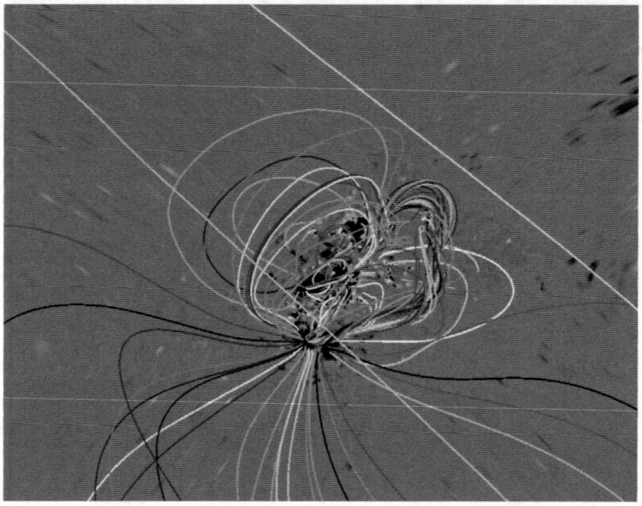

Figure 3. Structure at AR scale of the global solution shown in Figure 1. A twisted flux rope in equilibrium is neatly seen to be present.

4. Application to a composite magnetogram (SDO/HMI-SOLIS)

We have applied our method to the determination of the full Sun coronal magnetic configuration of February 14, 2011, 22:00 UT, using a composite magnetogram provided by two instruments. We give here only a preliminary account of our results, which are explained in details in a forthcoming paper.

During the first half of February 2011, the disk was characterized by the presence of the AR 11158, which was the sit of a very strong eruptive activity involving a CME and an X-class flare on February 15 at 01:40 UT. For our reconstruction with XTRAPOLS, we use as boundary conditions a vector magnetogram provided by SDO/HMI, embedded in a larger full disk vector magnetogram provided by the same instrument, finally embedded in a synoptic map provided by SOLIS. We also used a special processing, and set the current to zero but in the vicinity of AR 11158, where the accuracy of our mesh was high enough to take it into account. The calculations are done with a numerical resolution of 120x200x340 in spherical coordinates.

The solution allows us to catch the large scale magnetic structure (see Fig. 2), as well as the smaller AR scale (see Fig. 3). The structure of the AR can be very precisely characterized and one may note the presence of a twisted flux rope in equilibrium. This rope is possibly responsible for the energization of the AR and it may have stored the large amount of free magnetic energy that was later the source of the large scale eruptive event.

Acknowledgments

We are grateful to the referee for his careful reading of the paper. We thank the Centre National d'Etudes Spatiales (CNES) for support.

References

ALY, J.-J. & AMARI, T. 2007 Structure and evolution of the solar coronal magnetic field. *Geophys. Astrophys. Fluid Dyn.* **101**, 249–287.

AMARI, T. & ALY, J.-J. 2010 Observational constraints on well-posed reconstruction methods and the optimization-Grad-Rubin method. *Astron. Astrophys.* **522**, A52.

AMARI, T., ALY, J.J., CANOU, A. & MIKIC, Z. 2014 Reconstruction of the solar coronal magnetic field in spherical geometry. *Astron. Astrophys.* **553**, A43-A49.

AMARI, T., BOULMEZAOUD, T. Z. & ALY, J.-J. 2006 Well posed reconstruction of the solar coronal magnetic field. *Astron. Astrophys.* **446**, 691-705.

AMARI, T., BOULMEZAOUD, T. Z. & MIKIC, Z. 1999 An iterative method for the reconstruction of the solar coronal magnetic field. I. Method for regular solutions. *Astron. Astrophys.* **350**, 1051-1059.

BLEYBEL, A., AMARI, T., VAN DRIEL-GESZTELYI, L. & LEKA, K. D. 2002 Global budget for an eruptive active region. I. Equilibrium reconstruction approach. *Astron. Astrophys.* **395**, 685–695.

CANOU, A. & AMARI, T. 2010 A twisted flux rope as the magnetic structure of a filament in active region 10953 observed by Hinode. *Astrophys. J.* **715**, 1566–1574.

CANOU, A., AMARI, T., BOMMIER, V., SCHMIEDER, B., AULANIER, G. & LI, H. 2009 Evidence for a pre-eruptive twisted flux rope using the Themis vector magnetograph. *Astrophys. J.* **693**, L27–L30.

DEROSA, M. L., SCHRIJVER, C. J., BARNES, G., LEKA, K. D., LITES, B. W., ASCHWANDEN, M. J., AMARI, T., CANOU, A., MCTIERNAN, J. M., RÉGNIER, S., THALMANN, J. K., VALORI, G., WHEATLAND, M. S., WIEGELMANN, T., CHEUNG, M. C. M., CONLON, P. A., FUHRMANN, M., INHESTER, B. & TADESSE, T. 2009 A critical assessment of nonlinear force-free field modeling of the solar corona for active region 10953. *Astrophys. J.* **696**, 1780–1791.

DONATI, J.-F., JARDINE, M. M., GREGORY, S. G., PETIT, P., BOUVIER, J., DOUGADOS, C., MÉNARD, F., COLLIER CAMERON, A., HARRIES, T. J., JEFFERS, S. V. & PALETOU, F. 2007 Magnetic fields and accretion flows on the classical T Tauri star V2129 Oph. *Mon. Not. R. Astron. Soc.* **380**, 1297–1312.

GRAD, H. & RUBIN, H. 1958 Hydromagnetic equilibria and force-free fields. In *Proc. 2nd Intern. Conf. on Peaceful Uses of Atomic Energy*, vol. 31, pp. 190–197. United Nations, Geneva.

INHESTER, B. & WIEGELMANN, T. 2006 Nonlinear force-free magnetic field extrapolations: Comparison of the Grad Rubin and Wheatland Sturrock Roumeliotis algorithm. *Solar Phys.* **235**, 201-221.

LOW, B. C. & LOU, Y. Q. 1990 Modeling solar force-free magnetic fields. *Astrophys. J.* **352**, 343–352.

MACKAY, D. & YEATES, A. 2012 The Sun's global photospheric and coronal magnetic fields: observations and models. *Living Rev. Solar Phys.* **9** (6).

PETRIE, G. J. D., CANOU, A. & AMARI, T. 2011 Nonlinear force-free and potential-field models of active-region and global coronal fields during the whole heliosphere interval. *Solar Phys.* **274**, 163–194.

RÉGNIER, S. & AMARI, T. 2004 3D magnetic configuration of the Hα filament and X-ray sigmoid in NOAA AR 8151. *Astron. Astrophys.* **425**, 345-352.

RÉGNIER, S., AMARI, T. & KERSALÉ, E. 2002 3D Coronal magnetic field from vector magnetograms: non-constant-alpha force-free configuration of the active region NOAA 8151. *Astron. Astrophys.* **392**, 1119-1127.

RÉGNIER, S., AMARI, T. & CANFIELD, R. C. 2005 Self and mutual magnetic helicities in coronal magnetic configurations. *Astron. Astrophys.* **442**, 345–349.

SAKURAI, T. 1981 Calculation of force-free magnetic field with non-constant-alpha. *Solar Phys.* **69**, 343-359.

Journal of Physics: Conference Series **544** (2014) 012012 doi:10.1088/1742-6596/544/1/012012

Schrijver, C. J., DeRosa, M. L., Metcalf, T. R., Liu, Y., McTiernan, J., Régnier, S., Valori, G., Wheatland, M. S. & Wiegelmann, T. 2006 Nonlinear force-free modeling of coronal magnetic fields. Part I: A quantitative comparison of methods. *Solar Phys.* **235**, 161-190.

Valori, G., Kliem, B. & Keppens, R. 2005 Extrapolation of a non linear force-free field containing a highly twisted magnetic loop. *Astron. Astrophys.* **433**, 335-347.

Wheatland, M. 2007 Calculating and testing nonlinear force-free fields. *Solar Phys.* **245**, 251-262.

Wiegelmann, T. 2004 Optimization code with weighting function for the reconstruction of coronal magnetic fields. *Solar Phys.* **219**, 87-108.

Wiegelmann, T. 2007 Computing nonlinear force-free coronal magnetic fields in spherical geometry. *Solar Phys.* **240**, 227–239.

Wiegelmann, T. & Sakurai, T. 2012 Solar force-free magnetic fields. *Living Rev. Solar Phys.* **9** (5).

Tropical cyclogenesis: a numerical diagnosis based on helical flow organization

G.V. Levina[1,2] **& M.T. Montgomery**[3]

[1] Institute of Continuous Media Mechanics, Ural Branch of the Russian Academy of Sciences, Akademik Korolyov Street 1, Perm 614013, Russia
[2] Space Research Institute, Russian Academy of Sciences, Moscow, Russia
[3] Department of Meteorology, Naval Postgraduate School, Dyer Road, Monterey, CA 93943-5114, USA

E-mail: levina@icmm.ru, mtmontgo@nps.edu

Abstract. A numerical diagnosis for tropical cyclogenesis is proposed based on near-cloud-resolving atmospheric simulation. Calculation and analyses of helical and energetic characteristics together with hydro- and thermodynamic flow fields allow the diagnosis of tropical cyclogenesis as an event when the primary and secondary circulations become linked on system scales thereby making the nascent large-scale vortex helical. A key process of vertical vorticity generation from horizontal components and its amplification by special convective coherent structures - Vortical Hot Towers (VHTs) - is highlighted. The process is found to be a pathway for generation of a velocity field with linked vortex lines of horizontal and vertical vorticity on local and system scales. The role of VHTs as connectors of the primary and secondary circulation is emphasized.

1. Introduction

In our works of 2009–2012, a numerical approach was developed and applied for analysis of helical features of the vortex velocity field generated during tropical cyclone (TC) formation.

The main energy supply for tropical cyclones is provided from the ocean via latent and sensible heat fluxes (primarily latent fluxes). The supply is realized as cumulus convection. Recently, a long-standing idea on self-organization of convective processes in the tropical atmosphere under kinematically and thermodynamically favorable environmental conditions found a confirmation in a new scenario of tropical cyclogenesis proposed by Montgomery *et al.* (2006) and illustrated by high resolution (2-3 km horizontal grid spacing) direct numerical simulation by use of the Regional Atmospheric Modeling System (RAMS). The work demonstrated how a mesoscale tropical depression vortex could develop from cumulonimbus convection as a result of system-scale convergence and upscale vorticity growth.

The first investigation of TC formation from the perspective of helical features of atmospheric flows of different scales, which contributed to the organization of the cyclone (helical self-organization) was conducted by Levina & Montgomery (2010). Using the data of near-cloud-resolving numerical simulation (Montgomery *et al.*, 2006), helical characteristics of the velocity field were calculated and analyzed. It has been discovered that the process of hurricane formation is accompanied by the generation of nonzero and increasing helicity in moist convective atmospheric turbulence that implies a new topology of the flow when it is characterized by linked vortex lines (Moffatt, 1969). This means a break of the mirror symmetry of atmospheric turbulence when, following the theory of turbulence (Frisch, 1995), a large-scale vortex instability is possible.

Analysis of processes of vorticity and helicity generation on cloud convection and system scales was carried out in Levina & Montgomery (2011); Levina (2013) and revealed a new role of deep cumulonimbus rotating cores (Vortical Hot Towers – VHTs). While the work of Montgomery *et al.* (2006) showed that the convection could amplify pre-existing cyclonic vorticity and induce concentrations of vorticity much larger than that of the aggregative vortex, it was found by the foregoing studies that from the very first hours of TC formation, VHTs provided intense helicity generation on their local scales by linking horizontal and vertical components of vorticity. Further, a developing population of VHTs of different size and intensity (e.g., illustrated by Figure 1) created the secondary overturning circulation and sustained it during subsequent TC evolution. Moreover, VHTs linked the primary tangential and secondary overturning circulation on system scales (Figure 2) thereby providing the spiral structure and integrity of the evolving large-scale vortex. As soon as such linkage occurred, the nascent large vortex underwent a self-sustaining amplification process.

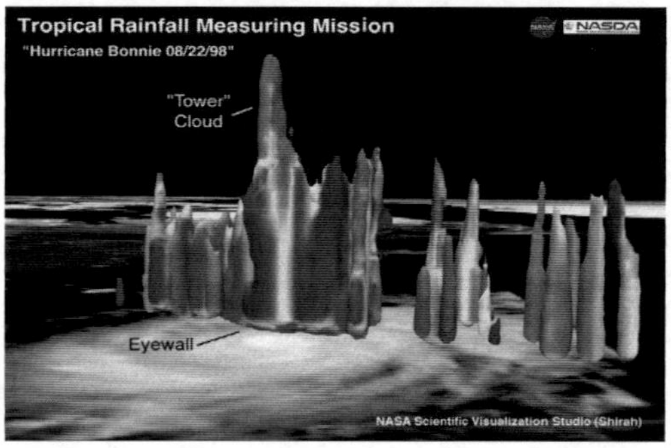

Figure 1. Hot Towers in Hurricane Bonnie 1998. Altitude of clouds is exaggerated. (Wikipedia, 2013).

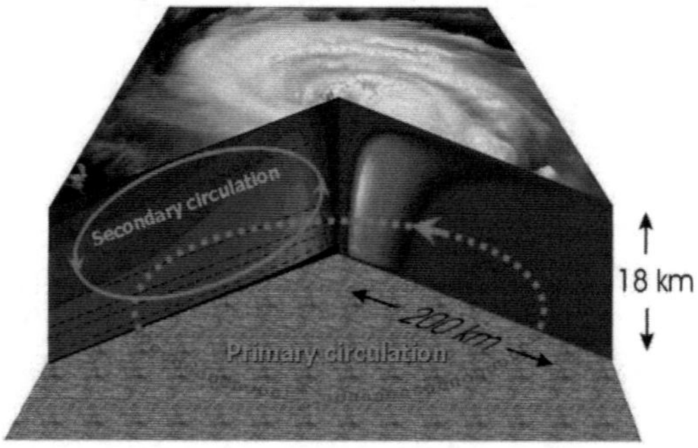

Figure 2. Scheme of the linkage of primary and secondary circulation. A composite TC is borrowed from (Emanuel, 2003).

In the present paper, we demonstrate how the VHTs work to form the secondary circulation, generate helicity and provide the linkage of the tangential and overturning circulation. By means of quantitative analysis, we substantiate that such linkage provides a positive energy feedback between these two circulations. The results are suggested to support the hypothesized model of a large-scale, helical-vortex instability that operates over the ocean in which sufficient moisture fluxes maintain convective instability. Based on this perspective we offer a numerical diagnosis for tropical cyclogenesis.

2. Helicity of the velocity field: a hypothesis on the turbulent vortex dynamo

Helicity of the velocity field is a pseudoscalar quantity defined as the scalar product of velocity $\mathbf{V}(\mathbf{r},t)$ and vorticity $\mathrm{curl}\mathbf{V}(\mathbf{r},t)$ vectors (Moffatt, 1969). More knowledge on helicity and helical turbulence can be gained in (Moffatt & Tsinober, 1992; Levina *et al.*, 2000; Levina, 2013) and references therein. The volume integral calculated in a specific space domain,

$$H = \int \mathbf{V}\cdot\mathrm{curl}\mathbf{V}d\mathbf{r}\,, \tag{1}$$

gives the helicity of vortex system, where $\mathbf{V}\cdot\mathrm{curl}\mathbf{V}$ is the helicity density of the flow. Both quantities are pseudoscalars, i.e., they change sign under change from a right-handed to a left-handed frame of reference (Moffatt & Tsinober, 1992).

A non-vanishing helicity implies the symmetry break of turbulence with respect to coordinate system reflections (Moffatt, 1978; Frisch, 1995). The mean helicity, like energy, is an inviscid quadratic constant of motion in barotropic fluids.

However, unlike energy the helicity can be both positive and negative. Its sign determines the predominance of the left-handed or the right-handed spiral motions in the examined flow. If we choose a right-handed Cartesian or orthogonal curvilinear frame for our further consideration, positive mean helicity will be generated in the moist atmosphere under the predominance of cyclonic updrafts and/or anticyclonic downdraft motions. Similarly, negative helicity will be generated for the case of anticyclonic updrafts and/or cyclonic downdraft flows.

Helicity is one of the most important characteristics for describing the structure of vortex fields. This quantity is a topological invariant, which measures the degree of linkage of the vortex lines (Moffatt, 1969, 1978; Moffatt & Tsinober, 1992; Frisch, 1995). The sources of helical turbulence are known to be the force fields of a pseudovector nature, such as magnetic or Coriolis force fields.

In the theory of turbulence, there exists a fundamental hypothesis about a small-scale helical turbulence that under certain conditions may evoke a large-scale instability governing the structure formation.

Thirty years ago in paper (Moiseev *et al.*, 1983), a hypothetical scenario for intensification and sustaining of large-scale vortex disturbances in the atmosphere due to energy transfer from small-scale helical convective turbulence – the helical-vortex instability or so called turbulent vortex dynamo – was proposed as a possible illustration for self-organization of turbulence with the broken mirror symmetry. The theoretical estimates obtained by substituting the specific atmospheric parameters in solutions (Moiseev *et al.*, 1983) were tested to describe tropical cyclone formation in the Earth's atmosphere as well as to explain the size and structure of large-scale long-lived vortex disturbances in Jovian atmosphere after the collision of comet Shoemaker-Levy 9 with Jupiter in July 1994. The theory showed a very good agreement with the characteristics of observed phenomena in the atmospheres of both planets. A summary of those results was given in a review work by Levina *et al.* (2000).

A proposed physical scenario for the helical self-organization (Moiseev *et al.*, 1983; Levina *et al.*, 2000) supposed an initial break of the mirror symmetry of turbulence, for example, due to a weak large-scale vortex disturbance. That is well consistent with the problem formulation in (Montgomery *et al.*, 2006), where simulations started with a pre-cursor mesoscale cyclonic vortex. The incipient process of helical self-organization would have to reveal itself in the non-zero mean helicity generation and consequent increasing of its level. Large helicity should suppress the energy flux to the scale of dissipation due to weakening of nonlinear interactions (Lilly, 1986), thereby favoring an energy accumulation in the inertial range, and then, its transfer to larger scale motions. In this connection, an inverse cascade or nonlocal energy transfer was expected to exist in developed helical turbulence. Under such circumstances, the flow patterns could exhibit a merging of small-scale turbulent cells and such an upscale growth would ultimately contribute to the organization of larger-scale structures.

As a sign that could help to precisely identify genesis of large-scale atmospheric vortex, one envisioned to have an emerging linkage of the system-scale primary (tangential) and secondary (overturning/transverse) circulation in a forming vortex structure. A schematic of such linkage is shown in Figure 2. The authors (Moiseev *et al.*, 1983; Levina *et al.*, 2000) anticipated that the linkage should result in a "helical" feedback between the system-scale circulations that would mean a mutual intensification of both circulations due to energy influx from small-scale moist convective helical turbulence. The initiation of such helical feedback would make the emerging large-scale vortex energy-self-sustaining. Thus the moment of time corresponding to the emergence of the feedback process could be considered as the genesis event for a tropical cyclone.

Journal of Physics: Conference Series **544** (2014) 012013 doi:10.1088/1742-6596/544/1/012013

3. Numerical approach to identify the helical-vortex instability in the atmosphere

A numerical approach for investigation of large-scale helical instability in the atmosphere was proposed in (Levina, 2006; Levina & Burylov, 2006) and first applied in a simpler case, namely, to simulate helical-vortex effects in laminar Rayleigh-Bénard convection by use of an additional helical force. Those simulations demonstrated effects, which were qualitatively new and appeared in a crisis manner. A nonzero helicity of the flow was generated, and after exceeding some its value, the large-scale vortex instability appeared. A key role in the process the vertical component of velocity played, and the corresponding evolutionary equation in the mathematical model (Levina, 2006; Levina & Burylov, 2006) closed a positive feedback loop. The instability evolution observed as an enlargement of horizontal scales of structures what happened by merging of helical vortex cells. The process was accompanied by flow intensification in newly forming larger vortices, similarly to that observed in atmospheric modeling (Montgomery *et al.*, 2006), which was based on hydro-thermodynamic equations for the atmosphere.

It should be particularly pointed out about a distinct enhancement of the heat transfer discovered in experiments (Levina, 2006; Levina & Burylov, 2006). The heat flux through a layer increased with an increase in mean helicity of the flow. A sharp intensification of heat transfer observed after exceeding the threshold of stability, and later during the instability evolution, after each merging of vortex structures. Thus, the effectiveness of heat transfer increased due to two factors, namely, increase in the mean helicity and decrease in a cell number of the convective system. Such energetic expedience suggests to us that a most effective "removal" of accumulating heat might be possibly one of the reasons for helical self-organization of convection in tropical cyclogenesis.

One of the most promising results of the applied approach was that it demonstrated how the onset of large-scale helical-vortex instability could be identified by examining helicity and kinetic energies of the tangential and overturning circulation.

In this paper, we are presenting a very similar analysis of helical and energetic characteristics based on high resolution atmospheric data and aimed at diagnosing TC genesis.

4. A key interaction between convection and vertical shear of horizontal wind

In paper (Montgomery *et al.*, 2006), a dipolar structure of the vorticity anomalies was found that was approximately colocated with the hot towers and an explanation was given of how such vertical vorticity anomalies might be produced. The following vorticity generation mechanism was proposed: The initial mesoscale convective vortex (MCV), with which simulation (Montgomery *et al.*, 2006) starts, provides an environment rich in horizontal and vertical vorticity. As the first updraft forms due to evolving convective instability, it tilts ambient horizontal vorticity into the vertical while at the same time stretching MCVgenerated vertical vorticity. As the updraft intensifies to become a hot tower, both ambient and tilting generated vertical vorticity is stretched even more, leading to a strong convectively generated vertical vorticity anomaly. At later times in the simulation(s) the convergence/stretching of near surface ($0 < z < 2$ km) vorticity by the convective plumes dominates the generation of vorticity by tilting processes.

A rudimentary understanding of the cloud-scale dynamics at early times can be obtained by examining the evolution of the updrafts (Montgomery *et al.*, 2006). Thus, the simulation in Experiment A2 "3 km", which is discussing in this paper, was initialized with a weak midlevel vortex elevated above the sea surface, with a maximal mean tangential wind at z = 4 km. At radii less than 75 km, the main vortex has a basic-state cyclonic tangential velocity field that increases in magnitude with height below z = 4 km and decreasing above. Ignoring buoyant effects for the time being, we can consider the horizontal vorticity profile at the initial time as being due solely to the vertical wind shear of the initial MCV. As sketched in Figure 3, this vertical shear profile will generate a radial vorticity profile that, when tilted upward by an updraft, generates negative (positive) relative vertical vorticity anomalies on the radially inward side of the updraft below (above) z = 4 km. As positive vertical vorticity is generated in the region of an updraft, vortex tube stretching further intensifies the positive vorticity anomaly. The corresponding VHT's vorticity dipole structure found in (Montgomery *et al.*, 2006) exhibited just this orientation and magnitude difference, supporting the proposed mechanism. Unlike the tilting term, however, the intensification of vertical vorticity by stretching can be exponential if the convergence is approximately constant during the parcel's ascent through the tower.

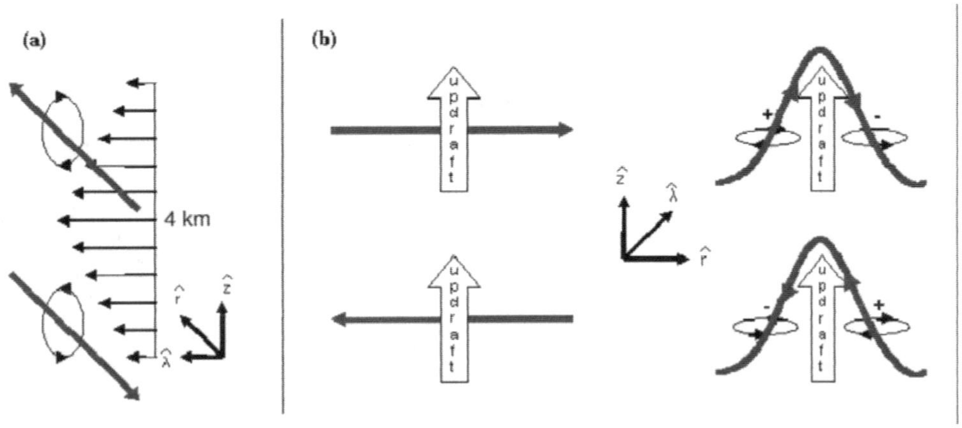

Figure 3. Schematic of vortex tilting within the initial MCV. Purple lines represent vortex filaments. (a) Radial vorticity generated by vertical shear profile of initial MCV. (b) Updraft tilts radial vortex filament upward, generating a vertical vorticity dipole with negative relative vorticity radially inward (outward) at heights below (above) z = 4 km. The figure is borrowed from (Montgomery *et al.*, 2006).

The described mechanism of horizontal vorticity transformation into the vertical one by a developing convective updraft was also interpreted in (Levina & Montgomery, 2011; Levina, 2013) as an effective way for generation of the linkage of vortex lines of horizontal and vertical components of vorticity, i.e., helicity generation. In work by Levina (2013), a comprehensive quantitative analysis was carried out for this process by using the data of six numerical experiments of Montgomery *et al.* (2006). Thus, it was found that maximal values of mean helicity generated by the interaction between the very first updraft and MCV within the initial two hours of those experiments were between $2.0 - 3.5 \times 10^{11}$ m^4 s^{-2}.

5. Numerical analysis: helical and energetic characteristics

Post-processing of the numerical simulation data of Montgomery *et al.* (2006) was carried out on the finest of three nested computational grids for subsequent times with a time increment of 10 minutes during 72 hours of numerical experiment. Characteristics were calculated in the computational domain of $276 \times 276 \times 20$ km^3, at first, in Cartesian co-ordinates (x, y, z) by use of uniform finite-difference grid with increments Δx, Δy, Δz. Throughout the post analysis the vertical increment was identical and equal to 500 m; the horizontal increments were $\Delta x = \Delta y =$ 3 km for numerical experiments discussed in this paper.

We also applied an analysis of system-scale dynamics from a traditional vortex-centric perspective when the Cartesian model data were transformed into a local cylindrical coordinate system. For these purposes we used the "Diagnostic Package" developed and described in (Montgomery *et al.* (2006), Appendix B).

For our investigation, let us choose a number of helicity and energy characteristics.

Between them are: helicity density field, i.e. helicity calculated in each grid point; mean helicity, i.e., helicity density integrated over the computational domain and averaged by a number of grid points, together with its integrated spatial contributions

$$\mathbf{V}\cdot\text{curl}\mathbf{V}\,, \quad < H >, \quad < H_{hor} >, \quad < H_{ver} > . \tag{2}$$

In real atmospheric conditions there always exists a definite preferred direction due to gravity. It makes reasonable to introduce in our investigation vertical and horizontal helicity, the latter as a sum of two spatial horizontal contributions. In the subsequent discussion we will use terms "total helicity", "horizontal helicity", and "vertical helicity" for $< H >$, $< H_{hor} >$, and $< H_{ver} >$, correspondingly. As it was analyzed and noted (Levina, 2013), non-zero total/horizontal helicity could be generated even when the vertical contribution of helicity vanished. However, that is only possible when the horizontal wind is changing with height, i.e., the occurrence of vertical shear of the horizontal wind. **Thus, non-zero horizontal helicity can be considered as a sign of existing or emerging shear flow. Non-zero vertical helicity, being a product of vertical velocity and vorticity, is an indicator of the presence of vortical convection in the examined area.**

To identify an emerging energy feedback between the primary and secondary circulation in the forming TC, we use the following set of kinetic energy characteristics:

$$E = \tfrac{1}{2}(\mathbf{V})^2\,, \qquad < E >, \qquad < E^P >, \qquad < E^S >, \tag{3}$$

which include E – kinetic energy density, and $< E >$ – averaged full energy in Cartesian coordinates; as well as separate averaged energy contributions calculated in cylindrical coordinates: of the primary tangential field, $< E^P >$, and secondary, $< E^S >$, formed by the radial and vertical components of velocity.

6. Diagnosis of TC genesis

Let us now examine how tropical cyclogenesis can be identified based on the new knowledge (Levina & Montgomery, 2010, 2011; Levina, 2013) on helical flow organization and pivotal role of VHTs in this process.

In the majority of numerical experiments (Montgomery *et al.*, 2006), a weak tangential circulation existed in low and mid troposphere (with a maximal tangential wind between 4 and 5 km altitude) from the very first beginning due to the initial MCV. The secondary circulation in six experiments examined in (Levina, 2013) appeared after several hours of flow development as a result of self-organization of vortical cloud convection or did not appear at all in unfavorable environment (Expt. C1, no initial MCV). The time interval needed for the secondary circulation to emerge depended on initial conditions. It should be noted that unlike initially organized tangential circulation, radial and vertical flows contributing to the secondary circulation are chaotic and weak enough during an initial time span. Nevertheless, such motions result in a small non-zero kinetic energy, $< E^S > \neq 0$. In order to diagnose an emerging secondary circulation, azimuthally mean fields of radial and vertical flow components will be analyzed as well as a number of other hydro- and thermodynamic mean fields.

Let us choose for diagnosing purposes, for instance, Experiment A2 "3km" of (Montgomery *et al.*, 2006). We will use kinetic energy and helicity characteristics (Fig. 4) as well as a set of other data gained from snapshots of spatial velocity, vorticity, helicity and temperature fields at horizontal and vertical cross-sections (not shown). To obtain azimuthal mean values the procedure applied in (Montgomery *et al.*, 2006)) was used.

The Expt. A2 started with kinetic energy of the primary tangential circulation, $< E^P >$, equal to 0.45×10^{16} m^5 s^{-2} (Fig. 4), entirely generated by the initial MCV, which had the maximum mean tangential wind 6.8 m s^{-1} at r = 75 km and z = 4 km. At the lower surface, the initial MCV had a relative vorticity of $O(1 - 2 \times 10^{-4}\ s^{-1})$. Initially, the secondary transverse circulation was absent that implied vanishing kinetic energy, $< E^S > = 0$. The maximal value of total helicity, $< H >$, generated by the initial conditions in this experiment was 3.3×10^{11} m^4 s^{-2} (Levina, 2013). This is an order of magnitude higher than 2.0×10^{10} m^4 s^{-2}, which was found insufficient for initiating the upscale organization in Expt. C1 (no initial MCV).

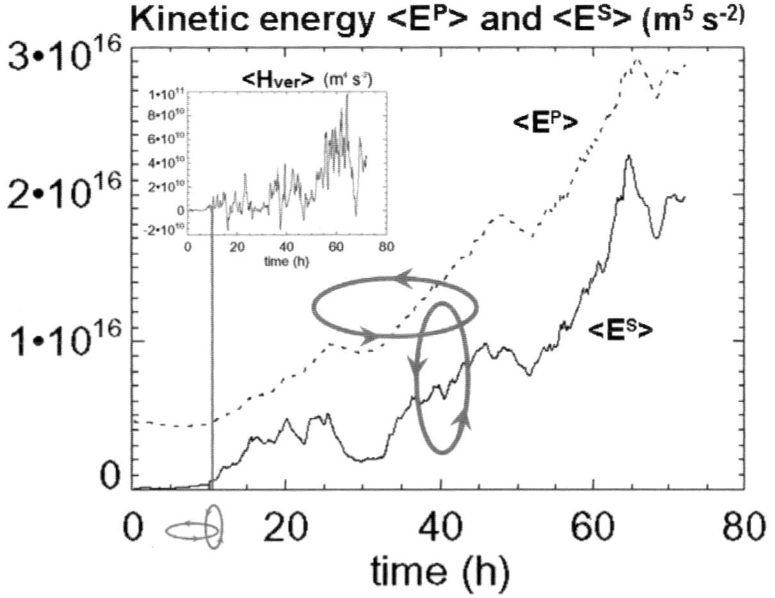

Figure 4. TC genesis at $t \sim 10$ h – the linkage of the primary and secondary circulation makes the forming vortex an integral helical system which undergoes self-sustaining amplification process. Vortical convection activity is illustrated by the vertical helicity in the upper panel.

During the first 10 hours, the tangential circulation is slightly weakening against its initial energy value (Fig. 4) and the maximal tangential wind undergoes a decrease from 6.8 m s^{-1} to 6.0 m s^{-1}, whilst $< E^S >$ becomes slowly increasing near t = 5 h. Our examination of the vertical velocity and vorticity fields confirmed a weak thermoconvective activity with no strong rotating updrafts at that time.

A few rotating convective flows become visible (with maximal vertical velocity about 2.5 m s^{-1} and cyclonic vertical relative vorticity 3.0×10^{-3} s^{-1} at z = 1 km) at t = 6 h. Near this time a process of merging of convective cells starts. This phenomenon was described in detail, analyzed quantitatively and interpreted in (Montgomery *et al.*, 2006) as a manifestation of upscale organization of atmospheric rotating moist convection. The process of merging is accompanied by not only an emergence of larger and stronger convective structures but also an increase in the background vorticity and helicity in adjacent areas. Convective activity is progressing during t = 6–9 h when a whole population of convective updrafts emerges, which interact with each other by merging, and result in more intense structures. At this time, one can observe a few cyclonically rotating convective cores - vortical hot towers (VHTs) of different horizontal and vertical sizes and intensity (the maximal vertical velocity and vorticity during these few hours are increasing up to 3.5 m s^{-1} and 5.0×10^{-3} s^{-1}, correspondingly, at z = 1 km). Within this time interval the most intense updrafts have 5–10 km in diameter and are growing gradually with time in height from 4–6 km up to 8–10 km.

The rotating convective cells start to generate (near t = 7–8 h) a non-vanishing and increasing third (vertical) contribution of helicity (Fig. 4, upper panel) and, thereby, a local linkage of vortex lines of horizontal and vertical flow components in a vicinity of each rotating updraft.

Each rotating convective structure contributes simultaneously to both the tangential and overturning circulation, namely, by its vertical vorticity to the former and by its vertical motion to the latter. Thus, such structure represents a natural link between the circulations on its local scale whilst their developed population should help link the primary and secondary circulation on the system scale.

To measure a degree of such linkage is just the main function of the quantity introduced above and known as "helicity".

Thus, in our case, an increase in the vertical contribution of helicity from zero up to approximately 5.0×10^9 m^4 s^{-2} within 6–9 h results in an amplification of horizontal/total helicity from 1.5×10^{11} m^4 s^{-2} up to 3.0×10^{11} m^4 s^{-2}. Just at this time, t = 6 h, a slight yet distinct increase in the kinetic energy $< E^S >$ starts.

Near t = 10 h dramatic changes in the flow intensity attract our attention – kinetic energy of the transverse circulation, $< E^S >$, increases sharply and soon after this, kinetic energy of the tangential circulation, $< E^P >$, becomes increasing as well (Fig. 4). This emerging mutual intensification of both circulations marks a critical point in a process of TC formation when **the vortex becomes self-sustaining. The time after which both $< E^S >$ and $< E^P >$ mutually increase may be considered a practical definition for the moment of tropical cyclogenesis**.

In order to identify reasons which are behind the observed phenomenon, let us search for explanations in corresponding flow structure and dynamics. Indeed, an intense helical updraft–VHT of about 14 km in height appears just at that time, t = 10 h (Fig. 5b,6-left).

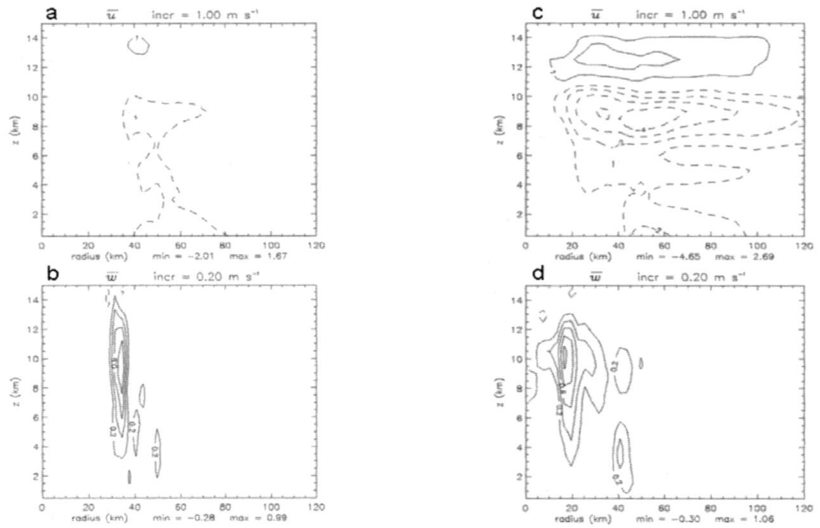

Figure 5. Azimuthal averages of radial wind, and vertical velocity at t = 10 h (a, b), and t = 12 h (c, d). The zero contour is omitted. Negative contours are dashed.

The VHT is found to be strong enough to generate a large transverse circulation, of tens of kilometers in horizontal directions and throughout the whole troposphere layer up to 14 km in height – Fig. 5a,b). This gives start to the formation of stable system-scale (hundreds of kilometers horizontally) secondary circulation during t = 10–12 h. In Fig. 5c,d plotted for

Journal of Physics: Conference Series **544** (2014) 012013 doi:10.1088/1742-6596/544/1/012013

t = 12 h, one can see the already formed transverse circulation with a pronounced low and middle level inflow, developed rising flow and upper level outflow. The circulation is sustained and linked with the primary circulation by the strong VHT and a number of smaller and less intense rotating convective flows (Fig. 6-right).

Since this point one can observe (Fig. 4) consequently increasing intensity of both the primary and secondary circulation with time, yet, interrupted periodically ($\sim$ 24 h) due to diurnal cycle. This indicates that a positive feedback has formed between the circulations and resulted in a developing large-scale vortex instability. The feedback is sustained by convective instability and vortical convection in the central region of the circulation. The convective instability there is maintained primarily by latent heat fluxes from the underlying sea surface, which need not increase with wind speed (Montgomery *et al.*, 2009; Montgomery & Smith, 2013). The existence of such rising warm flows suggests a release of potential energy that is converted into kinetic energy of developing large-scale helical vortex. The active feedback provides energy exchange between the primary and secondary circulation and their further mutual intensification. From a vorticity perspective, during the amplification of the system-scale circulation, the cyclonic vorticity consolidates via multiple diabatic vortex mergers and system-scale convergence of ambient and convectively-generated vorticity. The merging of cyclonic vorticity anomalies occurs also via the more familiar dry adiabatic vortex merger of convectively-generated vortex remnants.

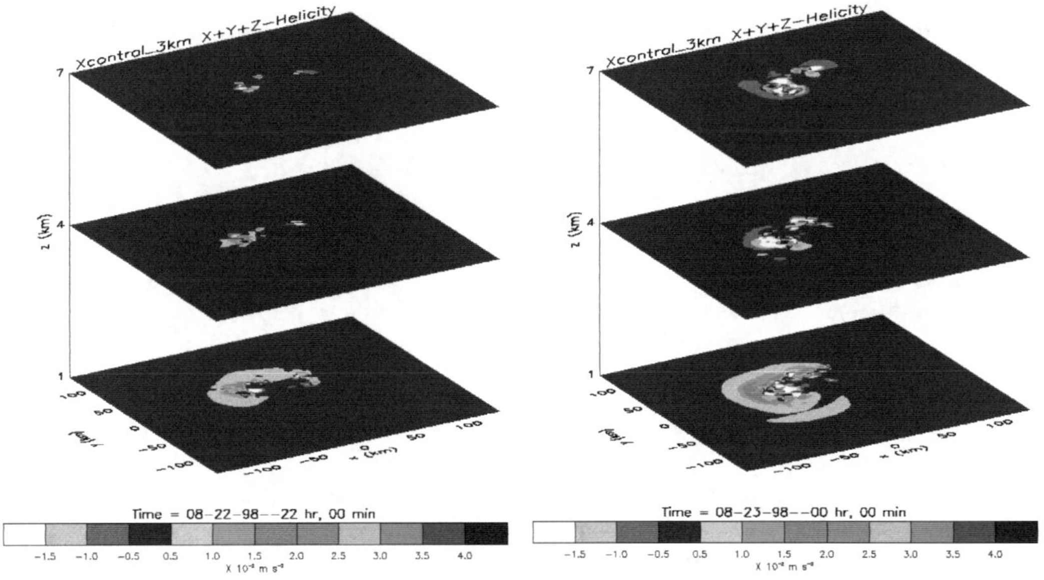

Figure 6. Helicity density ($\times$ 10^{-4} m s^{-2}) in three horizontal cross-sections of 276$\times$276 km^2 at $z = 1, 4, 7$ km, at $t = 10$ h (left) and $t = 12$ h (right).

The foregoing applied analysis suggests that diagnoses using helicity may help provide an answer to the important question of when will cyclogenesis commence given a favorable tropical environment? The findings here indicate that when the primary and secondary circulations become linked by vortical hot towers, the nascent vortex becomes helical. With adequate moisture fluxes from the underlying sea surface to maintain a degree of convective instability in the cumulus zone, the conditions are then set for a self-sustaining amplification process of the system-scale circulation.

Acknowledgments

G.L. is deeply grateful to Professor H.-K. Moffatt for his invitation to participate in the INI Programme on Topological Dynamics and important discussions, and thanks the co-organizers of the Programme and Isaac Newton Institute for hospitality. The work is supported by the Russian Academy of Sciences, Program of OEMMPU 12-T-1-1008, and the U.S. National Science Foundation, Grant ATM-0733380.

References

Emanuel, K. 2003 Tropical cyclones. *Annual Review of Earth and Planetary Sciences* **31**, 75–104.

Frisch, U. 1995 *Turbulence: the Legacy of A.N. Kolmogorov.* (Cambridge: Cambridge University Press).

Levina, G. V. 2006 Parameterization of helical turbulence in numerical models of intense atmospheric vortices. *Doklady Earth Sciences* **411** (9), 1417–1421.

Levina, G. V. 2013 Helical organization of tropical cyclones. *Preprint* NI13001-TOD. (Cambridge: Isaac Newton Institute for Mathematical Sciences, Preprint Series).

Levina, G. V. & Burylov, I. A. 2006 Numerical simulation of helical-vortex effects in Rayleigh-Bénard convection. *Nonlinear Processes in Geophysics* **13**, 205–222.

Levina, G. V., Moiseev, S. S. & Rutkevich, P. B. 2000 *Hydrodynamic alpha-effect in a convective system. (Advances in Fluid Mechanics Series* no 25*) Nonlinear Instability, Chaos and Turbulence*, vol 2, ed L Debnath and D N Riahi (Southampton, Boston: WITPress) chapter 4, pp 111–162.

Levina, G. V. & Montgomery, M. T. 2010 A first examination of the helical nature of tropical cyclogenesis. *Doklady Earth Sciences* **434** (1), 1285–1289.

Levina, G. V. & Montgomery, M. T. 2011 Helical scenario of tropical cyclone genesis and intensification. *J. Phys.: Conf. Series* **318** (7), 072012.

Lilly, D. 1986 The structure, energetics, and propagation of rotating convective storms. Part II: Helicity and storm stabilization. *Journal of Atmospheric Sciences* **43**, 126–140.

Moffatt, H.-K. 1969 The degree of knottedness of tangled vortex lines. *Journal of Fluid Mechanics* **35**, 117–129.

Moffatt, H.-K. 1978 *Magnetic Field Generation in Electrically Conducting Fluids.* (Cambridge: Cambridge University Press).

Moffatt, H.-K. & Tsinober, A. 1992 Helicity in laminar and turbulent flow. *Annual Review of Fluid Mechanics* **24**, 281–312.

Moiseev, S. S., Sagdeev, R. Z., Tur, A. V., Khomenko, G. A. & Shukurov, A. M. 1983 Physical mechanism of amplification of vortex disturbances in the atmosphere. *Soviet Physics Doklady (translation from Russian)* **28**, 925–928.

Montgomery, M. T., Nguyen, S. V., Smith, R. K. & Persing, J. 2009 Do tropical cyclones intensify by WISHE? *Quarterly Journal of the Royal Meteorological Society* **135**, 1697–1714.

Montgomery, M. T., Nicholls, M. E., Cram, T. A. & Saunders, A. B. 2006 A vortical hot tower route to tropical cyclogenesis. *Journal of Atmospheric Sciences* **63**, 355–386.

Montgomery, M. T. & Smith, R. K. 2013 Paradigms for tropical cyclone intensification. *Australian Oceanographic and Meteorological Journal, Bruce Morton Memorial Volume (accepted, March 5, 2013)* .

Wikipedia 2013 The Free Encyclopedia. www.en.wikipedia.org/wiki/Hot_tower.

Journal of Physics: Conference Series **544** (2014) 012014 doi:10.1088/1742-6596/544/1/012014

Higher order topological invariants from the Chern-Simons action

Roman V. Buniy[1], **Thomas W. Kephart**[2]

[1] Schmid College of Science, Chapman University, Orange, CA 92866, USA

[2] Department of Physics and Astronomy, Vanderbilt University, Nashville, TN 37235, USA

E-mail: roman.buniy@gmail.com,tom.kephart@gmail.com

Abstract. It is well known that for a field theory with Chern-Simons action, expectation values of Wilson line operators are topological invariants. The standard result is expressed in terms of the Gaussian linkings of closed curves defining the operators. We show how a judicious choice of Wilson lines leads to higher order topological linkings.

1. Introduction

Our understanding of knots and links in topological field theory began with the work of Polyakov (Polyakov, 1988) who showed that the expectation value of a Wilson line in Chern-Simons theory gives the Gaussian linking of the components of a link. This work was soon followed by Witten's (Witten, 1989) extensive exploration of expectation values of Wilson lines in non-abelian Chern-Simons theory which he showed is a natural framework for understanding Jones polynomials from knot theory. Perturbative expansion of the expectation values leads to higher order linking and the corresponding link polynomials (Guadagnini *et al.*, 1989). Our approach is somewhat different. Wilson loops in the abelian Chern-Simons theory also implicitly contains the requisite apparatus for the study of higher order linking, and we show here that this can be facilitated by a set of gauges (Buniy & Kephart, 2008*a*,*b*, 2006) tailored to the higher order linking problem. We begin with a discussion of intersection number and its relation to Gaussian linking. Next we review the minimal required background from Poincaré duality and de Rham theorem needed to carry out our calculations. We then introduce a special set of gauge potentials that allow us to arrive at our main result—higher order linking invariants at all orders. We conclude with detailed computations for several examples where linkings of various orders appear.

2. Intersection and linking numbers

Suppose C and C' are disjoint oriented closed curves in $\mathbb{R}^3$, and S and S' are surfaces such that $\partial S = C$ and $\partial S' = C'$. By smooth deformations of C which leave it in $\mathbb{R}^3\backslash C'$, we can change the number of points of intersection of C and S'. To find a quantity which is invariant under such deformations, we note that additional points that appear due to the deformations come in pairs, and the intersections of C and S' have opposite orientations for points in each pair (and likewise for disappearing pairs). If C and S' intersect transversely at a point P, we define the intersection index $I(C, S', P)$ to be equal 1 or -1 depending on the relative orientation of C and

Journal of Physics: Conference Series **544** (2014) 012014 doi:10.1088/1742-6596/544/1/012014

S' at P. We also define the intersection number $I(C, S')$ as the sum of the intersection indices over all points of intersection,

$$I(C, S') = \sum_{P \in C \cap S'} I(C, S', P). \tag{1}$$

It is clear that $I(C, S')$ is invariant under the above deformations as contributions due to additional pairs of points cancel.

One can also deform S' into $\tilde{S}'$ and notice that $I(C, S') = I(C, \tilde{S}')$, where $\partial S' = \partial \tilde{S}' = C'$. This means that the intersection number depends only on C and C'; for this reason we call it the linking number, $L(C, C') = I(C, S')$. Examination of relevant orientations gives $L(C, C') = L(C', C)$. The method of Green functions leads to an explicit Gauss expression for the linking number of the curves in terms of their parametrizations,

$$L(C, C') = (4\pi)^{-1} \int_{C \times C'} \sum_{a,b,c} \varepsilon_{abc} \frac{(x - x')^a}{\|x - x'\|^3} dx^b \wedge dx'^c. \tag{2}$$

3. Duality

From Poincaré duality and de Rham theorem (Griffiths & Harris, 1978; Hatcher, 2002), for a closed curve C in three dimensions, there exists a closed 2-form F such that for any 1-form B we have

$$\int_C B = \int_{\mathbb{R}^3} B \wedge F. \tag{3}$$

We call (C, F) a dual set. Since B is arbitrary, it is clear that F has support only on C, i.e. $\operatorname{supp} F = C$. Since $\mathbb{R}^3$ is simply connected, there exists a 1-form A such that $F = dA$. The Stokes theorem then gives [1]

$$\int_S dB = \int_{\mathbb{R}^3} dB \wedge A. \tag{4}$$

This means that there is a particular solution A such that

$$A(x) = \int_{y \in S} \delta(x - y) \sum_a dx^a * dy^a. \tag{5}$$

This is a singular gauge with $\operatorname{supp} A = S$. We can similarly introduce a dual set (C', F') such that $F' = dA'$ and for any 1-form B' we have

$$\int_{C'} B' = \int_{\mathbb{R}^3} B' \wedge F'. \tag{6}$$

Taking $B = A'$, $B' = A$ and using the Stokes theorem again, we find

$$\int_C A' = \int_{C'} A = \int_{\mathbb{R}^3} A \wedge dA' = L(C, C'). \tag{7}$$

Since $L(C, C')$ is a topological invariant, smooth deformations of C which leave it in $\mathbb{R}^3 \backslash C'$ should not change its value. This is possible if and only if $dA'|_{\mathbb{R}^3 \backslash C'} = 0$. If A' is exact, then $L(C, C') = 0$, and so a nontrivial case is when A' is a closed 1-form which is not exact.

Journal of Physics: Conference Series **544** (2014) 012014 doi:10.1088/1742-6596/544/1/012014

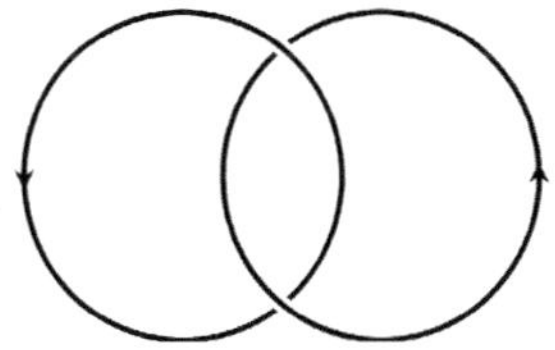

Figure 1. The Hopf link.

As an example with the first order linking, consider the Hopf link (Rolfsen, 1990; Kauffman, 2001) in Fig. 1. We use the singular gauge (5) to compute

$$L(C_1, C_2) = \int_{\mathbb{R}^3} dA_1 \wedge A_2 = 1. \tag{8}$$

This follows since we can choose $C_1 \cap S_2 = P$, where P is a point. The first order non-self linking is $\tilde{L}_1(C, C) = 2$. For $C = \cup_{1 \leq i \leq N} C_i$, we find

$$L_1(C, C) = \sum_{i,j} L(C_i, C_j), \tag{9}$$

which is in agreement with results in Refs. (Polyakov, 1988; Witten, 1989).

4. Second order fields

Suppose $\{C_i\}_{1 \leq i \leq N}$ are disjoint closed curves in $\mathbb{R}^3$ and let $F_i = dA_i$ be dual to C_i. We now construct a dual set (C_{ij}, F_{ij}) of the second order which satisfies

$$\int_{C_{ij}} B = \int_{\mathbb{R}^3} B \wedge F_{ij} \tag{10}$$

for any 1-form B and $i \neq j$. Since B is arbitrary, it follows that $\operatorname{supp} F_{ij} = C_{ij}$. The most general 2-form which can be expressed in terms of A_i and A_j is

$$F_{ij} = f_i dA_j - f_j dA_i + g A_i \wedge A_j, \tag{11}$$

where f_i, f_j and g are arbitrary functions. In order for F_{ij} to be closed, the functions have to satisfy certain conditions. A requirement $dF_{ij}|_{\mathbb{R}^3 \setminus (C_i \cup C_j)} = 0$ gives $dg|_{\mathbb{R}^3 \setminus (C_i \cup C_j)} = 0$ and without loss of generality we set $g = 1$. Requirements $dF_{ij}|_{C_i} = 0$ and $dF_{ij}|_{C_j} = 0$ then give

$$(df_j - A_j)|_{C_i} = 0, \tag{12}$$
$$(df_i - A_i)|_{C_j} = 0. \tag{13}$$

Integrating these conditions, we find a constraint $L(C_i, C_j) = 0$. This means that the second order field associated with a pair of closed curves can be defined only if the curves are unlinked.

Since $dF_{ij} = 0$, there exists a 1-form A_{ij} such that $F_{ij} = dA_{ij}$. We seek a solution in the form

$$A_{ij} = \tfrac{1}{2}(\gamma_i A_j - \gamma_j A_i). \tag{14}$$

[1] We assume that values of appropriate quantities vanish at infinity, so there is no contribution from boundary terms. Boundary terms also vanish if we use $\mathbb{S}^3$ instead of $\mathbb{R}^3$.

Journal of Physics: Conference Series **544** (2014) 012014 doi:10.1088/1742-6596/544/1/012014

From a requirement $F_{ij} = dA_{ij}$, the unknown functions γ_i and γ_j are found to satisfy

$$(d\gamma_i - A_i)|_{\mathbb{R}^3\backslash(C_i \cup C_j)} = 0, \tag{15}$$

$$d\gamma_i|_{C_i} = 0, \tag{16}$$

$$(d\gamma_i - 2A_i)|_{C_j} = 0, \tag{17}$$

$$(\gamma_i - 2f_i)|_{C_j} = 0 \tag{18}$$

and the same expressions with i and j interchanged. This implies $\gamma_i|_{\mathbb{R}^3\backslash(\cup_k C_k)} = \int_{\Gamma_i} A_i$, where Γ_i is a curve in $\mathbb{R}^3\backslash(\cup_k C_k)$; this means γ_i is a nonlocal quantity. If $S_i \cap C_j = \varnothing$, then $d\gamma_i|_{C_j} = 0$, and so $\gamma_i|_{C_j}$ is a constant. If $S_i \cap C_j \neq \varnothing$, then $S_i \cap S_j = \cup_m (S_i \cap S_j)_{(m)}$, where m labels disjoint segments of the intersection; for an example, see Fig. 2. Let $C_{j(m)}$ be the segment

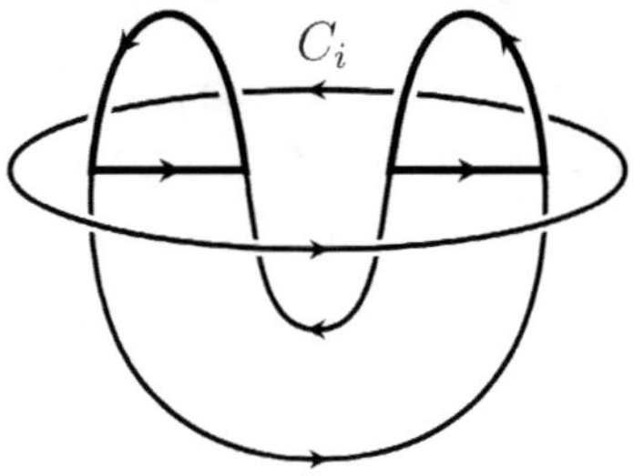

Figure 2. An example of the construction of the second order curve C_{ij}. Two components of C_{ij} are drawn with thick lines. S_i is a disk bounded by C_i, and the thick line segments in this disk are two disjoint segments of $S_i \cap S_j$.

of C_j which closes the curve $(S_i \cap S_j)_{(m)}$ and agrees with its orientation; this closed curve is $C_{ij(m)} = C_{j(m)} \cup (S_i \cap S_j)_{(m)}$. We define $C'_j = \cup_m C_{j(m)}$ and its complement in C_j is C''_j. It follows from the above relations that $\gamma_i|_{C'_j}$ and $\gamma_i|_{C''_j}$ are constants such that $\gamma_i|_{C'_j} - \gamma_i|_{C''_j} = 2$. Without loss of generality, we set $\gamma_i|_{C'_j} = 2$ and $\gamma_i|_{C''_j} = 0$. Using the definition of F_{ij}, the duality condition now becomes

$$\int_{C_{ij}} B = -\int_{C'_i} B + \int_{C'_j} B + \int_{S_i \cap S_j} B. \tag{19}$$

This implies $C_{ij} = C_i'^{-1} \cup C'_j \cup (S_i \cap S_j)$, which agrees with $\operatorname{supp} F_{ij} = C_{ij}$. See Fig. 2 for an example of the above construction. The Stokes theorem gives

$$\int_{S_{ij}} dA = \int_M dA \wedge A_{ij}, \tag{20}$$

where a surface S_{ij} is such that $\partial S_{ij} = C_{ij}$. This means $\operatorname{supp} A_{ij} = S_{ij}$.

The above construction requires $L(C_i, C_j) = 0$. Since the curves C_i and C_j are unlinked, the surfaces S_i and S_j can be chosen to be disjoint, in which case $dA_{ij}|_{\mathbb{R}^3\backslash(C_i \cup C_j)} = 0$. One can similarly proceed to construct higher order fields. For example, if $L(C_{ij}, C_k) = 0$ for all (i, j, k), then we can define the third order 1-forms A_{ijk}. In particular, they satisfy

$$dA_{ijk}|_{\mathbb{R}^3\backslash(C_i \cup C_j \cup C_k)} = (A_i \wedge A_{jk} + A_{ij} \wedge A_k)|_{\mathbb{R}^3\backslash(C_i \cup C_j \cup C_k)} \tag{21}$$

Journal of Physics: Conference Series **544** (2014) 012014 doi:10.1088/1742-6596/544/1/012014

for $i \neq j \neq k$.

If all linkings of order q, where $1 \leq q \leq p-1$, vanish, then we can similarly construct dual sets (C_{I_p}, dA_{I_p}), where $I_p = (i_1, i_2, \ldots, i_p)$. The quantities dA_{I_p} are related to what is known in algebraic topology as the Massey products of cohomology groups (Massey, 1959, 1968; Kraines, 1966; O'Neill, 1979; Fenn, 1983); see also Refs. (Monastyrsky & Retakh, 1986; Berger, 1990).

As an example with second order linking, consider the Borromean rings (Rolfsen, 1990; Kauffman, 2001) in Fig. 3. We use the singular gauge to compute

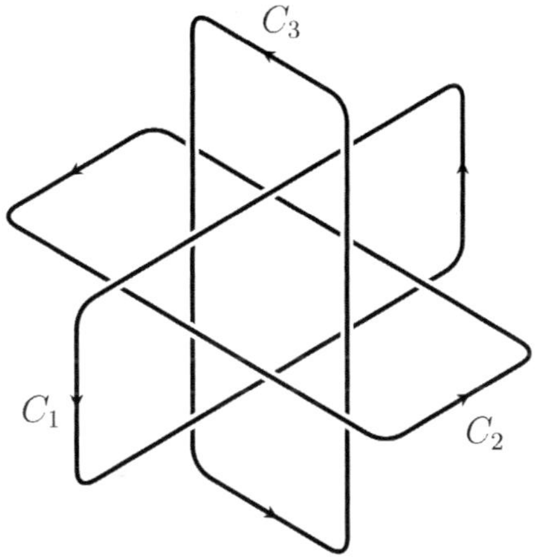

Figure 3. The Borromean rings.

$$L(C_i, C_j) = \int_{\mathbb{R}^3} dA_i \wedge A_j = 0 \tag{22}$$

for $i \neq j$. This follows since we can choose $C_i \cap S_j = \varnothing$ for $i \neq j$. We take $C = C_1 \cup C_2 \cup C_3$ and find that the first order non-self linking vanishes,

$$\tilde{L}_1(C, C) = 2L(C_1, C_2) + 2L(C_2, C_3) + 2L(C_3, C_1) = 0. \tag{23}$$

We define the second order curves $\{C_{ij}\}$ for $i \neq j$. Again we use the singular gauge to compute

$$L(C_{ij}, C_k) = \int_{\mathbb{R}^3} A_i \wedge A_j \wedge A_k - \tfrac{1}{2}\int_{C_i} \gamma_j A_k + \tfrac{1}{2}\int_{C_j} \gamma_i A_k = \varepsilon_{ijk}, \tag{24}$$

where (i, j, k) is a permutation of $(1, 2, 3)$. The integral over $\mathbb{R}^3$ equals ε_{ijk} since we can choose $S_1 \cap S_2 \cap S_3 = P$, where P is a point. The integrals over C_i and C_j vanish since we can choose S_k such that $C_i \cap S_k = \varnothing$ and $C_j \cap S_k = \varnothing$. Similarly,

$$\tilde{L}(C_{ij}, C_j) = \int_{\mathbb{R}^3} A_i \wedge A_j \wedge A_j - \tfrac{1}{2}\int_{C_i} \gamma_j A_j + \tfrac{1}{2}\int_{C_j} \gamma_i A_j = 0 \tag{25}$$

for any $i \neq j$. We take $C = C_1 \cup C_2 \cup C_3 \cup C_{12} \cup C_{23} \cup C_{31}$ and find that the second order non-self linking of C is

$$\tilde{L}_2(C, C) = 2L(C_{12}, C_3) + 2L(C_{23}, C_1) + 2L(C_{31}, C_2) = 6. \tag{26}$$

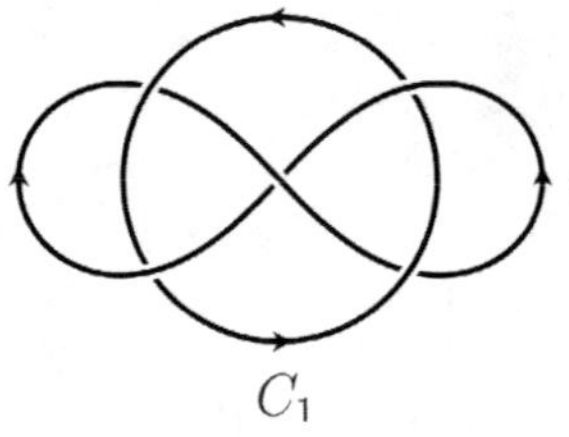

Figure 4. The Whitehead link.

As an example with the third order linking, consider the Whitehead link (Rolfsen, 1990; Kauffman, 2001; Monastyrsky & Retakh, 1986) in Fig. 4. We use the singular gauge to compute

$$L(C_1, C_2) = \int_{\mathbb{R}^3} dA_1 \wedge A_2 = 0. \tag{27}$$

This follows since we can choose $C_1 \cap S_2 = P_1 \cup P_2$, where P_1 and P_2 are points at which the contributions of $dA_1 \wedge A_2$ to the integral cancel due to the opposite orientations. We take $C = C_1 \cup C_2$ and find that the first order non-self linking vanishes, $\tilde{L}_1(C, C) = 0$.

We define the second order curve C_{12} and use the singular gauge to compute

$$L(C_{12}, C_k) = \int_{\mathbb{R}^3} A_1 \wedge A_2 \wedge A_k - \tfrac{1}{2} \int_{C_1} \gamma_2 A_k + \tfrac{1}{2} \int_{C_2} \gamma_1 A_k = 0 \tag{28}$$

for $k = 1$ or $k = 2$. We take $C = C_1 \cup C_2 \cup C_{12}$ and find that the second order non-self linking vanishes, $\tilde{L}_2(C, C) = 0$.

We use the singular gauge to compute the third order self linking,

$$L(C_{12}, C_{12}) = \int_{\mathbb{R}^3} A_1 \wedge A_2 \wedge A_{12} = 2. \tag{29}$$

This follows since S_{12} is the union of disks $S_{12(1)}$ and $S_{12(2)}$ such that $S_{12(1)} \cap S_{12(2)} = P$, where P is a point, and we can choose S_1 and S_2 such that $S_1 \cap S_2 \cap S_{12} = P$.

We define the third order curves C_{121} and C_{212}. We use the singular gauge to compute

$$L(C_{121}, C_2) = \int_{\mathbb{R}^3} (A_1 \wedge A_{21} + A_{12} \wedge A_1) \wedge A_2 = 2 \int_{\mathbb{R}^3} A_1 \wedge A_2 \wedge A_{12} = 4 \tag{30}$$

and similarly $L(C_{212}, C_1) = 4$. We take $C = C_1 \cup C_2 \cup C_{12} \cup C_{121} \cup C_{212}$ and find that the third order non-self linking of C is

$$\tilde{L}_3(C, C) = 2L(C_{121}, C_2) + 2L(C_{212}, C_1) = 16. \tag{31}$$

5. Path integral

We want to construct a field theory for which expectation values of observables are topological invariants of various orders. For this we need to specify the action and the observables. We choose the Chern-Simons action,

$$S(B) = \int_{\mathbb{R}^3} B \wedge dB, \tag{32}$$

Journal of Physics: Conference Series **544** (2014) 012014 doi:10.1088/1742-6596/544/1/012014

since it has all the necessary topological properties (Witten, 1989). First, it does not depend on the choice of metric. Second, it can be related to linking numbers in the following way. Suppose Γ_α is a closed curve on which dB takes a constant value dB_α. (In order for Γ_α to be closed, we may need to identify points at infinity by considering $\mathbb{S}^3$ instead of $\mathbb{R}^3$. We also avoid field configurations with sources, such as monopoles.) Since a union of curves $\Gamma = \cup_\alpha \Gamma_\alpha$ densely fills $\mathbb{R}^3$, an arbitrary 1-form B can be written as $B = \sum_\alpha B_\alpha$. The action becomes

$$S(B) = \sum_{\alpha,\beta} L(\Gamma_\alpha, \Gamma_\beta), \tag{33}$$

and so (apart from the choice of measure for Γ) we interpret it as the self-linking $L(\Gamma,\Gamma)$ of the set of closed field lines of dB.

If $C = \cup_\alpha C_\alpha$ is a union of disjoint closed curves, then an integral $\int_C B$ is invariant with respect to deformations of C if and only if $dB = 0$. Since $dB = 0$ is the classical equation of motion for $S(B)$, the integral is a topological observable (at least in the semiclassical limit). Proceeding as above, we find

$$\int_C B = \sum_{\alpha,\beta} L(\Gamma_\alpha, C_\beta), \tag{34}$$

which we interpret as the linking $L(\Gamma, C)$ of the sets of closed field lines of dB with C. Since the measure in the path integral is $\exp(iS(B))$, it is convenient to consider as an observable a Wilson loop operator

$$W(C,B) = \exp\Big(i \int_C B\Big). \tag{35}$$

We thus need to compute the expectation value

$$Z(C) = \int DB \exp(iS(B)) W(C,B). \tag{36}$$

Using duality, this becomes

$$Z(C) = \int DB \exp\Big(i \int_{\mathbb{R}^3} B \wedge d(B+A)\Big), \tag{37}$$

where dA is dual to C. The change of the variable $B = B' - \frac{1}{2}A$ gives four terms in the exponent. One term corresponds to the path integral of the measure without a Wilson loop, $Z(\varnothing)$, the two mixed terms combine into a boundary term, and the forth is the linking invariant. Ignoring the boundary term, we find

$$Z(C) = Z(\varnothing) \exp\big(-\tfrac{1}{4} i (L(C,C))\big). \tag{38}$$

A product of Wilson loop operators can be written as a single Wilson loop operator for the union of the corresponding loops,

$$\prod_p \prod_{I_p} W(C_{I_p}, B) = W(C,B), \tag{39}$$

where $C = \cup_p \cup_{I_p} C_{I_p}$ and C_{I_p} is a curve of order p. It follows that

$$L(C,C) = \sum_{p,q} \sum_{I_p, J_q} L(C_{I_p}, C_{J_q}). \tag{40}$$

Here $L(C_{I_p}, C_{J_q})$ is the first order linking of curves C_{I_p} and C_{J_q}. This quantity can also be interpreted as a linking of order $p+q-1$ of the curves $\{C_i\}_{1\leq i\leq N}$. In general, $L(C,C)$ is a sum of linkings of various orders,

$$L(C,C) = \sum_r L_r(C,C), \tag{41}$$

where the expressions for several lowest order terms are

$$L_1(C,C) = \sum_{i,j} L(C_i, C_j), \tag{42}$$

$$L_2(C,C) = \sum_{i,j,k} L(C_{ij}, C_k), \tag{43}$$

$$L_3(C,C) = L_{3,1}(C,C) + L_{3,2}(C,C), \tag{44}$$

$$L_{3,1}(C,C) = \sum_{i,j,k,l} L(C_{ijk}, C_l), \tag{45}$$

$$L_{3,2}(C,C) = \sum_{i,j,k,l} L(C_{ij}, C_{kl}). \tag{46}$$

6. Conclusions

In Chern-Simons theory the first order (Gaussian) linking of two curves C_1 and C_2 associated with the 1-forms A_1 and A_2 can be ascribed to the topological properties of the expectation value of the associated Wilson lines. Here we have generalized this idea to the case of higher order linking. First, if C_1 and C_2 are unlinked, then we can define a new second order 1-form A_{12} via Eq. (14). Unlike C_1 and C_2, the associated curve C_{12} is not fixed in space, but is however sufficient for our topological needs, which are to investigate its linking with other curves, since this property is invariant. If, for instance, we have a third curve C_3 unlinked with both C_1 and C_2, but linked with C_{12}, then we find nonzero second order linking, i.e., a nonvanishing second order topological invariant. The Borromean rings is the simplest topological configuration for which this invariant is nonzero. The argument generalizes to higher order invariants. The Whitehead link is the simplest example with a nonvanishing third order topological invariant. For curves $\{C_i\}_{1\leq i\leq N}$, with no linking of orders $1, 2, \ldots, p-1$ and nonzero linking of order p, the general results are obtained from the path integral of the expectation value of a product of Wilson loop operators which depend on curves of order p in the Chern-Simons gauge theory. One could apply these results to the investigation of higher order link polynomials.

Acknowledgments

We are grateful to the Isaac Newton institute for hospitality. RVB thanks Chapman University for support. The work of TWK was supported by U.S. DoE grant number DE-FG05-85ER40226 and Vanderbilt University College of Arts and Sciences.

References

BERGER, M. A. 1990 Third-order link integrals. *J. Phys. A: Math. Gen.* **23**, 2787.

BUNIY, R. V. & KEPHART, T. W. 2006 Higher order topological invariants from the Chern-Simons action. ArXiv:hep-th/0611336.

BUNIY, R. V. & KEPHART, T. W. 2008*a* A proposal for detecting second order topological quantum phase. *Phys. Lett.* **A372**, 2583–2586.

BUNIY, R. V. & KEPHART, T. W. 2008*b* Higher order topological actions. *Phys. Lett.* **A372**, 4775–4778.

FENN, R. A. 1983 **Techniques of Geometric Topology**. Cambridge University Press, Cambridge.

GRIFFITHS, P. & HARRIS, J. 1978 **Principles of Algebraic Geometry**. Wiley-Interscience, New York.

GUADAGNINI, E., MARTELLINI, M. & MINTCHEV, M. 1989 Perturbative Aspects of the Chern-Simons Field Theory. *Phys. Lett.* **B227**, 111.

HATCHER, A. 2002 **Algebraic Topology**. Cambridge University Press, Cambridge.

KAUFFMAN, L. 2001 **Knots and Physics**. World Scientific, Singapore.

KRAINES, D. 1966 Massey higher products. *Trans. Amer. Math. Soc.* **124**, 431.

MASSEY, W. S. 1959 Some higher order cohomology operations. In *Symp. Int. Topologia Algebraica*, p. 145. Mexico.

MASSEY, W. S. 1968 Higher order linking numbers. In *Proc. Conf. on Algebraic Topology*, p. 174. University of Illinois at Chicago.

MONASTYRSKY, M. I. & RETAKH, V. S. 1986 Topology of defects in condensed matter. *Commun. Math. Phys.* **103**, 445.

O'NEILL, E. J. 1979 Higher order Massey products and links. *Trans. Amer. Math. Soc.* **248**, 37.

POLYAKOV, A. M. 1988 Fermi-Bose Transmutations Induced by Gauge Fields. *Mod. Phys. Lett.* **A3**, 325.

ROLFSEN, D. 1990 **Knots and Links**. Publish or Parish, Houston.

WITTEN, E. 1989 Quantum Field Theory and the Jones Polynomial. *Commun. Math. Phys.* **121**, 351.

Journal of Physics: Conference Series **544** (2014) 012015 doi:10.1088/1742-6596/544/1/012015

On combinatorial properties of a higher asymptotic ergodic invariant of magnetic lines

P.M.Akhmet'ev[1]

[1] IZMIRAN,
Troitsk, Moscow region, Russia

E-mail: pmakhmet@izmiran.ru

Abstract. We investigate combinatorial properties of a higher invariant of magnetic lines, which is defined in the paper Akhmet'ev-1 (2013). Assume that a 3-component link $\mathbf{L}$ is modeled by a magnetic field $\mathbf{B}$, which is represented by 3 closed magnetic lines. Main Theorem relates the integral invariant $M(\mathbf{B})$ and a combinatorial invariant $\tilde{M}(\mathbf{L})$, defined from the Conway polynomial. As a corollary of Main Theorem, asymptotic properties for combinatorial links are proposed. The combinatorial invariant $\tilde{M}$ satisfies these asymptotic properties.

1. Introduction

Magnetic fields are investigated by the helicity integral $\chi_{\mathbf{B}}$, for details see the paper (Moffatt, 2001). The helicity integral is related with the Gauss integral for the linking number of two closed curves. The dimension of the helicity integral is $Gs^2 \cdot sm^4$. The idea of topological investigation of turbulent magnetic fields is assumed step-by-step generalization, using higher invariants of links instead of the linking number. Such an approach is possible because of a theorem by V.I.Arnol'd, see Arnol'd V.I. & Khesin B.A. (1998) Ch. 3, paragraph 4. This theorem provides the formula of the helicity integral as the asymptotic ergodic Hopf invariant. In the papers Akhmet'ev (2012), Akhmet'ev-2 (2013) the theorem by V.I.Arnol'd is generalized.

Briefly, the idea of the considered approach proposed by V.I.Arnol'd is following. Assume, for simplicity, that a magnetic field $\mathbf{B}$ is structured into a finite set $U = \cup_i U_i$ of magnetic tubes; definition of magnetic tubes is proposed in Berger M.A. & Field G.B. (1984). The case when a magnetic field $\mathbf{B}$ is inside the only magnetic tube $U = U_0$ is possible. Take the set of all magnetic lines of $\mathbf{B}$ (generally speaking, non-closed) and take all collections of k magnetic lines (non-trivial examples for $k \geq 4$ are unknown). Take an appropriate Vassiliev's invariant I of k-component links and apply this invariant formally for a k-collection of non-closed magnetic lines of the fixed length l. Take the asymptotic limit $I(+\infty) = \lim_{l \to +\infty} I(l)$ of the integrals, if this limit exists. The average value of $I(+\infty)$ over all k-collections of magnetic line is called an asymptotic ergodic invariant of $\mathbf{B}$.

In the paper Akhmet'ev (2012) a modification of the helicity invariant for $k = 2$ is introduced. This new invariant is called the quadratic helicity and is denoted by $\chi^{[2]}_{\mathbf{B}}$. The quadratic helicity is not a higher invariant: this invariant is derived from the square of the linking number, the dimension of the quadratic helicity invariant is $Gs^4 sm^4$.

A geometrical meaning of $\chi^{[2]}_{\mathbf{B}}$ is following. The asymptotic linking number of all pairs of magnetic lines determines the (absolutely) integrable function $\chi : U \times U \to \mathbb{R}$ on the 2-point configurations space (recall that an integrable function could be infinite-valued on a set of the zero-measure). This function is frozen-in for the ideal magnetic field and is called the helicity density. The helicity integral $\chi_{\mathbf{B}}$ is the mean value of the helicity density χ on $U \times U$. Moreover, the square χ^2 of the helicity density is an integrable function. The quadratic helicity $\chi_{\mathbf{B}}^{[2]}$ is the dispersion of χ. In the prove that $\chi_{\mathbf{B}}^{[2]}$ is invariant for ideal magnetic fields one replaces short paths of end-points of open long magnetic lines by pairs "'monopole–antimonopole"' as in Akhmet'ev (2012), Section 2.2.

Because the quadratic helicity is defined as an ergodic invariant, this invariant is continuously depended of dissipations of **B**. A priori this is not evident, because a C^{∞}–small perturbation of **B** changes totally a structure of magnetic lines. In the paper Akhmet'ev (2012), Theorem 2, shows that the quadratic helicity depends even with a Lipschitz coefficient (but not smoothly!) of the magnetic field **B**. The quadratic helicity could be explicitly estimated from the observed magnetic field **B** as in the paper Akhmet'ev (2012), inequality (5).

In the paper Arnol'd V.I. (2000) the following Problem 1984-12 is formulated: "To transform asymptotic ergodic definition of the Hopf invariant of divergence-free vector fields to a theory by S.P.Novikov, which generalized Whitehead product in homotopy group of spheres"'. As it is well-known, by Whitehead product in 3D is assumed the Massey integrals for Milnor's invariants of links. The Massey integrals is applied in the framework of MHD problems by Monastyrsky, M.V. & Retakh, V.S. (1986), for a detailed exposition and references see, f.ex. Mayer Ch. (2003). The Massey integrals are defined with restrictions. New aspects of the problem, related with configuration spaces, recently, is proposed in the paper De Turck, D. *et al.* (2011), Cohen, F.R. *et al.* (2012).

To apply the Massey integrals for magnetic lines the following problem arises. Massey integrals are not totally–defined: an integral of an order k is well defined, only if all integrals of the orders $\leq k-1$ are vanished. In particular, to define the simplest integral $\mu_{1,2,3}$ for 3-component link one should assume that all pairwise linking numbers between components are vanished (3 conditions). In the paper Akmetiev P. & Ruzmaikin A (1955) is observed that for the Sato-Levine invariant (this invariant for 2-component links is defined as the $\mu_{1,1,2,2}$ integral) the only linking number between components has to be vanished. This assumption is restrictive for applications. The Massey integrals are not asymptotic ergodic invariants.

In the paper Akhmetiev (2005) an invariant M for magnetic fields, which is structured into 3 disjoint magnetic tubes in introduced. This invariant is written as a totally well-defined Massey's integral. This is a higher integral: M is not a function of the pairwise integral linking coefficients of the magnetic tubes. In the paper Akhmet'ev-1 (2013) the integral M is well-defined for magnetic lines using a integral over a finite-type configuration space. In the paper Akhmet'ev-2 (2013) (Section 6, Theorem 14) is particularly proved the following conjecture, which is a positive solution of the Problem 1984-12 by V.I.Arnol'd.

Conjecture 1 The invariant M is well-defined as a higher asymptotic ergodic invariant of magnetic fields.

The most difficult part of the proof of the Conjecture is to rewrite peripheral terms of the integral (9) – (14) in an ergodic stile, analogously to the invariant $\chi_{\mathbf{B}}^{[2]}$, Akhmet'ev (2012).

The goal of the paper is to express the integral M in terms of Conway polynomial of links. Additionally, a combinatorial property of the invariant M namely, the formula (23), is deduced from Conjecture 1. This combinatorial formula (23) is explicitly proved in Akhmet'ev (2011) (Section 6 Theorem 8).

Journal of Physics: Conference Series **544** (2014) 012015 doi:10.1088/1742-6596/544/1/012015

2. The integral M

Let us recall the integral formula from Akhmetiev (2005) in the form Akhmet'ev-1 (2013). Assume that a magnetic field $\mathbf{B}$ is localized into 3 disjoint magnetic tubes

$$\mathbf{B} = \mathbf{B}_1 \cup \mathbf{B}_2 \cup \mathbf{B}_3,$$

each magnetic tube is presented by a thin solid torus $U_i \subset \mathbb{R}^3$, in this torus a prescribed coordinate system $U_i = D^2 \times S^1$ is well-defined, where $S^1 \subset \mathbb{C}$ is the standard circle. The integral magnetic flow of $\mathbf{B}_i$ is well-defined through a cross-section $D^2 \times pt \subset D^2 \times S^1$ of the magnetic tube U_i.

Denote by $\mathbf{A}_i$ the vector-potentials of $\mathbf{B}_i$, $i = 1, 2, 3$ in the corresponding magnetic tube:

$$\mathrm{rot}\mathbf{A}_i = \mathbf{B}_i.$$

Denote by $(1, 2), (2, 3), (3, 1)$ the integral linking coefficients, which are given by the integrals:

$$(i, j) = \int_{U_i} (\mathbf{A}_j, \mathbf{B}_i) dU_i = \int_{U_j} (\mathbf{A}_i, \mathbf{B}_j) dU_j, \quad i, j = 1, 2, 3, \quad i \neq j.$$

The restriction of the potential $(1, 3)(2, 3)\mathbf{A}_2 - (3, 2)(1, 2)\mathbf{A}_3$ to the magnetic tube U_1 coincides with the gradient of a function, which is well-defined up to an additive constant. This function is denoted by $\phi_1 : U_1 \to \mathbb{R}$, two analogous functions are denoted by $\phi_2 : U_2 \to \mathbb{R}$, $\phi_3 : U_3 \to \mathbb{R}$. Those functions are called scalar potentials.

We get $\varphi_{2,1} = (2, 1)\varphi_1^0 + \varphi_{j,i}^{var} + C_{2,1}$, where φ_1^0 is a linear multivalued function in $U_1 = D^2 \times S^1$ with respect to the second factor. The functions $\varphi_{2,1}^{var}$, $\varphi_{3,1}^{var}$ in U_1 satisfy the equation $\int_{U_1} \varphi_{2,1}^{var} dU_1 = 0$; $\int_{U_1} \varphi_{3,1}^{var} dU_1 = 0$.

With respect to the definition above, we get:

$$\phi_1 = (3, 1)\varphi_{2,1}^{var} - (1, 2)\varphi_{3,1}^{var} + C_1. \tag{1}$$

The functions ϕ_2, ϕ_3 are defined analogously:

$$\phi_2 = (1, 2)\varphi_{3,2}^{var} - (2, 3)\varphi_{1,2}^{var} + C_2, \tag{2}$$

$$\phi_3 = (2, 3)\varphi_{1,3}^{var} - (3, 3)\varphi_{2,3}^{var} + C_3. \tag{3}$$

The normalized constant C_1 is well-defined by the equation:

$$\begin{aligned} 2C_1 = \textstyle\int_{U_1} \varphi_{2,1}^{var}(\mathbf{grad}\varphi_{3,1}^{var}, \mathbf{B}_1) - (\mathbf{grad}\varphi_{2,1}^{var}, \mathbf{B}_1)\varphi_{3,1}^{var} dU_1 + \\ \tfrac{2}{3} \textstyle\int_{\mathbb{R}^3} \langle \mathbf{A}_1(x), \mathbf{A}_2(x), \mathbf{A}_3(x) \rangle d\mathbb{R}^3, \end{aligned} \tag{4}$$

$$\begin{aligned} 2C_2 = \textstyle\int_{U_2} \varphi_{3,2}^{var}(\mathbf{grad}\varphi_{1,2}^{var}, \mathbf{B}_2) - (\mathbf{grad}\varphi_{3,2}^{var}, \mathbf{B}_2)\varphi_{1,2}^{var} dU_2 + \\ \tfrac{2}{3} \textstyle\int_{\mathbb{R}^3} \langle \mathbf{A}_1(x), \mathbf{A}_2(x), \mathbf{A}_3(x) \rangle d\mathbb{R}^3, \end{aligned} \tag{5}$$

$$\begin{aligned} 2C_3 = \textstyle\int_{U_3} \varphi_{1,3}^{var}(\mathbf{grad}\varphi_{2,3}^{var}, \mathbf{B}_3) - (\mathbf{grad}\varphi_{1,3}^{var}, \mathbf{B}_3)\varphi_{2,3}^{var} dU_3 + \\ \tfrac{2}{3} \textstyle\int_{\mathbb{R}^3} \langle \mathbf{A}_1(x), \mathbf{A}_2(x), \mathbf{A}_3(x) \rangle d\mathbb{R}^3. \end{aligned} \tag{6}$$

In the construction it is sufficient to assume that the constants C_1, C_2, C_3 are well-defined using the only equation: the sum of the the equations (4),(5), (6).

Journal of Physics: Conference Series **544** (2014) 012015 doi:10.1088/1742-6596/544/1/012015

Using the functions ϕ_i by the equations (1) – (3) define a divergence-free vector function $\mathbf{F}$ by the formula:

$$\begin{aligned}\mathbf{F} = (1,3)(2,3)\mathbf{A}_1 \times \mathbf{A}_2 + (2,1)(3,1)\mathbf{A}_2 \times \mathbf{A}_3 + (3,2)(1,2)\mathbf{A}_3 \times \mathbf{A}_1 \\ -\phi_1\mathbf{B}_1(2,3) - \phi_2\mathbf{B}_2(3,1) - \phi_3\mathbf{B}_3(1,2).\end{aligned} \tag{7}$$

Let us define the invariant M as the the helicity integral of the vector-function (7) by the formula:

$$M(\mathbf{B}) = \int_{\mathbb{R}^3} (\mathbf{G},\mathbf{F}) d\mathbb{R}^3 + \dots, \tag{8}$$

where $\mathbf{G}$ is the vector-potential of $\mathbf{F}$, and $\dots$ in the formula (8) means the sum of the following 6 terms:

$$d_{1,1} = -(2,3)^2 \int (\mathbf{A}_1,\mathbf{B}_1)\phi_1^2 dU_1, \tag{9}$$

$$d_{2,2} = -(3,1)^2 \int (\mathbf{A}_2,\mathbf{B}_2)\phi_2^2 dU_2, \tag{10}$$

$$d_{3,3} = -(1,2)^2 \int (\mathbf{A}_3,\mathbf{B}_3)\phi_3^2 dU_3, \tag{11}$$

$$d_{1;3} = (2,3)(1,2) \int (\mathbf{A}_3,\mathbf{B}_1)\phi_1^2 dU_1, \tag{12}$$

$$d_{2;1} = (3,1)(2,3) \int (\mathbf{A}_1,\mathbf{B}_2)\phi_2^2 dU_2, \tag{13}$$

$$d_{3;2} = (1,2)(3,1) \int (\mathbf{A}_2,\mathbf{B}_3)\phi_3^2 dU_3. \tag{14}$$

In the paper Akhmetiev (2005) (the considered form of the integral is presented in Akhmet'ev-2 (2013), Theorem 2) the following result is proved:

2.1. *Theorem*

The expression (8), in which the scalar potentials, given by the formulas (1) - (3) with the gauge conditions (4) – (6), is the invariant (dimension $Gs^{12}sm^6$) of volume-preserved diffeomorphisms of the space. This invariant is not a function of pairwise integral integer coefficients of magnetic tubes.

Journal of Physics: Conference Series **544** (2014) 012015 doi:10.1088/1742-6596/544/1/012015

3. A combinatorial formula for the integral M

Let us consider a Conway polynomial for m–component link $\mathbf{L}$:

$$\nabla_{\mathbf{L}}(z) = z^{m-1}(c_0 + c_1 z^2 + \cdots + c_n z^{2n}). \tag{15}$$

We shall consider invariants of links, which are expressed from the first two coefficients c_0, c_1 of this polynomial (the Conway polynomial is applied to the link $\mathbf{L}$ itself and to the all proper sublinks of $\mathbf{L}$).

Let us start with the case $m = 2$ of 2-component links $\mathbf{L} = L_1 \cup L_2$. Recall results from the papers Melikhov (2003), Matveev S. & Polyak M. (2009), and Nikkuni, R. (2008). The simplest invariant of 2-component links, which is not expressed from pairwise linking coefficients of components, is called the generalized Sato-Levine invariant:

$$\beta(\mathbf{L}) = c_1(\mathbf{L}) - c_0(\mathbf{L})(c_1(L_1) + c_1(L_2)). \tag{16}$$

In this formula $c_0(\mathbf{L})$ coincides with linking coefficient $lk(L_1, L_2)$, $c_1(L_1)$; $c_1(L_2)$ are called Casson's invariants of the corresponding (knotted) component, Denote $lk(L_1, L_2) = c_0(\mathbf{L})$ by $(1, 2)$ for short.

Let us considered the case $m = 3$ of 3-component links $\mathbf{L} = L_1 \cup L_2 \cup L_3$. Define the Melikhov's invariant $\gamma(\mathbf{L})$, as in the papers Melikhov (2003), Matveev S. & Polyak M. (2009) by the following formula:

$$\gamma(\mathbf{L}) = c_1(\mathbf{L}) - \tag{17}$$

$$\begin{aligned}
&((1,2)(2,3) + (2,3)(3,1) + (3,1)(1,2))(c_1(L_1) + c_1(L_2) + c_1(L_3)) \\
&-((3,1) + (2,3))(c_1(L_1 \cup L_2) - (1,2)(c_1(L_1) + c_1(L_2))) \\
&-((1,2) + (3,1))(c_1(L_2 \cup L_3) - (2,3)(c_1(L_2) + c_1(L_3))) \\
&-((2,3) + (1,2))(c_1(L_3 \cup L_1) - (3,1)(c_1(L_3) + c_1(L_1))),
\end{aligned}$$

where (i, j) is the linking number of the pair of components L_i, L_j, $i, j = 1, 2, 3$, $i \neq j$, of the link $\mathbf{L}$.

Recall the formula by S.A.Melikhov of jumps of the invariant γ for homotopy of 3-component links with self-intersection and with no intersection between different components. Let $\mathbf{L}_{sing;3}$ be a singular 3-component link, the components L_1, L_2 are regular, the component $L_{sing,3}$ has the only self-intersection point. Denote by $\mathbf{L}_{+;3}$, $\mathbf{L}_{-;3}$ two 3-component links, the first 2 components of this links coincide with L_1, L_2, the third component $L_{3;+}$ of the link $\mathbf{L}_{3;+}$ is defined by means of the prescribed resolution of the self-intersection of the component $L_{sing,3}$, the third component $L_{3;-}$ of the link $\mathbf{L}_{-;3}$ is defined by means of the opposite resolution of the self-intersection of the component of the link $L_{sing,3}$. (The sign convention of resolutions of singular link is standard: a sign of the resolution is determined by the orientation of the basis, which is formed by the (non-ordered) pair of the tangent vectors and the vector from the (non-ordered) resolution points of the singular point.)

Let us define 4-component link $\mathbf{L}_{s;3}$ with the components $(L_1, L_2, L_{3+}, L_{3-})$. The first two components of the links $\mathbf{L}_{s;3}$, $\mathbf{L}_{sing;3}$ coincide, the components L_{3+}, L_{3-} of the link $\mathbf{L}_{s;3}$ are obtained by the orientation-preserved smoothing of the singular component $L_{sing,3}$ of the link $\mathbf{L}_{sing;3}$.

Denote by $(1, 3^+)$ $(1, 3^-)$ the linking coefficients of the component L_1 with the components L_{3+}, L_{3-} correspondingly. Define by $(2, 3^+)$ $(2, 3^-)$ the linking coefficients of the component L_2 with the components L_{3+}, L_{3-} correspondingly. Analogical denotations are well-defined after the replacement $1 \to 2$ of the numbers of the components.

Journal of Physics: Conference Series **544** (2014) 012015 doi:10.1088/1742-6596/544/1/012015

3.1. Lemma
The invariant $\gamma(\mathbf{L})$ of 3-component links satisfies the equations, which are obtained from the following equation by cyclic permutation of indexes:

$$\gamma(\mathbf{L}_{+;1}) - \gamma(\mathbf{L}_{-;1}) = (2,3)((2,1^+)(3,1^-) + (2,1^-)(3,1^+)), \tag{18}$$

Proof of Lemma 3.1 Lemma 3.1 is proved in Melikhov (2003), p.11.

Let $\mathbf{L} = (L_1 \cup L_2 \cup L_3)$ be an arbitrary 3-component link. Define the invariant $\tilde{M}(\mathbf{L})$ by the following formula:

$$\tilde{M}(\mathbf{L}) = (1,2)(2,3)(3,1)\gamma(\mathbf{L}) - \tag{19}$$

$$((1,2)^2(1,3)^2\beta(L_2 \cup L_3) + (2,3)^2(2,1)^2\beta(L_3 \cup L_1) + (2,3)^2(2,1)^2\beta(L_3 \cup L_1)).$$

The invariant $\tilde{M}$ is of the order 7 in the sense of V.A.Vassiliev.

3.2. Theorem
For homotopies of links with one self-intersection point on the corresponding component the invariant $\tilde{M}$ satisfies crossing change formulas, which is obtained from the following formula by cyclic permutation of indexes:

$$\tilde{M}(\mathbf{L}_{+;1}) - \tilde{M}(\mathbf{L}_{-;1}) = (1,2)(2,3)^2(3,1)((2,1^+)(3,1^-) + (2,1^-)(3,1^+)) \tag{20}$$

$$-(2,3)^2(1,2)^2(3,1^+)(3,1^-) - (3,1)^2(3,2)^2(2,1^+)(2,1^-),$$

Proof of Theorem 3.2
The crossing change formula for β is well known, see f.ex Matveev S. & Polyak M. (2009). Theorem follows from Lemma 3.1.

3.3. Main Theorem
Let $\mathbf{B} = \mathbf{B}_1 \cup \mathbf{B}_2 \cup \mathbf{B}_3$ a magnetic fields in 3 disjoint magnetic tubes $U_1 \cup U_2 \cup U_3$ with the central lines $\mathbf{L} = L_1 \cup L_2 \cup L_3$, and with the unit integral magnetic flows trough the cross-sections of the each magnetic tube. (Let us call in this case that $\mathbf{B}$ is a model of $\mathbf{L}$.) The integral invariant (8) is expressed from the combinatorial invariant (19) by he following equality:

$$M(\mathbf{B}) = \tilde{M}(\mathbf{L}) + P((1,2),(2,3),(3,1)), \tag{21}$$

where P is a polynomial of the pairwise linking numbers $(1,2)$, $(2,3)$, $(3,1)$ of the components of $\mathbf{L}$, $\deg(P) \leq 14$.

Proof of Main Theorem
Let $\mathbf{L}$ be an arbitrary 3-component link. In the case $(1,2) = (2,3) = (3,1) = 0$ we get $M(\mathbf{B}) = \tilde{M}(\mathbf{L}) = 0$.

Assume $(1,2)^2 + (2,3)^2 + (3,1)^2 \neq 0$. Let $k \in \mathbb{Z}$ be an arbitrary integer. Let us define the k-modification $\Psi_k : \mathbf{L} \mapsto \Psi_k(\mathbf{L})$ of $\mathbf{L}$, $k \in \mathbb{Z}$ (the corresponding modification of the magnetic tubes denote by $\Psi_k : \mathbf{B} \mapsto \Psi_k(\mathbf{B})$). Represents the link $\mathbf{L}$ in its isotopy class, such that there exists a ball $D^3 \subset \mathbb{R}^3$, which intersects $\mathbf{L}$ by 3 parallel segments, which belong to the different components. Each commutator is decomposed into 2 intersections of L_1 with L_2 with opposite signs. The commutator of the component L_1 with the component L_2, L_3 is called the "'delta move"', the delta move was introduced by Murakami and Nakanishi, see De Turck, D. *et al.* (2011) for a picture and a more detailed explanation of the commutator.

Apply k-times commutator, $k \in \mathbb{Z}$, of the segment on the component L_1 with the segments on the components L_2, L_3 inside D^3. Evidently, the links $\Psi_{k_1} \circ \Psi_{k_2}(\mathbf{L})$ and $\Psi_{k_1+k_2}(\mathbf{L})$ are isotopic.

3.4. Lemma
The integral invariant M, given by the formula (8), satisfies the crossing change formulas, as the formula (20), when the corresponding magnetic tube is changed by a homotopy with a self-intersection.

Proof of Lemma 3.4
It is sufficient to prove the lemma for a homotopy with the self-intersection of the tube U_1 (of the component L_1). The idea of the proof is in the paper by Akhmet'ev P.M. & Kunakovskaya O.V. (2009), where the invariant β is investigated.

Denote by V, W the two short segments of the magnetic tube U_1 in a small ball $D^3 \subset \mathbb{R}^3$, inside this ball the crossing of the homotopy is. Without loss of a generality we may assume that the scalar potential in U_1 satisfies the equation $\phi_1|_{V_1} = 0$. Then for the restriction of ϕ_1 to the second segment of the magnetic tube we have: $\phi_1|_W = (1,2)(3,1^+) - (1,3)(2,1^+)$ (obviously, $(1,2)(3,1^-) - (1,3)(2,1^-) = -(1,2)(3,1^+) + (1,3)(2,1^+)$). The gauge (4)-(6) remains fixed by the homotopy, the only term (9) gives a contribution to the crossing change formula by

$$\delta(M) = (2,3)^2[(1,2)(3,1^+) - (3,1)(2,1^+)]^2.$$

From the formula (20) (in the case $(2,3) \neq 0$) we get:

$$(2,3)^{-2}\delta(\tilde{M}) = (1,2)(3,1)[(2,1^+)(3,1^-)+(2,1^-)(3,1^+)]-(1,2)^2(3,1^+)(3,1^-)-(3,1)^2(2,1^+)(2,1^-).$$

Using the identities $(3,1^+) + (3,1^-) = (3,1)$, $(2,1^+) + (2,1^-) = (1,2)$, we get the proof of Lemma 3.4.

From Lemma 3.4 the difference $M(\mathbf{B}) - \tilde{M}(\mathbf{L})$ is fixed by a singular homotopy with no intersections between different magnetic tubes. By Akhmet'ev-2 (2013), Lemma 11, the invariant M is derived from a finite-type configuration space. Let us prove that the function $F(k) : k \mapsto M(\Psi_k(\mathbf{B}))$ is polynomial.

Denote the m-difference $\nabla^m F(k)$, $k \in \mathbb{Z}$, by the inductive formula:

$$\nabla^m F(k) = \nabla(\nabla^{m-1} F(k)) = \nabla^{m-1} F(k) - \nabla^{m-1} F(k-1), \quad i \in \mathbb{Z}.$$

The m-difference $\nabla^m F(k)$ is represented as following. Define the collection $\aleph = \aleph(\varepsilon)$ of 2^m links, which is defined as the modifications of the magnetic tube $\Psi_k(\mathbf{B})$ by a sequence of m copies of the delta-move, the delta-move with a number i, $1 \leq i \leq m$, in the sequence is included, or excluded, correspondingly.

Let us consider an algebraic sum of values of $M(\alpha)$ over a collection $\aleph$, $\alpha \in \aleph$. The algebraic sum is defined as following: $\sum_{\alpha \in \aleph} \pm M(\alpha)$, with sign $+1$ (-1), if the corresponding collection α consists of an even (odd) number of delta-moves, correspondingly. Obviously, $\nabla^m F(k) = \sum_{\alpha \in \aleph} \pm M(\alpha)$.

Analogously to Akhmet'ev-2 (2013), Theorem 6, there exists a positive m, such that for arbitrary $\mathbf{B}$, k the following identity is satisfied:

$$\nabla^m F(k) = 0.$$

From this equitation we derive that $F(k)$ is a polynomial of a degree $\le m$.

The invariant $\tilde{M}$ is a finite type invariant in the sense of V.A.Vassiliev, from the definition we get $\tilde{F}(k) : k \mapsto \tilde{M}(\Psi_k(\mathbf{L}))$ is polynomial. By the Gusarov's Theorem, see Baader, S. & Marché, J. (2008), Matveev S. & Polyak M. (2009) for details and references, $\tilde{M}$ is the algebraic sum of representations of a suitable collection of Gauss diagrams in the diagram of $\Psi_k(\mathbf{L})$. Therefore, for a sufficiently great m (where m is less or equals to the order of M, which is the maximal numbers of arrows of the Gauss diagrams) we get

$$\nabla^m \tilde{F}(k) = 0.$$

Therefore $F(k) - \tilde{F}(k) : k \mapsto M(\Psi_k(\mathbf{B})) - \tilde{M}(\Psi_k(\mathbf{L}))$ is a polynomial. Assume that $k \equiv gcd\{(1,2),(2,3),(3,1)\}$. Then the link $\Psi_k(\mathbf{L})$ is related with $\mathbf{L}$ by a homotopy with self-intersection of components and with no intersection between different components, see f.ex. Cohen, F.R. *et al.* (2012). Therefore for a such k we get $M(\Psi_k(\mathbf{B})) - \tilde{M}(\Psi_k(\mathbf{L})) = 0$, and therefore $M = \tilde{M} + C$, where C is a polynomial, which depends only of the pairwise linking numbers. The degree of C satisfy the equality: $\deg(C) \le 14$, because $\tilde{M}$ is a finite-type invariant in the sense of V.A.Vassiliev of the degree 7, and the assumption $\deg(C) > 14$ contradicts to the Property (22) below.

Main Theorem is proved.

4. Asymptotic properties for combinatorial links

Let $\mathbf{K}$ be an arbitrary m-component framed link with a framed marked component (K, ξ), $K \subset \mathbf{K}$. The framing ξ determines the coordinate system on the boundary of a tubular neighborhood of the component K, such a coordinate system is represented by a family of parallels and meridians.

For an arbitrary integer $r \in \mathbb{Z}$ let us define the following two operations:

$$A^r : (\mathbf{K}; K, \xi) \to \mathbf{L},$$

where $\mathbf{L}$ is m-component link, and

$$B^r : (\mathbf{K}; K, \xi) \to \mathbf{L},$$

where $\mathbf{L}$ is $(m+1)$-component link.

Denote non-marked components of $\mathbf{K}$ by K_i, $i = 2, \dots, m$. By the transformation A^r, all the non-marked components of the link $\mathbf{K}$ remain unchanged, $K_i = L_i$, $i = 2, \dots, m$, unless the first marked component K, which is transformed into the standard $(r,1)$–time winding L_1 along (K, ξ). The component L_1 passes r times along the parallel and 1–time along the meridian on the boundary of the thin regular tubular neighborhood of the component K. In the case $r = 0$ we get the link $\mathbf{L}$ with a small non-knotted component L_1.

Denote non-marked components of $\mathbf{K}$ by K_i, $i = 3, \dots, m+1$. By the transformation B^r, all the non-marked components of the link $\mathbf{K}$ remain unchanged, $K_i = L_i$, $i = 3, \dots, m+1$, unless

Journal of Physics: Conference Series **544** (2014) 012015 doi:10.1088/1742-6596/544/1/012015

the first marked component K, which is transformed into the pair of closed parallel standard $(r, 1)$–time windings $L_1 \cup L_2$ along (K, ξ). In the case $r = 0$ we get the link $\mathbf{L}$ with two small parallel non-knotted components L_1, L_2.

Let $\mathbf{L}$ be a m-component link with the marked component $L \subset \mathbf{L}$, which is represented as $\mathbf{L} = A^r(\mathbf{K}; K, \xi)$ for some $r \in \mathbb{Z}$. let us define the following operations

$$C : \mathbf{L} \to \mathbf{L}^{tw},$$

where $\mathbf{L}^{tw}$ is an m-component link, which is obtained from $\mathbf{L}$ by an arbitrary homotopy (which is sufficiently small in C^1-topology), inside a small regular neighborhood V_K of the marked component $K \subset \mathbf{K}$, $L \subset V_K$, with self-intersection points of the component L. To get a definition of C, which is not ambiguous, we say that $\mathbf{L}^{tw}$ is obtained from $\mathbf{L}$ by the modification, using a prescribed colored r–component braid. Recall, an arbitrary colored braid is a composition of the elementary braids $\sigma_{i,i+1}$, $1 \le i < i+1 \le r$, in $\sigma_{i,i+1}$ the 2-strings with numbers i, $i+1$ are twisted.

4.1. Definition
Let us say that a finite-type invariant I for $(m+1)$-component links is an asymptotic invariant of the degree s, $s \in \mathbb{N}$, if the following equations (22) -(24) are satisfied:

$$I(A^r(\mathbf{K}; K, \xi)) = r^s I(\mathbf{K}) + o(r^s), \tag{22}$$

where $o(r^s)$ is a polynomial of a variable r of the degree less then s, coefficients of this polynomial depend only on the isotopy class of the framed link $(\mathbf{K}; K, \xi)$.

$$I(B^r((\mathbf{K}; K, \xi)) = r^{2s} I_0 + o(r^{2s}), \tag{23}$$

where $I_0 \in \mathbb{Z}$ is an integer, $o(r^{2s})$ is a polynomial of r of the degree less then $2s$, coefficients of this polynomial depend only on the isotopy class of the framed link $(\mathbf{K}; K, \xi)$.

$$I(\mathbf{L}) = I(\mathbf{L}^{tw}), \tag{24}$$

where $\mathbf{L}^{tw} = C(\mathbf{L})$.

4.2. Theorem
The combinatorial invariant M, determined by the right side of the formula (21), is an asymptotic invariant in the sense of Definition 4.1 for $m + 1 = 3$, $s = 4$.

Proof of Theorem 4.2
Let us replace the combinatorial invariant M by the integral invariant M using Theorem 3.3, and prove the analogous equations for the integral invariant M.

4.3. Lemma
The integral invariant M, given by (8), depends of the integral flows of the magnetic field thought the cross-sections of the magnetic tubes and of a position of the magnetic tubes in the space, and depends no of magnetic fields (with prescribed integral flows) inside the magnetic tubes U_i, $i = 1, 2, 3$.

Proof of Lemma 4.3
Assume that magnetic fields $\mathbf{B}$, $\mathbf{B}'$ coincide in U_2, U_3, but, probably, are different in U_1. Take the gauge with the trivial scalar potential (1) inside U_1. We may assume that $\mathbf{A}_2 = \mathbf{A}'_2$, $\mathbf{A}_3 = \mathbf{A}'_3$, and $\mathbf{A}_2, \mathbf{A}_3$ are collinear inside U_1. Moreover, $\mathbf{A}_1$ and $\mathbf{A}'_1$ coincide outside of U_1. In the presented gauge all terms of $M(\mathbf{B})$, $M(\mathbf{B}')$ coincide correspondingly. Lemma 4.3 is proved.

The equations (24) is a corollary of Lemma 4.3, which is applied for the magnetic fields $\mathbf{B}$ and $\mathbf{B}'$, which are models of $\mathbf{L}$ and $C(\mathbf{L})$ correspondingly.

To proof the equation (22) apply the scaling $\mathbf{B}_1 \mapsto r\mathbf{B}_1$ to the formula $M(\mathbf{B})$. By dimension arguments $r^4 M(\mathbf{B}_1 \cup \mathbf{B}_2 \cup \mathbf{B}_3) = M(r\mathbf{B}_1 \cup \mathbf{B}_2 \cup \mathbf{B}_3)$. By Lemma 4.3 $M(r\mathbf{B}_1 \cup \mathbf{B}_2 \cup \mathbf{B}_3) = M(\mathbf{B}'_1 \cup \mathbf{B}_2 \cup \mathbf{B}_3)$, where $\mathbf{B}_1$ is a model for the component $K \subset \mathbf{K}$, $\mathbf{B}'_1$ is a model for the first component of the link $A^r(\mathbf{K}; K, \xi)$.

Let us prove the equation (23). Take an arbitrary 2-component link $\mathbf{K}$, where (K, ξ) is a framed component of $\mathbf{K}$ inside U_1. Take the magnetic field $\mathbf{B} = \mathbf{B}_{1,2}(r) \cup \mathbf{B}_3$, which is a model for the link $B^r(\mathbf{K}; K, \xi)$. This magnetic field is defined by the two parallel magnetic lines $\mathbf{B}_{1,2}(r) = \mathbf{B}_1(r) \cup \mathbf{B}_2(r)$ on the boundary ∂U_1, the third magnetic line $\mathbf{B}_3$, which is a model for $K_3 \subset \mathbf{K}$. Denote the magnetic field $r^{-1}\mathbf{B}_{1,2}(r)$ by $\mathbf{B}^r_{1,2}$ and take the limit $r \to +\infty$ of the links $\mathbf{B}^r_{1,2}$. This limit is an integrable vector field on ∂U_1, denoted by $\mathbf{B}^{+\infty}_{1,2}$. The constant I_0 is defined by $I_0 = M(\mathbf{B}^{+\infty}_{1,2} \cup \mathbf{B}_3)$. It r_1, r_2 are sufficiently large, $\mathbf{B}^{r_1}_{1,2}$, $\mathbf{B}^{r_2}_{1,2}$ are closed as integrable magnetic fields. The equation (23) is a corollary of Conjecture 1. Theorem 4.2 is proved.

Remark
The combinatorial invariant M is not a function of pairwise linking coefficients. By Conjecture 1, the invariant M is ergodic. For a higher ergodic invariant by Baader, S. & Marché, J. (2008), Theorem 1, the following inequality is satisfied:

$$2d(M) > (m+1)s,$$

where $d(M)$ is the order of the combinatorial invariant M in the sense of V.A.Vassiliev (recall, $m+1=3$, $s=4$). By the formula (21) we get $d(M) \geq 7$, because for a combinatorial invariant $\tilde{M}$, defined by the formula (19), we get $d(\tilde{M}) = 7$.

5. Conclusion Remarks

It is well known that Hopf invariants in stable homotopy groups of spheres relates with E^1-term of the Adams spectral sequence. The helicity integral detects the Hopf bundle $S^3 \to S^2$, which is the main example of a mapping with the Hopf invariant 1. By Browder (1969) the Arf-Kervaire invariants, which are the "'quadratic constructions"' on the Hopf invariants, are related with E^2-term of the Adams spectral sequence. Main Theorem of the paper determines a connection of a higher helicity integral M with the Arf-invariants: the invariants γ, β are the integer lifts of the Arf-invariants of 3- and 2-component links. Integer lifts of elements of homotopy groups of the 2-sphere into finite-type invariants of spherical braids is the main subject of the survey by Cohen, F.R. & Wu, J. (2004). From an author's opinion, the problem by V.I.Arnol'd should be investigated using this algebra.

Acknowledgments
The author is grateful to E. A. Kudryavtseva, L.P. Plakhta, S.S.Podkoritov, Jason Cantarella, Rafal Komendarczyk, R. Jason Parsley, and Clayton Shonkwiler for discussions. This work was supported in part by the Russian Foundation of Basic Research Grant No. 15-02-01407A.

Journal of Physics: Conference Series **544** (2014) 012015 doi:10.1088/1742-6596/544/1/012015

References

ARNOL'D, V.I. 2000 Arnol'd problems. *Moscow, Fasis.*

ARNOL'D, V.I., KHESIN, B.A. 1998 Topological methods in hydrodynamics. *Applied Mathematical Sciences, Springer* vol. 125.

AKHMETIEV, P., RUZMAIKIN, A. 1995 A fourth-order topological invariant of magnetic or vortex lines. *Journal of Geometry and Physics* **15** (2), 95-101.

AKHMETIEV, P.M. 2005 On a new integral formula for an invariant of 3-component oriented links. *Journal of Geometry and Physics* **53** (2), 180-196.

AKHMET'EV, P.M. & KUNAKOVSKAYA, O.V. 2009 Integral Formula for a Generalized SatoLevine Invariant in Magnetic Hydrodynamics. *Mat. Zametki* **85** (4), 524537.

AKHMET'EV, P.M. 2011 On asymptotic higher analogs of the magnetic helicity invariant in MHD [v2]. ARXIV:1105.5876.

AKHMET'EV, P.M. 2012 Quadratic magnetic helisity and magnetic energy. *Proc. Steklov Math. Inst.* **278** 16-28.

AKHMET'EV, P.M. 2013-1 On a higher integral invariant for closed magnetic lines *Journal of Geometry and Physics* **74** 381391.

AKHMET'EV, P.M. 2013-2 On asymptotic higher analogs of the magnetic helicity invariant in MHD [v4]. ARXIV:1105.5876.

BAADER, S. & MARCHÉ, J. 2008 Asymptotic Vassiliev invariants for vector fields arXiv:0810.3870.

BERGER, M.A. & FIELD, G.B. 1984 The topological properties of magnetic helicity. *J. Fluid Mech.* **147**, 133-148.

BROWDER, W. The Kervaire Invariant of Framed Manifolds and its Generalization *The Annals of Mathematics, Second Series* 1969 **90** 1, 157-186

COHEN, F.R. & WU, J. 2004 On braids, free groups and the loop space of the 2-sphere. *Progress in Mathematic Techniques, Algebraic Topology: Categorical Decompositions* **215** 93-105.

COHEN, F.R., & KOMENDARCZYK, R. & SHONKWILLER C. 2012 *On homotopy Brunnian links and the κ-invariant* arXiv:1208.4587v2

MATVEEV, S. & POLYAK, M. 2009 A simple formula for the Casson-Walker invariant. *Journal of Knot Theory and Its Ramifications* **18** 6, 841864; arXiv:0811.0606.

MAYER, CH. 2003 Topological Link Invariants of Magnetic Fields, *PhD Thesis, university at Bochum.*

MELIKHOV, S.A. 2003 Colored finite type invariants and a multi-variable analogue of the Conway polynomial, Preprint. arXiv:math/0312007v2

MOFFATT, H.K. 2001 The topology of turbulence. Lesieur, M. (ed.) et al., Ecole de Physique des Houches UJF INPG Grenoble, a NATO Advanced Study Institute. Les Houches session LXXIV 31 July 1 September 2000, New trends in turbulence. *Turbulence: nouveaux aspects. Berlin: Springer; Les Ulis: EDP Sciences* 321-340.

MONASTYRSKY, M.V. & RETAKH, V.S. 1986 Topology of linked defects in condensed matter *Commun. Math. Phys.* **103**, 445-459.

NIKKUNI, R. 2008 Homotopy on spatial graphs and Generalized Sato-Levine Invariants, arXiv:0710.3627v2.

DENNIS DETURCK & ET HERMAN GLUCK & RAFAL KOMENDARCZYK & PAUL MELVIN & CLAYTON SHONKWILER & DAVID SHEA VELA-VICK 2011 Triple linking numbers, ambiguous Hopf invariants and integral formulas for three-component links *arXiv:0901.1612.*

Journal of Physics: Conference Series **544** (2014) 012016 doi:10.1088/1742-6596/544/1/012016

Higher-order topological phase: generalizations of the Aharonov-Bohm and Josephson effects

Roman V. Buniy
Chapman University, Schmid College of Science, Orange, CA 92866
E-mail: roman.buniy@gmail.com

Thomas W. Kephart
Department of Physics and Astronomy, Vanderbilt University, Nashville, TN 37235
E-mail: tom.kephart@gmail.com

Abstract.
In the magnetic Aharonov-Bohm effect a semiclassical path of a charged particle is gaussian linked with a magnetic flux tube and one observes an interference phase angle proportional to the magnitude of the enclosed flux. We first generalize to the case of knotted paths. Then we argue that quantum mechanical wave functions with semiclassical paths that have higher order linking to multiple magnetic flux tubes have interference angle proportional to the product of the fluxes. Similar results hold for generalizing the Josephson effect.

1. Introduction

The magnetic Aharonov-Bohm effect [Aharonov & Bohm, 1959] results when a charged particle travels around a closed path in a region of vanishing magnetic field but nonvanishing vector potential. The magnetic flux is confined to a region where the particle is excluded, but the wave function of the particle nonetheless responds to the vector potential and an interference pattern is detected, see Fig. 1.

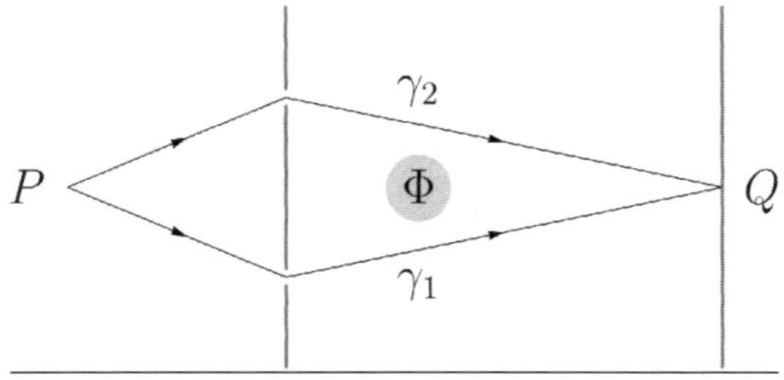

Figure 1. A plane projection of the standard magnetic Aharonov-Bohm effect apparatus. The electrons travel along semiclassical paths γ_1 and γ_2 from the source P to the detection screen Q. The electrons do not penetrate the interior of the solenoid which carries the flux Φ.

Let C be the center line of a magnetic solenoid carrying flux Φ and having vector potential $\boldsymbol{A}$. For an electron moving along a semiclassical path γ, the wave function is $\psi(\boldsymbol{A}) = \psi(0)\exp\left(i\xi\int_\gamma \boldsymbol{A}\cdot d\boldsymbol{x}\right)$, where $\xi = e(\hbar c)^{-1}$. For a standard magnetic Aharonov-Bohm effect apparatus in Fig. 1, the total amplitude for paths γ_1 and γ_2 is $\psi(\boldsymbol{A}) = e^{i\theta}(\psi_1(0) + e^{-i\phi}\psi_2(0))$. The quantity θ is an overall irrelevant phase and the quantity $\phi = \xi\oint_\gamma \boldsymbol{A}\cdot d\boldsymbol{x}$, where $\gamma = \gamma_1\gamma_2^{-1}$, measures the relative phase shift between paths γ_1 and γ_2. Applying the Stokes theorem, we find $\phi = \frac{e\Phi}{\hbar c}$.

The ratio of the phase to the flux is proportional to the gaussian linking of the curves γ and C, which is symmetric with respect to the curves. We will later use this duality in the more complicated case of knotted curves.

For a fixed closed curve C, any possible path of an electron belongs to $C' = \mathbb{R}^3\backslash C$, the complement of C in $\mathbb{R}^3$. The set of all such paths form a group Γ under the operation of multiplication of paths. This group is $\Gamma = \pi_1(C')$, the first homotopy group [Hatcher, 2002] of C', also called the fundamental group of C'.

Topologically, C bounds a disk S. If we continuously deform this disk, the signed intersection number N with γ does not change. This can be seen by noting that during deformations of S, new intersection points of γ and S appear in pairs, and the two points in each pair have intersections of opposite signs. This means that N depends only on the topological class to which $\gamma \in \Gamma$ belongs. It follows from the above that $\Gamma \cong \mathbb{Z}$. In the dual picture, we need $\Gamma_* = \pi_1(\mathbb{R}^3\backslash\gamma)$, the fundamental group of the complement of γ, and it similarly follows that $\Gamma_* \cong \mathbb{Z}$. Now let us generalize these ideas to orientable surfaces which are bounded by knots.

2. Aharonov-Bohm effect for knotted solenoids

We consider the case of a closed self-knotted curve C. For each such C, there are surfaces which are bounded by C. If such a surface is connected and orientable, then it is called a Seifert surface of C. A well-known theorem [Rolfsen, 1976] states that there is a Seifert surface for every knot. In general, for a given C, there can be more than one nonequivalent Seifert surface [Alford, 1970; Trotter, 1975; Lyon, 1974; Rolfsen, 1976]. However, the signed number of intersections of a closed curve γ and S is independent of the choice of S. Hence our results will be independent of the choice of a Seifert surface. A Seifert surface for the trefoil knot is shown in Fig. 2.

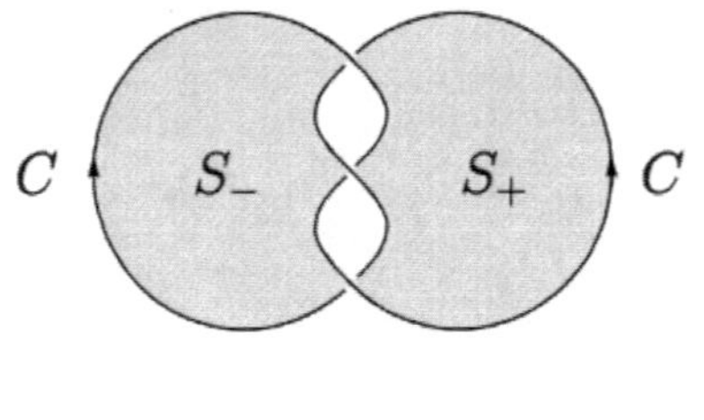

Figure 2. The trefoil knot C and its Seifert surface S. S_+ and S_- are the two sides of the orientable S. If a semiclassical path of an electron intersects S, then the Aharonov-Bohm effect results.

A generalization of the Stokes theorem for knotted closed curves [Kephart, 1985] states that $\oint_\gamma \boldsymbol{A}\cdot d\boldsymbol{x} = N\Phi$, where N is the signed intersection number of γ and S, and S is a Seifert surface of the knot C. In the dual picture with the unknotted solenoid γ with flux Φ_* and a knotted semiclassical path of an electron C, we have $\oint_C \boldsymbol{A}_*\cdot d\boldsymbol{x} = N_*\Phi_*$, where N_* is the signed intersection number of C and σ, and σ is a Seifert surface of γ. As in the case of unknotted closed curves, the numbers N and N_* are equal since they both again represent the gaussian

Table 1. Phase components along the path C_3.

path segment Γ	γ_1	γ_2	$K^{-1}\phi_{12}(\Gamma)$
1	δ_1	δ_2	0
a_1	$\delta_1+\Phi_1$	δ_2	$-\frac{1}{2}\Phi_1\delta_2$
a_1a_2	$\delta_1+\Phi_1$	$\delta_2+\Phi_2$	$\frac{1}{2}[(\delta_1+\Phi_1)\Phi_2-\Phi_1\delta_2]$
$a_1a_2a_1^{-1}$	δ_1	$\delta_2+\Phi_2$	$\frac{1}{2}[(\delta_1+\Phi_1)\Phi_2-\Phi_1\delta_2+\Phi_1(\delta_2+\Phi_2)]$
$C_3=a_1a_2a_1^{-1}a_2^{-1}$	δ_1	δ_2	$\Phi_1\Phi_2$

linking of the curves γ and C. In both cases we find a generalized magnetic Aharonov-Bohm effect with phase proportional to $N\Phi$ and $N_*\Phi_*$ respectively [J. C. Kimball, 2004; Buniy & Kephart, 2009].

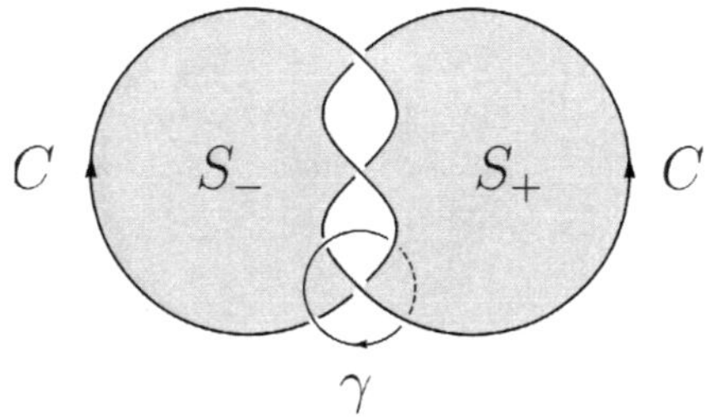

Figure 3. An example of an electron path γ which does not intersect the Seifert surface S but nevertheless links with the knot C.

Higher order linking of particle paths and knotted solenoids is possible as shown in Fig. 3. However, to study the case of higher order linking it is better to first look at the simpler example of linking of unknots rather than of knots.

3. Generalization of the Aharonov-Bohm effect for higher order linking

Now let us consider the Borromean ring configuration consisting of three rings C_1, C_2, and C_3 that are pairwise unlinked but linked as a triple as shown in Fig. 4. We follow the semiclassical path of a particle around the circuit C_3 where we now take C_1 and C_2 to be a pair of unlinked solenoids. If, and only if, the particle path has no net first order linking with either solenoid C_1 or C_2, will we then define a gauge [Massey, 1959, 1968; Berger, 1990] that describes the higher order linking. That gauge is

$$\boldsymbol{A}_{12}=\tfrac{1}{2}(\gamma_1\boldsymbol{A}_2-\gamma_2\boldsymbol{A}_1),$$

where subscripts 1 and 2 refer to the solenoids along C_1 and C_2, and

$$\gamma_k=\delta_k+\int_\Gamma \boldsymbol{A}_k\cdot d\boldsymbol{x}.$$

Here $\boldsymbol{A}_1$ and $\boldsymbol{A}_2$ are the vector potentials due to the two solenoids, and Γ is the path that will run along C_3. A sketch of an experimental setup is shown in Fig. 6.

We want to calculate the overall phase difference $K^{-1}\phi_{12}=\int_{C_3}\boldsymbol{A}_{12}\cdot d\boldsymbol{x}$. (For a discussion of the normalization factor K^{-1} multiplying ϕ_{12} see [Buniy & Kephart, 2008*a*].) Table 1 follows the path step by step through the experimental setup along path C_3 using the gauge $\boldsymbol{A}_{12}$. The

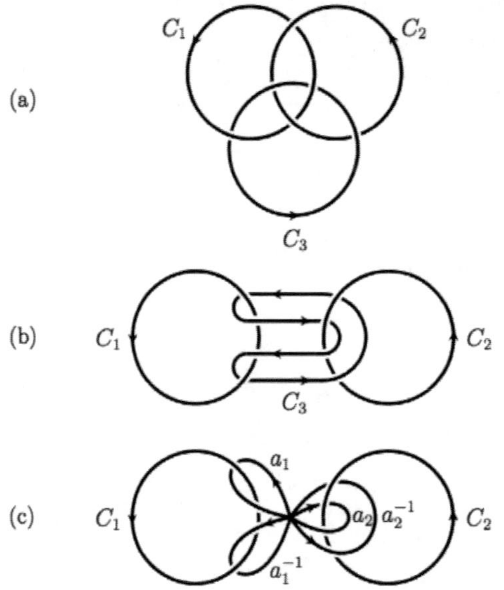

Figure 4. (a) The Borromean rings. The topological significance of this arrangement is that while no pair of loops is linked, the triple of loops is linked. This is the simplest configuration with zero first order (gaussian) linking and nonzero second order linking. (b) To obtain this figure from (a), pull C_1 and C_2 apart for flexible C_3. (c) To obtain this figure from (b), pinch C_3 at the point x_0 so that it becomes a commutator loop $C_3 = a_1 a_2 a_1^{-1} a_2^{-1}$.

first column labels the current positions on the path, the next two columns are the cumulative values of γ_1 and γ_2 at these points, and the last column gives the value of $\phi_{12}(\Gamma)$ at these points. In the third row we have used a_1 to take us from x_0 around C_1 and back to x_0. In the process γ_1 has increased by Φ_1 since $\int_{\Gamma=a_1} \boldsymbol{A}_1 \cdot d\boldsymbol{x} = \Phi_1$ while γ_2 stays fixed since $\int_{\Gamma=a_1} \boldsymbol{A}_2 \cdot d\boldsymbol{x} = 0$. Hence we have

$$K^{-1}\phi_{12}(a_1) = \tfrac{1}{2}\gamma_1 \int_{\Gamma=a_1} \boldsymbol{A}_2 \cdot d\boldsymbol{x} - \tfrac{1}{2}\gamma_2 \int_{\Gamma=a_1} \boldsymbol{A}_1 \cdot d\boldsymbol{x} = -\tfrac{1}{2}\Phi_1\delta_2. \tag{1}$$

From here it is obvious how to generate the remaining entries in the table. An alternative representation of this information is given in Fig. 5. The path C_3 begins at the initial position (δ_1, δ_2). We first use a_1 to travel to $(\delta_1 + \Phi_1, \delta_2)$, picking up an area $\delta_2\Phi_1$, which corresponds to a contribution of $-\frac{1}{2}\delta_2\Phi_1$ to $K^{-1}\phi_{12}$ (see Eq. (1)). Next, a_2 takes us to $(\delta_1 + \Phi_1, \delta_2 + \Phi_2)$ and it generates a contribution $\frac{1}{2}(\delta_1 + \Phi_1)\Phi_2$. Next, a_1^{-1} takes us to $(\delta_1, \delta_2 + \Phi_2)$ and contributes $\frac{1}{2}\Phi_1(\delta_2 + \Phi_2)$. Finally, a_2^{-1} returns us to (δ_1, δ_2) and contributes $-\frac{1}{2}\delta_1\Phi_2$, for a total phase of $K^{-1}\phi_{12} = \Phi_1\Phi_2$ for traversing the full loop C_3. The last row of Table 1 (or the full loop in Fig. 4c) gives the final result for the full path when $\Gamma = C_3$. We find

$$\phi_{12}(C_3) = \frac{e^2}{\hbar^2 c^2}\Phi_1\Phi_2 \tag{2}$$

once physical units have been restored and where the normalization has been fixed by requiring single valued results as discussed in [Buniy & Kephart, 2008*a*]. We see that the phase is proportional to the product of fluxes. Higher order generalizations can be found in [Buniy

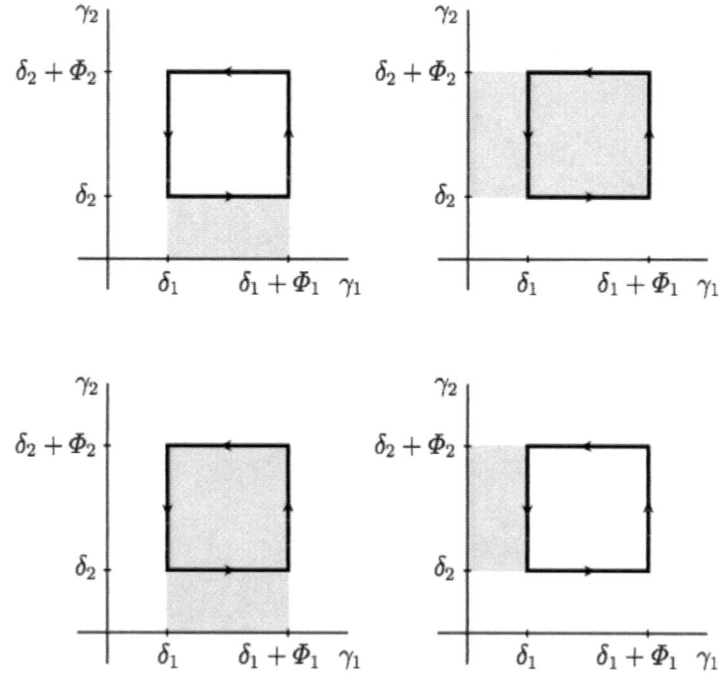

Figure 5. Graphic representation of the information in Table 1.

& Kephart, 2008*b*], while further discussion of commutator paths can be found in [Magnus *et al.*, 1966; Schulman, 1971].

4. Higher order generalization of the Josephson effect

Solving the Schrödinger equation for a superconductor as arranged in Fig. 7 with current flowing from P to Q along the two paths C' and C'' through Josephson junctions J' and J'' respectively, where the loop $C = C'C''^{-1}$ is linked with the solenoid C_1 carrying flux Φ_1 leads to a total current from the point P to the point Q

$$I = I_0 \sin\left(\tfrac{1}{2}(\theta' + \theta'')\right) \cos\left(\pi\Phi_1/\Phi_0\right).$$

For a fixed value of Φ_1, the corresponding maximal total current is

$$I_{\max} = I_0 |\cos\left(\pi\Phi_1/\Phi_0\right)|,$$

which itself has maxima when $\Phi_1 = n\Phi_0$, $n \in \mathbb{Z}$. This is the Josephson effect [Josephson, 1962; Bloch, 1970; Jaklevic *et al.*, 1964; Feynman *et al.*, 1963].

The Josephson analog of the Borromean ring generalization of the Aharonov-Bohm effect [Buniy & Kephart, 2010] is gotten from the arrangement shown in Fig. 8. Again due to the interference phase angle being proportional to the product of fluxes, we find on solving the Schrödinger equation the total current from point P to point Q is

$$I = I_0 \sin\left(\tfrac{1}{2}(\theta' + \theta'')\right) \cos\left(2\pi^2 k_2 \Phi_1\Phi_2/\Phi_0^2\right).$$

Journal of Physics: Conference Series **544** (2014) 012016 doi:10.1088/1742-6596/544/1/012016

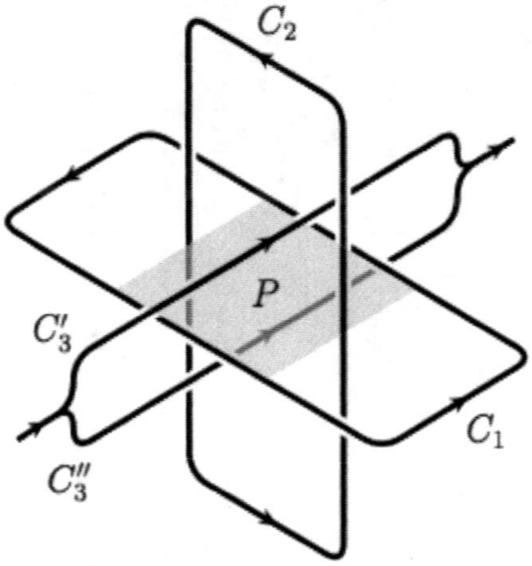

Figure 6. Shown is a schematic of a Borromean ring arrangement to detect the second order phase ϕ_{12}, where C_1 and C_2 are magnetic solenoids (leads not shown) carring flux Φ_1 and Φ_2, and C_3' and C_3'' correspond to two topologically distinct paths and are parts of the closed path $C_3 = C_3' C_3''^{-1}$ of a charged particle path starting from the source and ending at the screen. To prevent second order (gaussian) linking of the wave function with the solenoids one would install a rectangular plate P in the plane of C_1 that covers the area between the two sides of C_1 and fills the region between the sides of C_2. For particle wave packets that do not spread much beyond the center of the region containing the plate, only second order linking will be detected at the screen.

For fixed values of Φ_1 and Φ_2, the maximal total current flowing in the superconductor is

$$I_{\max} = I_0 |\cos{(2\pi^2 k_2 \Phi_1 \Phi_2 / \Phi_0^2)}|.$$

5. Summary

First order (Gaussian) linking leads to interference with phase proportional to enclosed flux in the case of the Aharonov-Bohm effect. Higher order linking also leads to interference, but with phases proportional to products of fluxes from different solenoids. Even though path components a_1 and a_2 in the above example do not commute, the phase is still abelian as required [Schulman, 1971]. Our analysis needs nothing more than quantum mechanics and a judicious choice of gauge, and our conclusions are easily testable with tabletop experiments using known techniques.

Acknowledgments

We thank the Isaac Newton institute for their hospitality. RVB thanks Chapman University for support. The work of TWK was supported by U.S. DoE grant number DE-FG05-85ER40226.

References

Aharonov, Y. & Bohm, D. 1959 Significance of electromagnetic potentials in the quantum theory. *Phys.Rev.* **115**, 485–491.

Alford, W. R. 1970 *Ann. of Math* **91**, 419.

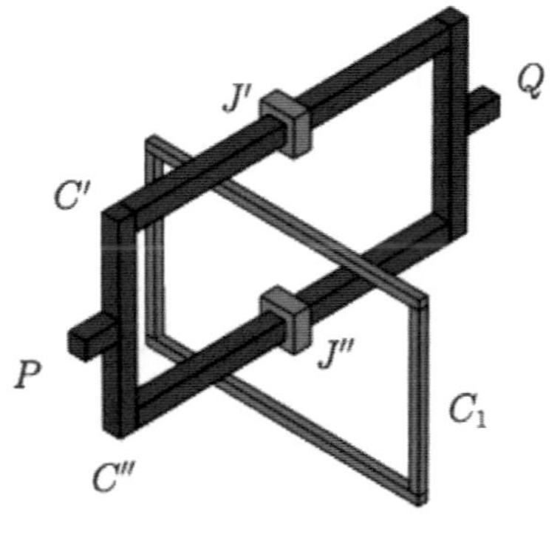

Figure 7. A diagram of an experimental setup for the detection of the Josephson effect. C' and C'' are paths from the point P to the point Q through the superconductors with the Josephson junctions J' and J'' and the total current I from P to Q. C_1 is the magnetic solenoid carrying flux Φ_1. The Josephson effect (for a review see Ref. [Golubov *et al.*, 2004]) is due to the first order (gaussian) linking of the closed curves $C = C'C''^{-1}$ and C_1.

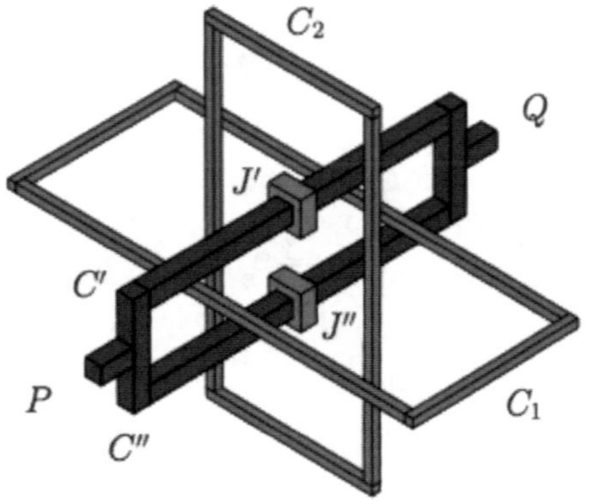

Figure 8. A diagram of an experimental setup for the detection of the second order Josephson effect. C_1 and C_2 are the magnetic solenoids. C' and C'' are paths from the point P to the point Q through the superconductors connected by Josephson junctions J' and J''. The total current from P to Q is I. C_1 and C_2 are magnetic solenoids carrying fluxes Φ_1 and Φ_2. The second order Josephson effect is due to the second order linking of the set of three closed curves $C = C'C''^{-1}$, C_1 and C_2.

Berger, M. A. 1990 Third-order link integrals. *J. Phys. A* **23**.

Bloch, F. 1970 Josephson effect in a superconducting ring. *Phys. Rev.* **B2**.

Buniy, R. V. & Kephart, T. W. 2008*a* A Proposal for detecting second order topological quantum phase. *Phys.Lett.* **A372**, 2583–2586.

Buniy, R. V. & Kephart, T. W. 2008*b* Higher order topological actions. *Phys.Lett.* **A372**, 4775–4778.

Journal of Physics: Conference Series **544** (2014) 012016 doi:10.1088/1742-6596/544/1/012016

BUNIY, R. V. & KEPHART, T. W. 2009 The Aharonov-Bohm effect for a knotted magnetic solenoid. *Phys.Lett.* **A373**, 919–921.

BUNIY, R. V. & KEPHART, T. W. 2010 Higher order Josephson effects. *J.Phys.* **A43**, 225301.

FEYNMAN, R. P., LEIGHTON, R. B. & SANDS, M. 1963 **The Feynman Lectures on Physics**. See volume III, chapter 21. Addison-Wesley.

GOLUBOV, A. A., KUPRIYANOV, M. Y. & IL'ICHEV, E. 2004 The current-phase relation in Josephson junctions. *Rev.Mod.Phys.* **76**, 411–469.

HATCHER, A. 2002 **Algebraic Topology**. Cambridge University Press, Cambridge, New York.

J. C. KIMBALL, H. L. F. 2004 Aharonov-Bohm effects in entangled molecules. *Phys. Rev. Lett.* **93**, 093001.

JAKLEVIC, R. C., LAMBE, J., SILVER, A. H. & MERCEREAU, J. E. 1964 Quantum interference effects in Josephson tunneling. *Phys. Rev. Lett.* **12** (7), 159–160.

JOSEPHSON, B. D. 1962 Possible new effects in superconductive tunnelling. *Phys.Lett.* **1**, 251–253.

KEPHART, T. W. 1985 Fluxoid quantization in knotted superconductors. *Phys. Rev.* **B32**.

LYON, H. C. 1974 *Proc. Amer. Math. Soc.* **43**, 449.

MAGNUS, W., KARRASS, A. & SOLITAR, D. 1966 **Combinatorial Group Theory: Presentations of Groups in Terms of Generators and Relations**. Interscience, New York, N. Y.

MASSEY, W. S. 1959 Some higher order cohomology operations. In *Symp. Int. Topologia Algebraica, Mexico*, p. 145.

MASSEY, W. S. 1968 Higher order linking numbers. In *Proc. Conf. on Algebraic Topology, Chicago University of Illinois at Chicago*, p. 174.

ROLFSEN, D. 1976 **Knots and Links**. Publish or Perish, Wilmington, DE.

SCHULMAN, L. S. 1971 Approximate topologies. *J.Math.Phys.* **12**, 304.

TROTTER, H. F. 1975 *Ann. of Math. Studies* **84**, 51.

Journal of Physics: Conference Series **544** (2014) 012017 doi:10.1088/1742-6596/544/1/012017

Random walks and polygons in tight confinement

Y. Diao[1], **C. Ernst**[2] **& U. Ziegler**[2]

[1] Department of Mathematics and Statistics,
University of North Carolina Charlotte, Charlotte, NC 28223, USA
[2] Department of Mathematics and Computer Science,
Western Kentucky University, Bowling Green, KY 42101, USA

E-mail: yuanandiao@uncc.edu, claus.ernst@wku.edu, uta.ziegler@wku.edu

Abstract. We discuss the effect of confinement on the topology and geometry of tightly confined random walks and polygons. Here the walks and polygons are confined in a sphere of radius $R \geq 1/2$ and the polygons are equilateral with n edges of unit length. We illustrate numerically that for a fixed length of random polygons the knotting probability increases to one as the radius decreases to $1/2$. We also demonstrate that for random polygons (walks) the curvature increases to πn ($\pi(n-1)$) as the radius approaches $1/2$ and that the torsion decreases to $\approx \pi n/3$ ($\approx \pi(n-1)/3$). In addition we show the effect of length and confinement on the average crossing number of a random polygon.

1. Introduction

Many systems and processes in nature do not exist under ideal conditions, but are subject to constraints imposed by the environment in which they occur or the manner in which they are created. This article investigates polygonal random walks and random polygons that are constrained in a small volume. The study of such objects is motivated by physical entities that exist or are created within a small volume and that contain self-entanglements (knots or links). Such topologically and space constrained systems can be found in areas of biology, chemistry and physics. The two examples described below demonstrate the importance of studying such systems. Some macromolecular self-assembly processes pack DNA molecules in a highly condensed manner. For example, in the case of a human cell, the genome contains about 2 meters of DNA, all folded into the nucleus, which is a million times smaller in diameter. The ratio between the volume occupied by a given genome and the volume occupied by a random walk of the same length as the genome (if the genome were to be modeled by a random walk), generally ranges from 10^2 to 10^4 (Holmes & Cozzarelli, 2000). This process of compacting DNA molecules is often called DNA condensation and affects virtually every biological process involving DNA molecules (Horn & Peterson, 2000). Despite the important role that DNA packing mechanisms play in all living organisms and the efforts spent studying them, only some details are known about the workings of DNA packaging motors (Zhang *et al.*, 2012). The general packing mechanisms of double-stranded DNA (dsDNA) remain largely unknown even in the simplest organisms such as viruses like bacteriophages where DNA molecules are packed in a viral capsid (a bacteriophage is a virus that infects and replicates within bacteria). In the case of bacteriophages, the organization of the condensed DNA facilitates the process of DNA packing and provides stability to the capsid while packaged and also facilitates the release of DNA upon infection. Important topological aspects of DNA packing are detected in DNA extracted from bacteriophage P4. These extracted circular DNA molecules are non-trivially knotted with very high probability (Liu *et al.*, 1981*a*,*b*). In particular, quantitative analysis of the knots extracted from bacteriophage P4 revealed that the DNA inside P4 is likely to be chirally organized (Arsuaga *et al.*, 2005).

Another motivation to study random walks and polygons arises in physics. In theoretical physics, quantum chromodynamics (QCD) is the theory of the actions of the strong force (color force), a fundamental force describing the interactions between quarks and gluons which make up hadrons (such as the proton, neutron or pion). The theory is an important part of the standard model of particle physics. A huge body of experimental evidence for QCD has been gathered over the years. In high-energy particle collision, the center of the collision forms a small superheated volume of hot plasma which contains quarks, antiquark and gluons. In a model proposed by physicists (Buniy & Kephart, 2003; Buniy *et al.*, 2013) knots and links play a central role. As the plasma cools the particles rehadronize to form mesons and baryons, and in this process chromoelectric flux tubes can form that may close into knots and links. The spectrum of tight knotted and linked flux tubes in QCD are the observed f_J states seen in high-energy experiments (Buniy *et al.*, 2013). Confinement arises because the knotted flux tubes are confined in the volume of the superheated plasma.

In both examples, one or more (the latter in the case of the quantum flux) tube- or chainlike, self-closing (circular) structures are created or packed inside a confining volume. This creates structures that contain non-trivial self-entanglements (expressed as knots and links). The packing process cannot be observed directly, and experiments (and simulations) provide only scattered snapshots and evidence of the activity. A potential packing mechanism under investigation may only be valid if its simulation produces results that are (statistically) consistent with those observed in experiments. The data reported in this article presents observable packing results for a random packing mechanism. This article is structured as follows: In Section 2 we provide the necessary background terminology in topology and geometry. In Section 3 we give a brief introduction to the generation of random walks and polygons. In Sections 4.1, 4.2 and 4.3 we discuss the numeric data on curvature, torsion, average crossing number and knotting probabilities. Finally, in section 5 we describe future work and state some open questions.

2. Background

The random rope-like physical structures such as the ones encountered in the above examples are modeled in this study as equilateral random polygons confined inside a sphere of a fixed radius R. This section provides important background information about topics related to topology and geometry. Many of these terms can be found in any standard book in knot theory such as (Adams, 2004; Burde & Zieschang, 2002; Cromwell, 2004). The topology of a polygon is expressed by the different knots that can be formed in the polygon. Mathematically, a knot is a just simple closed curve in $\mathbb{R}^3$. Intuitively, two knots are considered topologically equivalent if one can continuously deform one to the other without breaking the curve or causing self-intersections. A *knot type* is defined as the set of all topologically equivalent knots. We use the symbol K when referring to a particular member of the knot type $\mathcal{K}$. A *knot diagram* D is the projection of a knot K onto a plane. D is *regular* if no more than two segments of K cross at the same point in the projection. At each intersection one of the strands of the knot is below or above the other strand. An intersection in a regular projection is called a *crossing*. The minimum number of crossings taken over all regular projections of all members from a knot type $\mathcal{K}$ is called the *crossing number* of the knot type $\mathcal{K}$ and is denoted by $Cr(\mathcal{K})$. A diagram of $\mathcal{K}$ that realizes the crossing number of $\mathcal{K}$ is called a *minimal diagram*. Figure 1 shows two knot diagrams that look very alike but are not. A *knot invariant* of a knot type $\mathcal{K}$ is a quantity Q (such as a number or a polynomial or a group) that can be computed (usually using a knot diagram) with the following property: If K and L are two knots of the same knot type $\mathcal{K}$ then $Q(K) = Q(L)$. Thus if for two knots K and L the knot invariant is different ($Q(K) \neq Q(L)$) then K and L represent different knot types. However the converse of this statement is not true.

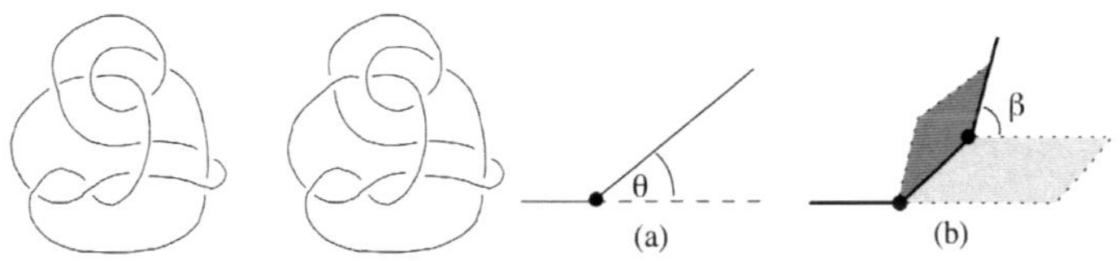

Figure 1. On the left: Two non-minimal knot diagrams: The left one simplifies to the eight-crossing knot 8_9 in the knot table (Cromwell, 2004)) while the right one simplifies to the unknot. On the right: (a) and (b): The angles used to define the total curvature and total torsion of a polygon

Different packing mechanisms affect the overall geometric properties of the random walks and polygons (to be defined in Section 3.1). Note that these geometric quantities are not topological invariants, their value depends on the particular chosen polygonal representative. However these geometric quantities are well suited to describe a collection of random walks and polygons by taking the average value over all members of the collection. Below we define some simple geometric quantities that we discuss in this article. Any two consecutive edges of a given polygon define an angle θ as shown in Figure 1(a). Notice that $0 \leq \theta \leq \pi$. The *total curvature* of the polygon is defined as the sum of all these angles. Any two consecutive edges of a given polygon define a plane when the two edges are not co-linear. The two planes defined by three consecutive edges define a torsion angle β such that $0 \leq \beta \leq \pi$ as shown in Figure 1(b). The *total torsion* of the polygon is defined as the sum of all these torsion angles. The *average crossing number* (or ACN) of a polygon is the average number of crossings over all possible diagrams (or projection) of the polygon. (The ACN should not be confused with the crossing number of a knot type.)

3. Equilateral Random Walks/Polygons

Equilateral random walks/polygons are the simplest and most fundamental type of random walks/polygons which are composed of freely jointed segments of equal length.

3.1. Unconfined Random Walks/Polygons

Let $\vec{r}_k$, $k = 1, 2, ..., n$ be n independent random unit vectors uniformly distributed on the unit sphere centered at the origin O. An equilateral random walk W_n of length n is then defined as the piece-wise linear path whose vertices are given by $X_0 = O$, $X_k = \vec{r}_1 + \vec{r}_2 + \cdots + \vec{r}_k$ ($1 \leq k \leq n$). An equilateral random polygon of length n is then defined as an equilateral random walk of length n subjected to the condition $X_n = O$. Equilateral random walks/polygons consist of individual segments which have no thickness. Such a random walk/polygon is also known as an ideal random walk/polygon and it was already used to model the behavior of chain polymers/closed chain polymers under the so-called theta conditions (where polymer segments that are not in direct contact neither attract nor repel each other) in the middle of last century by Flory and de Gennes (Flory, 1953; Gennes, 1979). This is a subject that had been extensively studied. Much is known about the overall behavior of equilateral random walks/polygons. For example, the overall dimensions such as the average end-to-end distance or the average radius of gyration is known to scale with the number of segments n as $\sqrt{n}$ (Doi & Edwards, 1986; Flory, 1953; Gennes, 1979; Zirbel & Millett, 2012), the average crossing number grows as $O(n \ln n)$ (Diao *et al.*, 2003) and the average squared writhe is believed (with numerical and partial analytical evidence) to grow as $O(n)$ (Portillo *et al.*, 2011). Moreover, there are several generating methods that are well tested and efficient in generating random equilateral polygons such as the hedgehog method (Dobay *et al.*, 2003), the crankshaft method (Klenin *et al.*, 1988; Millett, 2000), or the generalized hedgehog method (Varela *et al.*, 2009). These methods are all well tested and believed to generate non-correlated samples of equilateral random polygons. In the case of the generalized hedgehog method (developed by one of the authors), an equilateral random polygon of length n can be generated in $O(n)$ time (Varela *et al.*, 2009).

3.2. Confined Equilateral Random Walks/Polygons

Unlike the unconfined case, generating confined equilateral random polygons is more difficult. In this project and all the related references cited here, the confining space (volume) is modeled by a sphere S_R of radius $R > .5$. (A sphere S of radius $R < .5$ cannot contain any segments of unit length.) However in a more realistic setting, the walks/polygons behave as "worm-like chains" (meaning that a certain stiffness has to be applied to the joints of the segments), which requires a larger R. If one uses an accept/reject approach and one of the methods mentioned in section 3.1, namely accepting only those un-confined random walks/polygons generated that are contained in S, then the computation time grows exponentially with the length of the walks/polygons. Recently the authors have developed several methods to efficiently generate confined equilateral random polygons using probability density functions at each step of the generation process (Diao *et al.*, 2011, 2012*a*,*b*). These methods have several advantages: first, they can generate random polygons efficiently even in tight confinement, and secondly they allow a theoretical analysis on geometric properties of the polygons generated.

At a first glance, a confined equilateral random walk/polygon just means an equilateral random walk/polygon subjected to the condition that it is within the confining sphere S and the matter seems to be just how to apply this condition in the generating process (from a numerical analysis point of view). However, closer examinations reveal that the matter is more complicated. First, there is no reason why the walk or the polygon has to be rooted at the center of S or at any particular point in S. This is different from the case of the un-confined equilateral random walks/polygons where all walks/polygons are rooted at the origin O (since one can always move the starting point of the walk/polygon to the origin through translation without changing the relative relations of the edges along the walk/polygon). In general, one has to choose the starting point of the random walk/polygon according to a certain probability distribution of the vertices. Second, once the starting point of the random walk/polygon is chosen, the random walk/polygon is an equilateral random walk/polygon subject to the condition that the entire walk/polygon has to stay within S. It turns out that there are different ways to impose this condition leading to random walks/polygons with different probability distributions for the distances of vertices to the center of the sphere.

Despite its apparent importance with many potential applications, the confined equilateral random walks/polygons have received little attention in the past. Consequently, there have been no systematic study on this subject until the recent results by the authors (Diao *et al.*, 2011, 2012*a*,*b*). The following are two different ways to impose (define) the confining condition on a random walk/polygon that have been discussed in their work.

Let n be the length of the equilateral walk/polygon, and let X_0 be the starting point of the walk/polygon with X_1, X_2, ..., X_{n-1}, X_n being the other vertices on the walk/polygon, labeled according to their order along the walk/polygon. In the case of the polygon, $X_n = X_0$. Assume that the distribution of X_0 is given and that X_0 has been chosen according to this distribution. The random walk/polygon is defined in an iterative manner in both definitions.

Condition R. Assuming that X_k has been selected, then the next vertex X_{k+1} is selected subject to the conditions that X_k, X_{k+1}, X_{k+2}, ..., X_n form an equilateral random walk with fixed end point X_k and X_{k+1} is within the confining sphere S (In the case of a polygon we must

Journal of Physics: Conference Series **544** (2014) 012017 doi:10.1088/1742-6596/544/1/012017

also have $X_n = X_0$) .

Condition A. Assuming that X_k has been selected, then the next vertex X_{k+1} is selected subject to the conditions that X_k, X_{k+1}, X_{k+2}, ..., X_n form an equilateral random walk with fixed end point X_k such that this random walk is entirely contained in S. (In the case of a polygon we must also have $X_n = X_0$) .

Both conditions can be used to generate confined equilateral random walks/polygons and generate the un-confined equilateral random walks/polygons once the confining condition is removed (by setting $R = \infty$). Condition A is equivalent to the following (Diao *et al.*, 2012*a*): Suppose we generate random polygons ignoring the confinement condition and starting at the origin. Out of all the polygons generated we now only keep those that lie in confinement. This is often called an absorbent boundary, any polygon that penetrates the boundary is absorbed (i.e. it vanishes). The letter R is referring to the fact that the confining sphere acts in a way similar to a reflective surface: a polygons hitting the boundary is reflected back into the confining sphere. The two conditions lead to different distributions of random walks/polygons. This is illustrated by Figure 2 which shows an example of two different vertex density distributions generated by the two conditions in the special case of $R = 1.5$, $n = 20$ and $X_0 = O$. More specifically Figure 2 shows the distances of each vertex X_i to the origin O for $3 \leq i \leq n-3$ and the histogram has been normalized to represent a probability density function. The lighter (yellow) shading corresponds to Definition R and darker (blue) shading corresponds to Definition A.

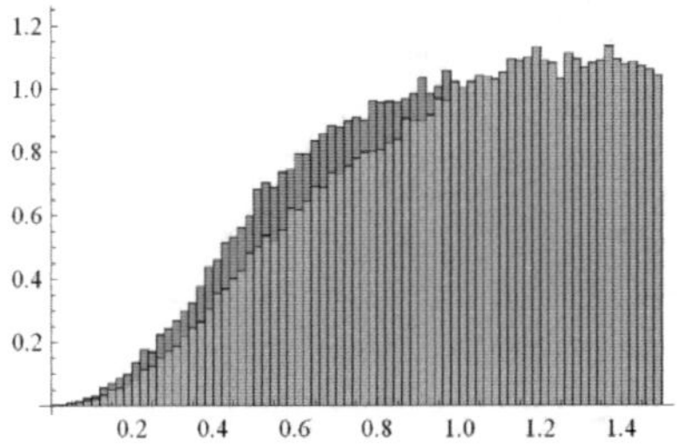

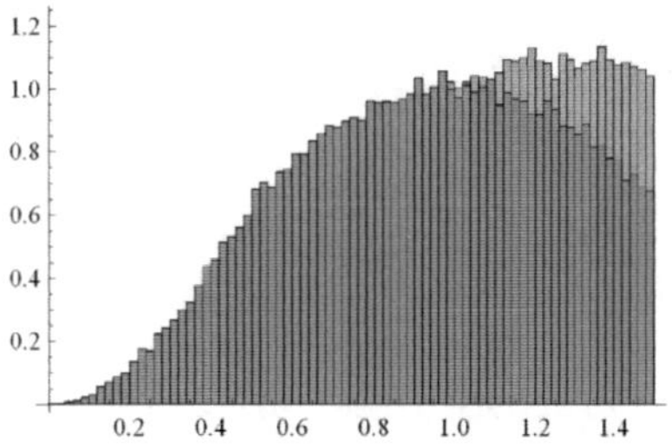

Figure 2. Probability density functions for the distances to the center of the sphere for the vertices of 20-step polygons rooted at the center of the confining sphere with $R = 1.5$. The lighter shading is for Definition R and the darker for Definition A.

4. The Effect of confinement

4.1. Mean Total Curvature and Torsion

It is quite intuitive that confined random walks (polygons) have expected total curvature values larger than their unconfined counterparts since the confinement condition forces the random walks (polygons) to make sharper and more frequent turns on average. In the extreme case the confinement sphere has a diameter close to the unit edge length, and the random walk (polygon) has to make a near 180 degree turn at each step hence its average total curvature would be close to $(n-1)\pi$ ($n\pi$) where n is the number of edges. On the other hand, if the confinement radius is very large relative to the edge length, then the walk (or polygon) behaves like it is unconfined. In this case the average curvature per edge approaches $\pi/2$ as $n \to \infty$ for both the random walks and the random polygons. However the random polygons have slightly larger mean total curvatures than the random walks. For large n, the mean total curvature of the unconfined random polygons of n edges is approximately $n\pi/2 + 3\pi/8$, while the mean total curvature of the unconfined random walks of n edges is precisely $(n-1)\pi/2$ (Grosberg, 2008). These two simple cases demonstrate that the extreme confinement condition can almost double the expected curvature. The left side of Figure 3 is a numerical comparison the mean curvatures per edge between a random walk and a random polygon with $n = 60$. Clearly we can see that the polygons have a slightly larger curvature due to the correction term of $3\pi/8$. In fact, expressed per edge this term is $\pi/160 \approx .02$. Lowering the squares representing the polygons by this amount would show the two data sets matching almost perfectly. (For the radii where we have both data points the mean difference between the two sets is $\approx .014$.)

It is much less intuitive why a confined random walk (or polygon) would have an expected torsion smaller than that of its unconfined counterpart. Again let us look at the two extreme cases. If the confinement radius is very large relative to the edge length, then the walk (or polygon) behaves as an unconfined one. In this case the average torsion per edge approaches $\pi/2$ as $n \to \infty$ (for both the random walks and the random polygons), though a random polygon has a slightly smaller average torsion than a random walk does. As $n \to \infty$, the average total torsion of an unconfined random polygon approaches $n\pi/2 - 3\pi/8$ (Grosberg, 2008). In the case of tight confinement, our numerical results show that the average torsion per edge is smaller than $\pi/2$ per edge. In fact, our numerical results strongly support the following conjecture:

Conjecture: As $n \to \infty$ and $R \to 1/2$, the average torsion (per edge) of a random walk (polygon) of n edges approaches $\pi/3$.

See the right side of Figure 3. Here we see that the polygons have a slightly smaller torsion due to the correction term $-3\pi/8$. As above, expressed per edge this term is $\pi/160 \approx .02$ and raising the squares representing the polygons by this amount would show the two data sets matching almost perfectly. (For the radii where we have both data points the mean difference between the two sets is $\approx .02$.) At this point a rigorous proof of this conjecture is elusive, and certainly seems difficult. The authors believe that a careful analysis of an integral expression of the average torsion might be possible for the case of a random walk. However for a random polygon this seems extremely difficult if not impossible. As a final remark we want to point out that the data shown in Figure 3 suggest that random walks and polygons in confinement have very similar values with respect to their mean curvature and torsion values. This is a little misleading. In fact, the length of the polygon also plays a role here. If the random polygons are relatively short with respect of the confinement condition then it might only barely reach the boundary and the confinement plays no role. If the random polygon is very long then we know that the vertex distribution of the random polygon (generated using Condition A) does not match the vertex distribution of the random walk, see Figure 2 and thus there might be a difference between the torsion and curvature values of random walks and random polygons that does not get smaller as the length of the walks and polygons increases. Finally, we should also point out that when R is close to $1/2$, the confined random polygons have to have even length (unless the polygon is very long): the equilateral triangle is the simplest example of a random polygon of odd length that cannot be confined in a sphere of radius close to $1/2$.

4.2. Mean Average Crossing Number (ACN)

A standard measure of the geometric complexity of random polygons is their mean average crossing number (mean ACN) . Figure 4 shows the mean ACN of polygons ranging in length from 10 to 90 with radii of confinement ranging from $R = 1$ to $R = 4.5$, with a clear indication on how tight confinement leads to a drastic increase in the mean ACN. In (Diao *et al.*, 2011; Arsuaga *et al.*, 2009; Diao *et al.*, 2012*a*) the mean ACN is modeled by a best fit-function of the form $y = a(R)n^2 + b(R)n\log(n)$, where $a(R)$ and $b(R)$ are constants that depend on the radius of confinement R. Using this model we fit the data to a function of the type

$$f(x, y) = (a + \frac{b}{x^2})y^2 + (c + \frac{d}{x^2})y\log(y),$$

where the x variable models the confinement radius R and the y variable models the polygon length. We found that values of $a \approx 0.0005$, $b \approx 0.12$, $c \approx 0.12$ and $d \approx 0.29$ give an excellent fit

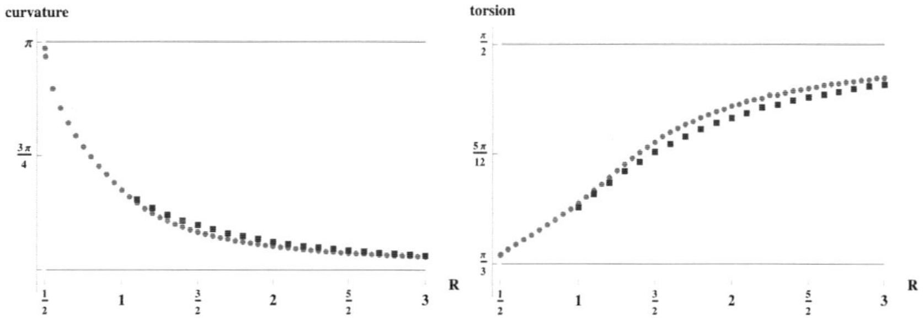

Figure 3. Curvarture (on the left) and torsion (on the right) per edge: The squares represent random polygons of length 60 and the dots represent a single random walk of length one million. The confinement radii (on the x-axis) vary from 1/2 to 3. The random polygons are generated under Condition A using the methods of (Diao *et al.*, 2012*a*): Each data point for the polygons is based on a sample of size 10, 000 while each data point for the random walks is based on a single walk with one million edges generated under Condition R.

with an R^2-value of 0.999978 and a mean difference between the best-fit function and the data of less than 0.35.

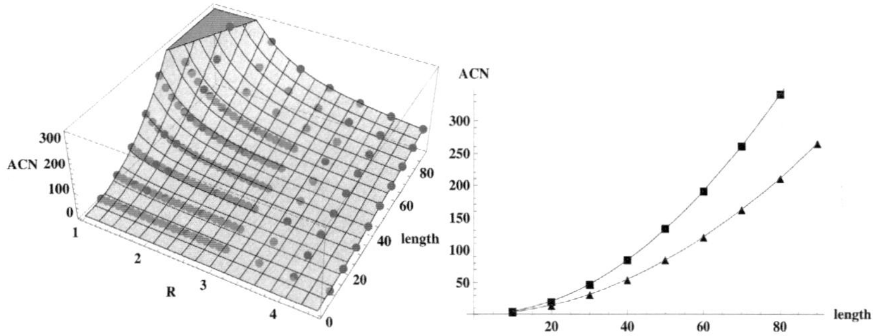

Figure 4. The mean ACN in a 3D plot (on the left) and two slices through the surface for radii $R = 1.5$ (■) and $R = 2$ (▲). Each data point is obtained from a sample of 10, 000 polygons.

One interesting question here is whether the mean ACNs of knotted and unknotted random polygons are different. A positive answer to this question implies that one may use the mean ACN of certain random polygons to make inferences about their topological properties such as their knotting probabilities (which is addressed in the next section). Figure 5 shows the numerical estimation of the ratio (for each fixed length and radius) between the mean ACN of the knotted random polygons and the mean ACN of all random polygons. As expected this ratio is greater than one for all data points. Furthermore, the shorter the polygons and the larger the confining radii, the larger the ratios. We interpret this data as follows: In shorter polygons with a larger confinement radius where knotting is less likely, the ACN of a knotted polygon is higher than the ACN of an unknotted polygon on the average because the topological complexity imposes an increased geometric complexity. The ACN ratio for unconfined random polygons is larger than 1 due to the same reason. With increased length and/or decreased confinement radius, the probability of knotting increases as does the number of crossings in diagrams of unknotted polygons and hence the knotting complexity no longer results in a much increased

ACN, causing a much smaller difference and a smaller ratio. To illustrate this we give a few examples: For length ten and radius one we get a ratio of ≈ 1.27 which increases to ≈ 2.18 for radius three. For length forty and radius one we get a ratio of ≈ 1.002 which increases to ≈ 1.17 for radius three. The noisy results for polygons of length ten at higher confinement radii are due to the fact that there were very few knotted polygons in these ranges. Much larger sample sizes may be needed in order to reduce the noise.

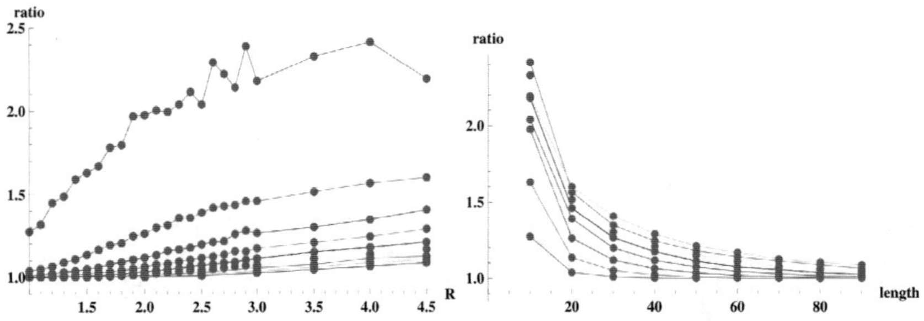

Figure 5. The ratio of the average crossing number of knotted random polygon over the average crossing number of all random polygons. Left: The radius R is on the x-axis and the there are nine curves representing the lengths of the polygons ranging from 10 (on the top) to 90 (at the bottom). Right: The lengths of the polygons is on the x-axis and there are eight curves for various values of R from 1 (bottom) to 4.5 (top) in increments of 1/2.

4.3. Knotting Probabilities

The famous Frisch-Wasserman-Delbrück conjecture states that sufficiently long ring polymers should be knotted with probability one (Frisch & Wasserman, 1961; Delbruck, 1962). This conjecture has been proved for several models of polymer rings including the (unconfined) equilateral random polygons (Diao, 1995). While it is very intuitive and plausible that this conjecture is also true for confined (equilateral) random polygons, no one has been able to prove it to date. The left of Figure 6 shows the numerical estimations of the knotting probabilities using the same random polygon sample used in Figure 3. The nine fitting curves represent polygons of lengths ranging from ten to ninety with increments of ten. All the random polygons shown are generated under Condition A (Diao *et al.*, 2012*a*) and are rooted at the origin (thus the confinement radius cannot be less than one). Knotting was determined by a calculation of the HOMFLY polynomial. For long polygons and small radii R this polynomial computation becomes impossible due to the complexity of the calculation. The unknotting probabilities for $R = 1$ and $R = 1.5$, $n = 10$, 20, 30, 40 and 50 are shown at the right side of Figure 6. At length $n = 50$, most polygons are knotted at $R = 1$ and $R = 1.5$. This shows how drastically the confinement can increase the knotting probability: only about 20% of unconfined equilateral random polygons at length 50 are knotted and this percentage increases to only 50% at length 500 (Millett & Rawdon, 2005).

5. Conclusions

In this paper, we demonstrated, using numerical simulations, the effect of volume confinement on several important geometric measures of random polygons including the mean total curvature, mean total torsion and mean total ACN. We also investigated the effect of volume confinement to the knotting probabilities of random polygons of various lengths. Our results show that confinement leads to increased mean total curvature, but to a decreased mean total torsion.

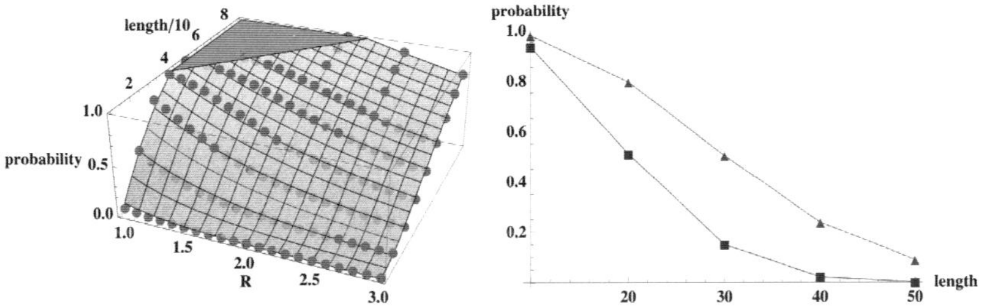

Figure 6. Left: the knotting probability with the confinement radii (x-axis) vary from 1 to 3 and the length (on the y-axis) varies from 10 to 90; Right: the unknotting probability for confinement radii $R = 1$ (■) and $R = 1.5$ (▲) and length $n = 10$, 20, 30, 40 and 50. Each data point is based on a sample of $10,000$ polygons.

Our study on the overall mean ACN matches well with previously published results, but further reveals that for a fixed length, the confinement also elevates the ratio between the mean ACN of knotted polygons and the mean ACN of all polygons. As we pointed out in the introduction section, generating confined equilateral random polygons itself is a difficult problem. There are different ways to define and generate such confined random polygons. How these different definitions and generating methods affect these geometric or topological properties of the polygons is a question for a future study.

Let us end this paper with some open questions. Although our numerical results seem to favor a positive answer for each of them, a theoretical proof for any of these will not be easy.

1. For any fixed confinement radius R, the knotting probability of a random polygon of length n goes to 1 as $n \to \infty$;

2. For any suitably large n, the knotting probability of a confined random polygon of length n goes to 1 as $R \to 1/2$;

3. The total torsion of a confined random walk of n edges goes to $(n-1)\pi/3$ as $R \to 1/2$;

4. If n is even, then the total torsion of a confined random polygon of n edges goes to $n\pi/3$ as $R \to 1/2$.

Acknowledgments
We want to thank Eric Rawdon for his work in determining the HOMFLY polynomial for many of our polygons and determining their knottypes.

References

Adams, C. 2004 **The Knot Book**. American Mathematical Society.

Arsuaga, J., Borgo, B., Diao, Y. & Scharein, R. 2009 The growth of the mean average crossing number of equilateral polygons in confinement. *J. Phys. A: Math. Theor.* **42**, 465202.

Arsuaga, J., Vazquez, M., McGuirk, P., Trigueros, S., Sumners, D. W. & Roca, J. 2005 Dna knots reveal a chiral organization of dna in phage capsids. *Proc. Natl. Acad. Sci. USA* **102**, 9165–9169.

Buniy, R. V., Cantarella, J., Kephart, T. W. & Rawdon, E. 2013 The tight knot spectrum in qcd. *preprint* .

Buniy, R. V. & Kephart, T. W. 2003 A model of glueballs. *Physics Letters B* **576**, 127–134.

Burde, G. & Zieschang, H. 2002 **Knots, Revised edition**. de Gruyter.

Cromwell, P. 2004 **Knots and Links**. Cambridge University Press.

DELBRUCK, M. 1962 Mathematical problems in the biological sciences. *Proceeding of the Symposium on Applied Mathematics, Vol. 14 (American Mathematical Society, Providence, RI, 1962)* **83**.

DIAO, Y. 1995 The knotting of equilateral polygons in $\mathbb{R}^3$. *J. of Knot Theory and its Ramifications* **4**.

DIAO, Y., DOBAY, A., KUSNER, R., MILLETT, K. & STASIAK, A. 2003 The average crossing number of equilateral random polygons. *J. Phys. A: Math. Theor.* **36**.

DIAO, Y., ERNST, C., MONTEMAYOR, A. & ZIEGLER, U. 2011 Generating equilateral random polygons in confinement. *J. Phys. A: Math. Theor.* **44**.

DIAO, Y., ERNST, C., MONTEMAYOR, A. & ZIEGLER, U. 2012*a* Generating equilateral random polygons in confinement ii. *J. Phys. A: Math. Theor.* **45**, 275203.

DIAO, Y., ERNST, C., MONTEMAYOR, A. & ZIEGLER, U. 2012*b* Generating equilateral random polygons in confinement iii. *J. Phys. A: Math. Theor.* **45**, 465003.

DOBAY, A., DUBOCHET, J., MILLETT, K., SOTTAS, P.-E. & STASIAK, A. 2003 Scaling behavior of random knots. *Proc Natl Acad Sci USA* **100**.

DOI, M. & EDWARDS, S. 1986 **The theory of polymer dynamics**. Oxford University Press.

FLORY, P. 1953 **Principles of Polymer Chemistry**. Cornell University Press.

FRISCH, H. L. & WASSERMAN, E. 1961 Chemical topology. *J. Amer. Chem. Soc.* **83**.

GENNES, P. G. D. 1979 **Scaling Concepts in Polymer Physics**. Cornell University Press.

GROSBERG, A. Y. 2008 Total curvature and total torsion of a freely jointed circular polymer with $n >> 1$ segments. *Macromolecules* **41**.

HOLMES, V. F. & COZZARELLI, N. R. 2000 Closing the ring: links between smc proteins and chromosome partitioning, condensation, and supercoiling. *Proc. Natl. Acad. Sci. USA* **97**, 1322–1324.

HORN, P. J. & PETERSON, C. L. 2000 Chromatin higher order folding: Wrapping up transcription. *Science* **297**, 1824–1827.

KLENIN, K. V., VOLOGODSKII, A. V., ANSHELEVICH, V. V., DYKHNE, A. M. & FRANK-KAMENETSKII, M. D. 1988 Effect of excluded volume on topological properties of circular dna. *J. Biomolec. Str. and Dyn.* **5**.

LIU, L. F., DAVIS, J. L. & CALENDAR, R. 1981*a* Novel topologically knotted dna from bacteriophage p4 capsids: studies with dna topoisomerases. *Nucleic Acids Res.* **9**, 3979–3989.

LIU, L. F., PERKOCHA, L., CALENDAR, R. & WANG, J. C. 1981*b* Knotted dna from bacteriophage capsids. *Proc. Natl. Acad. Sci. USA* **78**, 5498–5502.

MILLETT, K. 2000 Monte carlo explorations of polygonal knot spaces. *Knots in Hellas'98* .

MILLETT, K. & RAWDON, E. 2005 Universal characteristics of polygonal knot probabilities. *Jorge Alberto Calvo, Kenneth C. Millett, Eric J. Rawdon, and Andrzej Stasiak, editors Physical and Numerical Models in Knot Theory, Ser. Knots Everything, River Edge, NJ, World Sci. Publishing.* **36**, 247–274.

PORTILLO, J., ARSUAGA, J., DIAO, Y., SCHAREIN, R. & VAZQUEZ, M. 2011 On the mean and variance of the writhe of random polygons. *J. Phys. A: Math. Theor.* **44**.

VARELA, K., HINSON, K., ARSUAGA, J. & DIAO, Y. 2009 A fast ergodic algorithm for generating ensembles of equilateral random polygons. *J. Phys. A: Math. Theor.* **42**.

ZHANG, H., SCHWARTZ, C., DONATIS, G. M. D. & GUO, P. 2012 "push through one-way valve" mechanism of viral dna packaging. *Adv Virus Res.* **83**, 415–465.

ZIRBEL, L. & MILLETT, K. 2012 Characteristics of shape and knotting in ideal rings. *J. Phys. A: Math. Theor.* **45**.

Journal of Physics: Conference Series **544** (2014) 012018 doi:10.1088/1742-6596/544/1/012018

How averaged Menger curvatures control regularity and topology of curves and surfaces

P. Strzelecki[1], **H. von der Mosel**[2]

[1] University of Warsaw, Institute of Mathematics, ul. Banacha 2, 02–097 Warsaw, Poland
[2] RWTH Aachen University, Templergraben 55, 52062 Aachen, Germany

E-mail: `pawelst@mimuw.edu.pl`, `heiko@instmath.rwth-aachen.de`

Abstract. We demonstrate how integrated Menger curvatures can replace the nonsmooth ropelength functional to control regularity and topology of knotted curves and surfaces.

1. Introduction

Classic local curvature κ_γ of a closed smooth curve $\gamma \subset \mathbb{R}^3$ is a function that is defined pointwise: at the point $x = \gamma(s)$ one has $\kappa_\gamma(x) = |\gamma''(s)|$ via the arclength or unit-speed parametrization $\gamma : \mathbb{R}/\mathbb{Z} \to \mathbb{R}^3$ with $|\gamma'(t)| = 1$ for all $t \in \mathbb{R}/\mathbb{Z}$. Alternatively and in more geometric terms, the best local approximation of γ by circles near the curve point x is given by the osculating circle that touches γ at the point x. The radius $r = r(x)$ of this osculating circle is the *local radius of curvature* whose inverse equals the local curvature $\kappa_\gamma(x)$ of γ at x. Likewise for surfaces or higher-dimensional submanifolds, classic curvature functions such as mean or Gauß curvature are local, that is, they take single surface points as arguments. In contrast to that, Menger (1930) considered functions on curves that depend on more than one curve point. The most prominent example of such a *multipoint function* is the circumcircle radius $R(x, y, z)$ of three points $x, y, z \in \gamma$, which tends to the local radius of curvature $r(x)$ in the coalescent limit $y, z \to x$ if γ is smooth. This and the fact that $R(x, y, z)$ may be expressed solely in terms of the mutual distances of the three points x, y, and z, led Menger to the idea of developing a purely metric geometry based on such multipoint functions in contrast to classic differential geometry. Instead of following that thread here let us add one further simple observation: in order to define $R(x, y, z)$ along γ it is by no means necessary to assume any smoothness of γ, which turns out to be useful from the variational point of view. Indeed, quite often one has to enlarge the classic function spaces to a wider class of nonsmooth functions, such as Sobolev spaces, to actually find energy minimizers in the calculus of variations, or to find weak solutions of partial differential equations.

With that in mind and motivated by computational issues in the numerical simulation of macromolecules with self-contact, Gonzalez & Maddocks (1999) reconsidered the circumcircle radius as a means to define an analytically tractable notion of *thickness* for curves. Indeed, their idea was to search for the smallest possible circumcircle radius among all triples of curve points. This way one may end up with an osculating circle at a point of highest local curvature of the curve, or, on the other hand, with a limiting circle whose radius is much smaller than the local curvature at any of the curve strands it touches, thereby reflecting the fact that two (or more) strands of the same curve are close to each other in the ambient space; see Figure 1. This interplay between local and global

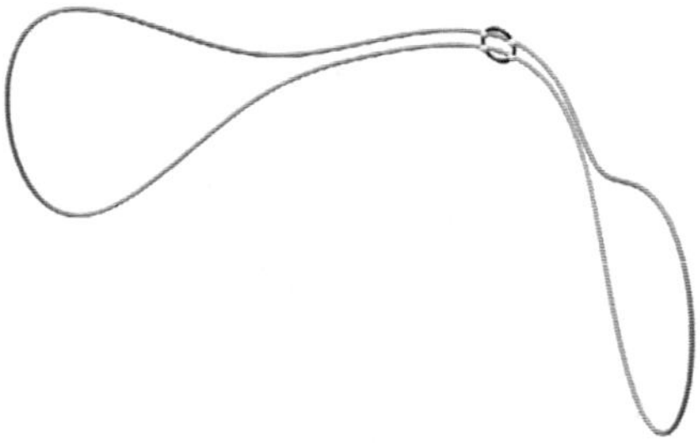

Figure 1. $1/R$ can be high at many locations where classic local curvature remains moderate.

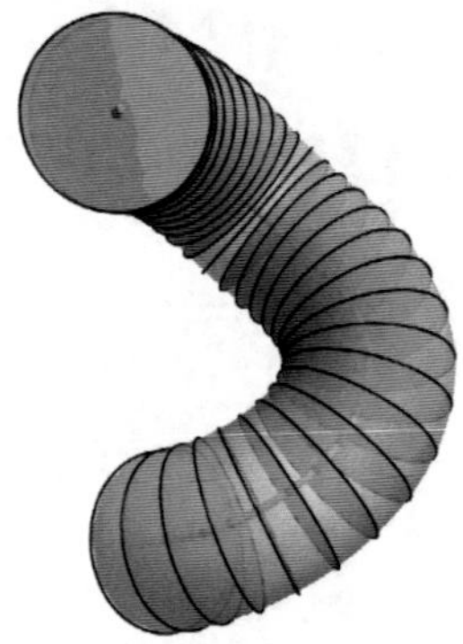

Figure 2. The tubular neighbourhood of a $C^{1,1}$-curve

control by means of a pointwise minimization of $R(x, y, z)$ is reflected in the following theorem proved independently by Cantarella *et al.* (2002) and Gonzalez *et al.* (2002).

Theorem 1.1. *An arclength parametrized closed curve* γ *with*

$$\triangle[\gamma] := \inf_{x \neq y \neq z \neq x} R(x, y, z) > 0 \tag{1}$$

is embedded and continuously differentiable, with a Lipschitz continuous tangent vector, and local curvature κ_γ *defined and bounded almost everywhere by* $1/\triangle[\gamma]$.

Even more is true: the quantity $\triangle[\gamma]$ can justifiably be called *thickness* of γ since there is a uniform tube around the curve, consisting of the disjoint union of normal disks centered on the curve and with uniform radius $\triangle[\gamma]$, so that in particular the nearest-point projection onto γ is well-defined and unique within that tubular neighbourhood; see Figure 2. In that sense, positive thickness $\triangle[\gamma]$ serves as an *exact excluded volume constraint* for γ. Theorem 1.1 may be viewed as the foundation for the analytic investigation of *ideal knots* as minimizers of *ropelength*, i.e. the quotient of length and thickness, see Cantarella *et al.* (2002, 2006, 2011, 2012); Gonzalez *et al.* (2002); Gonzalez & de la Llave (2003); Schuricht & von der Mosel (2004), as well as for elastic rods with self-contact Schuricht & von der Mosel (2003), or packing problems as in Gerlach & von der Mosel (2011*b*,*a*). One possible generalization of thickness to surfaces including an existence theory for area-minimizing thick surfaces with prescribed genus or in given isotopy classes was published by Strzelecki & von der Mosel (2005, 2006).

The analytic drawback of thickness or ropelength is that it is a nonsmooth functional due to the minimization over triples of points; see (1). Consequently, techniques from nonsmooth analysis are required to derive variational equations as a set of necessary conditions for minimizers of ropelength. In the light of this, it is not too surprising that up to now the optimal regularity of ideal knots is not known[1], and there is no rigorous analytic treatment of a suitable gradient flow for ropelength. So, it is

[1] Cantarella *et al.* (2002) have constructed ideal links with more than one component that are only $C^{1,1}$, and there is numerical evidence that the ideal trefoil knot may have jumps in local curvature.

Journal of Physics: Conference Series **544** (2014) 012018 doi:10.1088/1742-6596/544/1/012018

natural to ask for suitable relaxations of the ropelength functional without loosing too many features that qualify ropelength as a *knot energy* separating different knot classes by infinite energy barriers.

In this short note we will address one possible relaxation of ropelength $1/\triangle[\gamma]$ (keeping unit length of all curves γ fixed), by replacing each pointwise maximization by averaging. That is, we consider the *integral Menger curvature*[2]

$$\mathscr{M}_p(\gamma) := \int_\gamma \int_\gamma \int_\gamma \frac{1}{R(x,y,z)^p}\, d\mathscr{H}^1(x) d\mathscr{H}^1(y) d\mathscr{H}^1(z) \tag{2}$$

for a suitable range of integrabilities $p > 0$. It is clear that one may loose a lot of control over the shape and regularity of a finite energy curve, since even for large positive p the circumcircle R may be arbitrarily small at certain triples of curve points. The only hope is that such sets of triples cannot have large measure. We discuss in this note which analytic and topological effects can still be observed for curves with finite integral Menger curvature – at least for p above the scale invariant case $p = 3$. To address a wider audience we will make an effort to survey our results in a more informal way avoiding rigorous mathematical proofs, but we will try to explain some of the underlying (mostly) geometric ideas and principles. We would like to invite the reader who is interested in the detailed proofs to consult our respective papers (partly joint work with S. Kolasiński or M. Szumańska), explicit references will be given.

In Section 2 we discuss integral Menger curvature and its regularizing and topological effects on closed curves (Theorem 2.1), including a short sketch on how to obtain initial higher regularity and a weak form of an excluded volume constraint, which we refer to as the *diamond property*. This suffices to identify $\mathscr{M}_p$ as a valuable knot energy: in each knot class we find minimizing representatives, the energy bounds the stick number and the average crossing number, thus the number of isotopy types, and the energy separates knot space in terms of Hausdorff-distance; see Theorem 2.2. In Section 3 we describe how to generalize integral Menger curvature to higher dimensions and co-dimensions and present some of our more recent results such as a geometric variant of the Morrey-Sobolev embedding theorem (Theorem 3.1), the existence of $\mathscr{M}_p$-minimizing submanifolds in given isotopy classes, and a finiteness theorem on the isotopy types of Lipschitz submanifolds with unit area with uniformly bounded integral Menger curvature (Theorem 3.2). Apart from the technically more demanding proofs in higher dimensions that cannot even be sketched in the present note, there is one crucial conceptual difference to one-dimensional sets, the so-called *Ahlfors regularity*, that we do address at the end of Section 3, for simplicity for two-dimensional surfaces in $\mathbb{R}^3$: there is absolutely no control how much measure of a closed surface is contained in a small ball that is centered on the surface, it depends heavily on the individual surface; see Figure 3. For closed curves, on the other hand, it is clear that the length of the curve within a small ball centered on the curve exceeds the ball's radius. We present a new algorithm to find "voluminous" tetrahedra on a general closed surface, which may be of independent interest; see Theorem 3.3. These voluminous tetrahedra can then be used to control Ahlfors regularity of finite energy surfaces purely in terms of the energy value, which is the essential first step to even start regularity considerations for integral Menger curvature in higher dimensions.

To conclude this introduction let us point out an interesting connection to measure theoretic results in harmonic analysis where integral Menger curvature was used to prove Vitushkin's conjecture on the solution of the famous Painlevé problem on the removability of singularities for bounded analytic functions in the complex plane[3]. Léger (1999) proved that one dimensional Borel sets $E \subset \mathbb{R}^n$

[2] As already suggested in Gonzalez & Maddocks (1999) as a potential knot energy; see also Sullivan (2002). Banavar *et al.* (2003) discuss its possible relevance for modeling self-avoidance since no regularization is necessary in contrast to repulsive potentials such as O'Hara's energies, in particular the Möbius energy; see O'Hara (1991, 1992*a*,*b*).

[3] For more details in that context; see, e.g., the surveys David (1999), Tolsa (2006) or Mattila (2004).

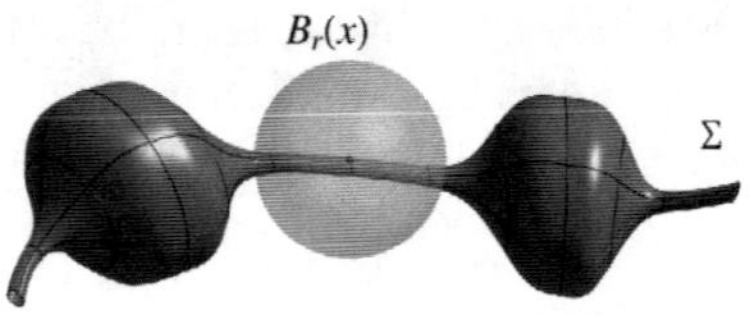

Figure 3. For fixed radius r, the measure of the surface in $B_r(x)$ can be made arbitrarily small just by making the tube thinner.

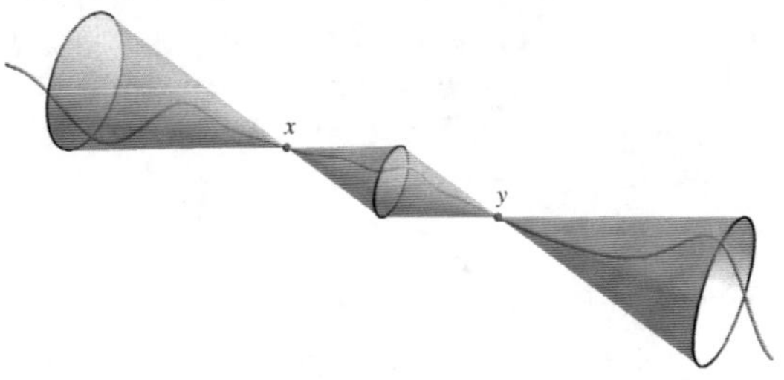

Figure 4. In a small scale, all arcs of a finite energy curve are trapped in double cones.

with $\mathscr{M}_2(E) < \infty$ are 1-rectifiable in the sense of geometric measure theory. That is, such sets can (apart from a set of measure zero) be covered by a countable union of graphs of Lipschitz continuous functions on the real line. Notice that the set can still consist of many disconnected little bits and pieces, nevertheless this result yields a dramatic increase of regularity even though $p = 2$ is below the scale-invariant exponent. This, on the one hand, serves as additional motivation to prove classic higher regularity, starting with closed rectifiable curves already on a higher regularity level and with $p \geq 3$ as we describe in the present note. On the other hand, it opens more terrain for future research possibly bridging the gap between the different ranges of integrability: polygons are shown to have finite integral Menger curvature $\mathscr{M}_p$ exactly for $p \in (0, 3)$ (Scholtes (2011)), so classic differentiability is not to be expected in this range, but the situation remains unsettled for the scale-invariant case $p = 3$.

2. Integral Menger curvature for curves

To fix notation, we denote by $\mathcal{C}$ the class of all *unit loops*, i.e., closed curves $\gamma : \mathbb{R}/\mathbb{Z} \to \mathbb{R}^3$ parametrized by arclength with $\gamma(0) = 0$, and with $\mathscr{H}^1(\gamma) = 1$ to exclude undesirable parametrizations that cover subarcs or all of γ more than once. Since we are mainly interested in the various shapes of different representatives of knot classes this is not a serious restriction. Moreover, integral Menger curvature does not penalize multiple coverings as can be seen, e.g., for the twice covered semicircle: the integrand of $\mathscr{M}_p$ is simply constant along the semicircle which gives finite energy, the image is a one-dimensional manifold with boundary but the parametrization is not smooth at the two endpoints since the tangent vector performs an abrupt turn by $180°$.

The first theorem is fundamental, it reflects how much control on topology and regularity finite integral Menger curvature imposes on a unit loop.

Theorem 2.1 (Strzelecki *et al.* (2010)). *All unit loops $\gamma \in \mathcal{C}$ with $\mathscr{M}_p(\gamma) \leq E < \infty$ for $p \geq 3$ are homeomorphic to a circle. Moreover, if $p > 3$, then γ is of class $C^{1,1-(3/p)}$ with a uniform estimate of the corresponding $C^{1,1-(3/p)}$-norm solely in terms of E and p.*

So, self-intersections are excluded for $p \geq 3$, and for $p > 3$ one has a Hölder continuous tangent vector with the Hölder exponent $1 - (3/p)$. This exponent is optimal as one can either see by specific examples constructed in Szumańska (2009); Kolasiński & Szumańska (2013), or by the more recent characterization of finite energy curves in terms of fractional Sobolev spaces which exactly embed into $C^{1,1-(3/p)}$ and not into any better classic function space; see Blatt (2013); Blatt & Kolasiński (2012). For the reader who is more familiar with classic Sobolev spaces we would like to offer an interpretation of Theorem 2.1 as a *geometric variant of the Morrey-Sobolev embedding theorem*: the integrand $1/R$ is a very weak form of curvature integrated on a three-dimensional domain $\gamma \times \gamma \times \gamma$,

Journal of Physics: Conference Series **544** (2014) 012018 doi:10.1088/1742-6596/544/1/012018

so it is comparable to the classic Morrey-Sobolev embedding theorem that guarantees $C^{1,1-(3/p)}$-regularity for functions whose second derivatives are p-integrable on a three-dimensional domain. Theorem 2.1 is a generalization of Theorem 1.1 for curves of finite ropelength. Notice that ropelength $1/\triangle[\gamma]$ is the limit of $\mathscr{M}_p^{1/p}(\gamma)$ as $p \to \infty$ for any unit loop $\gamma \in \mathcal{C}$, so that one might think of taking the formal limit in the regularity statement as well, to obtain the limiting $C^{1,1}$-regularity for finite ropelength.

For the proof of Theorem 2.1 one establishes first a geometric rigidity of the curve that serves as a weak form of an excluded volume constraint, and provides an intermediate Hölder regularity of the tangent with non-optimal exponent. After that, one proves in a more technical second step the optimal regularity by slicing and iteration arguments. Let us very briefly sketch the more geometric idea behind the first step, in which one tries to control the *beta numbers* introduced in the context of harmonic analysis in Jones (1990) and defined for $x \in \gamma$ and $d > 0$ as

$$\beta_\gamma(x,d) := \inf\Big\{ \sup_{y\in\gamma\cap B_d(x)} \frac{\operatorname{dist}(y,L)}{d} : L \text{ is a straight line through } x \Big\}. \tag{3}$$

The beta numbers measure how well the curve can be approximated locally by straight lines, or in other words, $\beta_\gamma(x,d)\cdot d$ is the width of the thinnest cylinder centered at $x \in \gamma$ that contains the portion of the curve γ in the ball $B_d(x)$. If the beta number happens to be small then the set is locally fairly flat. In the present situation the following holds:

Claim. There is a universal constant $c_0(p) > 0$ depending only on p such that for any $\epsilon \le 0.001$ and $d < \operatorname{diam}\gamma$ satisfying the balance condition

$$\epsilon^{6+p} d^{3-p} \ge c_0(p)\mathscr{M}_p(\gamma) \tag{4}$$

one has $\beta_\gamma(x,d) \le \epsilon$.

Applying this for ϵ with equality in the balance condition one observes that $\beta_\gamma(x,d) \lesssim d^\kappa$ for $\kappa = (p-3)/(p+6)$ which means that the height of the cylinder confining γ within the ball $B_d(x)$ is bounded by $d^{\kappa+1}$. Keeping the curve point x fixed, one can repeat the same reasoning on every scale $d/2^n$ for $n = 1,2,\ldots$ to obtain a sequence of thinner and thinner cylinders centered at $x \in \gamma$ and containing the curve in the balls $B_{d/2^n}(x)$. The height of the previous cylinder bounds the maximal tilt angle of the following cylinder, and all these angles add up like a geometric series to a total tilt angle $\lesssim d^\kappa$. Consequently, taking the limit $n \to \infty$, one has established the following geometric rigidity: $\gamma \cap B_d(x)$ is contained in a double cone with vertex x, and cone axis containing the segment $x-y$ for any curve point $y \in \partial B_{d/2}(x)$, and with cone angle $\lesssim d^\kappa$. Reversing the roles of x and y one actually finds locally near x and y the curve γ being trapped in the intersection of two such double cones; see Figure 4. This immediately implies the intermediate Hölder regularity of the tangent γ' with the non-optimal exponent κ, since one can argue first at all points x and y of differentiability, so that the corresponding unit tangents differ by at most the cone angle $\lesssim d^\kappa$. This uniform estimate on the oscillation of tangents, that exist almost everywhere, can then be uniquely extended to all of the parameter domain to find that the tangent exists everywhere and satisfies the same oscillation estimate. In order to establish the claim one assumes to the contrary that any cone with axis through $x, y \in \gamma$ with $|x-y| < d$ misses some point $z \in B_d(x)$, so that the triangle with vertices x, y, z has height greater than ϵd; see Figures 5 and 6. Elementary geometry yields the inequality

$$\frac{1}{R(x,y,z)} = 2\frac{\text{height}}{|x-z|\cdot|y-z|} \gtrsim \frac{\epsilon d}{d^2} = \frac{\epsilon}{d}. \tag{5}$$

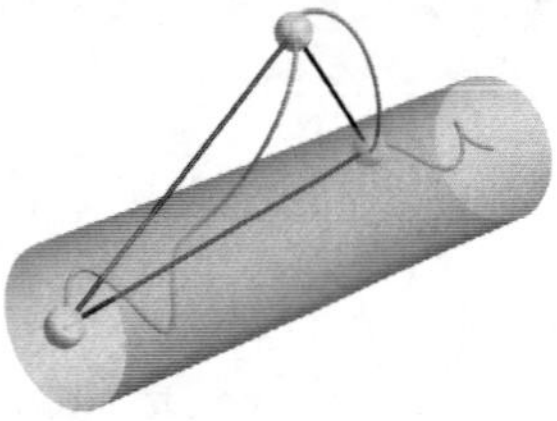

Figure 5. The beta number is large: the curve does not fit into a narrow cylinder.

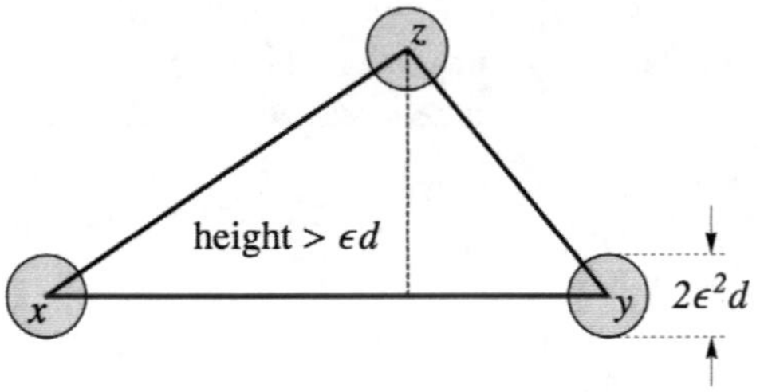

Figure 6. Small perturbations of vertices do not change $1/R$ significantly.

This estimate alone is useless since it is an estimate for just *one* triple of points which could be ignored completely by the energy (in contrast to ropelength!), but there are enough triples of points close by with roughly the same estimate, as long as those triples are in sufficiently small balls (say of radius $\epsilon^2 d$) about x, y, and z, respectively; see Figure 6. Indeed, any curve that hits the center of a ball and that leaves the ball has one dimensional Hausdorff measure at least as large as the radius of the ball:

$$\mathscr{H}^1(\gamma \cap B_{\epsilon^2 d}) \geq \epsilon^2 d. \tag{6}$$

This property is commonly referred to as *uniform (lower) Ahlfors regularity*, a fact not shared by higher dimensional objects such as two-dimensional surfaces, and we will make an effort in Section 3 to explain an algorithm which implies uniform Ahlfors regularity for general finite energy surfaces.

Returning to the proof of the claim we can now see how to obtain the desired contradiction by estimating the total energy from below by the contribution of the three little portions of γ within the balls of radius $\epsilon^2 d$ about x, y, and z, and finally using the balance condition (4):

$$\mathscr{M}_p(\gamma) > \int_{B_{\epsilon^2 d}(x)\cap\gamma} \int_{B_{\epsilon^2 d}(y)\cap\gamma} \int_{B_{\epsilon^2 d}(z)\cap\gamma} \frac{1}{R^p} \overset{(5)(6)}{\gtrsim} (\epsilon^2 d)^3 \cdot \left(\frac{\epsilon}{d}\right)^p = \epsilon^{6+p} d^{3-p} \overset{(4)}{\geq} c_0(p)\mathscr{M}_p(\gamma), \tag{7}$$

which is absurd for an appropriately chosen constant $c_0(p)$. Thus, the initial step in the proof of Theorem 2.1 is completed.

Let us recall that the curve is actually locally trapped in the intersection of two double cones with vertices at curve points on a sufficiently small scale $d(E)$ that depends only on the energy value $\mathscr{M}_p(\gamma) \leq E$. Since this property holds true on this and every smaller scale, no matter where the two points lie on the curve, we have established what we call the *diamond property*:

For each two points $x, y \in \gamma$ with $|x-y| \leq d(E)$ the portion of γ within the intersection of balls, $B_{2|x-y|}(x) \cap B_{2|x-y|}(y)$ is contained in the intersection of the two double cones with opening angle $\lesssim |x-y|^\kappa$, common axes through x and y, and with vertices at x and y, respectively; see Figure 4. In addition, in each cross section perpendicular to the common cone axis, one finds exactly one curve point γ.

One can use this property to discover that γ carries along a whole necklace of such double cones, in our terms *diamonds*, that have mutually disjoint interiors. This necklace serves as a weak form of an excluded volume constraint, since the curve is restricted to the diamonds in a whole neighbourhood of the curve. This holds for every partition of the curve into small arcs, as long as each such arc is sufficiently small in diameter, i.e. smaller than the scale $d(E)$. However, let us point out that there is no unique nearest-point projection, and we do not have the disjoint union of uniformly sized normal disks about the curve, as finite ropelength would imply. Still, the diamond property gives the curve enough rigidity to control local bending, and to allow for topological results that qualify integral Menger curvature as a valuable knot energy.

Theorem 2.2 (Strzelecki *et al.* (2013)). *Let $p > 3$, then $\mathscr{M}_p$ satisfies the following conditions:*

(i) *$\mathscr{M}_p$ is* charge, *i.e., it blows up along sequences $\{\gamma_i\} \subset \mathcal{C}$ that converge uniformly to some curve γ with self-intersections. Moreover, $\mathscr{M}_p$ is* tight *in the sense that it blows up along sequences for which small nontrivially knotted parts pull tight in the limit.*

(ii) *In each knot class $[K]$ one finds a representative $\gamma_K \in \mathcal{C} \cap [K]$ that minimizes $\mathscr{M}_p$, i.e., $\inf_{\gamma\in[K]\cap\mathcal{C}} \mathscr{M}_p(\gamma) = \mathscr{M}_p(\gamma_K)$.*

(iii) *$\mathscr{M}_p$ bounds the stick number and the crossing number, hence it bounds the number of knot classes that can be represented under a given energy level.*

(iv) *Any two finite energy curves $\gamma_1, \gamma_2 \in \mathcal{C}$ that are closer in Hausdorff-distance than a constant depending only on p and on $\max\{\mathscr{M}_p(\gamma_1), \mathscr{M}_p(\gamma_2)\}$ represent the same knot class.*

Let us comment on some of the aspects and on typical arguments that enter the proof of this result. For example, as to part (i), if $\gamma_i \to \gamma$ for some sequence with $\mathscr{M}_p(\gamma_i) \leq E$ for all i, then we may extract a subsequence converging to γ in C^1 because of the uniform $C^{1,1-(3/p)}$-a-priori estimates depending only on E. These estimates for the embedded curves γ_i yield a uniform bi-Lipschitz estimate which carries over to the limit curve γ. Hence γ cannot have a double point. A similar compactness argument, or alternatively, the rigidity imposed by the diamond property, shows that $\mathscr{M}_p$ is also tight, since each cross section of any diamond contains only one curve point, which prevents small knotted subarcs to pull tight. The improvement to C^1-convergence guarantees the same knot class in the limit, which yields the existence of energy minimizers in each given knot class in part (ii) by the direct method in the calculus of variations. The stick number is the minimal amount of straight segments (sticks) needed to construct a polygonal representative of a given knot class. To bound that number which is a knot invariant we can show that for a given curve $\gamma \in \mathcal{C}$ with $\mathscr{M}_p(\gamma) \leq E < \infty$, any inscribed polygon with edge length smaller than a constant $C = C(p, E)$ is automatically isotopic to γ. To that end we cover γ with a necklace of diamonds connecting the polygonal vertices, so that the axial segments of the diamonds coincide with the polygonal edges. Then we construct a suitable ambient isotopy between γ and the polygon by keeping everything constant outside the diamonds, while suitably mapping homeomorphically each of the distinct cross section of each diamond onto itself (keeping the boundaries fixed), whilst moving in each cross section the one and only curve point towards the axis of the double cone, i.e., onto the respective polygonal edge. The *minimal crossing number* of a representative γ of a knot class $[K]$ is the minimal number of self-crossings that one obtains when looking at all planar projections of γ, and the *crossing number* of $[K]$ is then the smallest possible such minimal crossing number if one looks at all representatives of $[K]$. To bound the latter we bound the minimal crossing number of a given representative $\gamma \in [K]$ by bounding its *average crossing number* $\mathrm{acn}(\gamma)$, which is the average value of crossing numbers upon taking all possible planar projections into account. Due to the practical integral representation; see Freedman *et al.* (1994),

$$\mathrm{acn}(\gamma) = \frac{1}{4\pi} \int\int_{[0,1]^2} \frac{|\det(\gamma'(s), \gamma'(t), \gamma(s) - \gamma(t))|}{|\gamma(s) - \gamma(t)|^3}\, dsdt, \tag{8}$$

and following ideas of Buck & Simon (1997) for thick knots, we can use the diamond property as a weaker from of excluded volume constraint to estimate this double integral. The local estimate of this integral on pairs of parameters that are sufficiently close to each other is a result of the geometry in one diamond: the respective tangents and the chord $\gamma(s) - \gamma(t)$ for such parameters span a fairly flat parallelepiped, so that the integrand is sufficiently small. For mutually distant parameters, on the other hand, one can use the necklace of diamonds as an excluded volume constraint to compute a worst case scenario of closest possible packing since the numerator in (8) as the dominating term needs to be controlled. Both, stick number and the crossing number, are knot invariants, and one knows from knot theory how bounds on either of these transfer to bounds on the number of possible knot types representable under these bounds, which concludes part (iii). For part (iv) one uses inscribed polygons of the same knot type as in part (iii) for each of the two curves, and shows by virtue of so called Δ-moves the combinatorial equivalence of these inscribed polygons to guarantee the same knot type of the two curves.

3. Integral Menger curvature in higher dimensions

Léger (1999) proposed to replace the integrand $1/R$ in $\mathscr{M}_p$ by the expression $\mathcal{L}(\xi, x, y, z) := \frac{\operatorname{dist}(\xi, \langle x,y,z\rangle)}{|\xi-x||\xi-y||\xi-z|}$ for two-dimensional sets, where $\langle x, y, z\rangle$ denotes the 2-plane spanned by x, y, and z. This suggestion might or might not work to generalize his proof of rectifiability, but it is certainly not the right choice in our context, since even the unit sphere $\mathbb{S}^2$ would have infinite energy

$$\int_{\mathbb{S}^2}\int_{\mathbb{S}^2}\int_{\mathbb{S}^2}\int_{\mathbb{S}^2} \mathcal{L}^p(\xi, x, y, z)\, d\mathscr{H}^2(\xi) d\mathscr{H}^2(x) d\mathscr{H}^2(y) d\mathscr{H}^2(z) \tag{9}$$

for sufficiently large p. The most obvious generalization of the circumcircle radius of three curve points would be the circumsphere radius of four points on a surface, which would provide a constant integrand equal to 1 for the unit sphere $\mathbb{S}^2$. But even that choice is too singular for our purposes, since there are smooth surfaces, e.g. with straight nodal lines such as the graph of the smooth function $f(x, y) := xy$, where the inverse of the circumsphere radius as a possible integrand is not bounded. This led us to the idea to use a very similar but less singular integrand defined on tetrahedra $T = (\xi, x, y, z)$ of surface points, namely $\text{Volume}(T)/(\text{Area}(T) \cdot (\operatorname{diam} T)^2)$ (Strzelecki & von der Mosel (2011)), where $\text{Area}(T)$ is the sum of all facet areas of T, or even simpler (Kolasiński (2011, 2012)), and for all dimensions $k < n$ and $(k + 1)$-dimensional simplices $T = (x_1, \ldots, x_{k+2}) \in \mathbb{R}^n$:

$$\mathcal{K}(T) := \text{Volume}(T)/(\operatorname{diam} T)^{k+2} \quad \text{if } T \text{ is non-flat}, \ \mathcal{K}(T) := 0 \text{ otherwise.} \tag{10}$$

The scale invariant case for integral Menger curvature $\mathscr{M}_p(\Sigma)$ on k-dimensional subsets $\Sigma^k \subset \mathbb{R}^n$,

$$\mathscr{M}_p(\Sigma) := \underbrace{\int_\Sigma \cdots \int_\Sigma}_{k+2 \text{ integrations}} \mathcal{K}^p(T)\, d\mathscr{H}^k \cdots d\mathscr{H}^k, \tag{11}$$

would be $p = k(k + 2)$, which corresponds to $k + 2$ integrations with respect to the k-dimensional Hausdorff measure $\mathscr{H}^k$. Again, for exponents above the scale invariant case we obtain a *geometric Morrey-Sobolev embedding theorem* (Strzelecki & von der Mosel (2011); Kolasiński (2011, 2012)):

Theorem 3.1. *If an admissible k-dimensional set $\Sigma^k \subset \mathbb{R}^n$ satisfies $\mathscr{M}_p(\Sigma) \leq E < \infty$ for some $p > k(k + 2)$, then Σ is a submanifold of class $C^{1,1-(k(k+2))/p}$. In addition, there is a length scale $R_0 = R_0(E, p)$ depending only on E and p, such that for each $x \in \Sigma$ and $0 < r < R_0$ the intersection $\Sigma \cap B_r(x)$ equals, up to isometry,* Graph*$f \cap B_r(0)$, where* Graph*f is the graph of a function $f \in C^{1,1-(k(k+2))/p}(\mathbb{R}^k, \mathbb{R}^{n-k})$ with $\|f\|_{C^{1,1-(k(k+2))/p}} \lesssim \mathscr{M}_p(\Sigma)^{1/p}$.*

We will briefly comment on the notion of "admissible sets" later on, for the moment one may assume that Σ is a compact, connected Lipschitz submanifold of $\mathbb{R}^n$ without boundary, but much more general sets are in fact admissible. As for curves, the integrand $\mathcal{K}(T)$ is a form of weak discrete curvature, and it is integrated on a $k(k+2)$ dimensional domain, so that Theorem 3.1 corresponds to the classic Morrey-Sobolev theorem for functions whose second weak derivatives are p-integrable on a $k(k+2)$-dimensional domain. But our result says more: the uniform control over the "patch size" of the local graph representation purely in terms of the integrability p and the energy value E allows for nice topological applications, stated here for two-dimensional[4] surfaces in $\mathbb{R}^3$ (Strzelecki & von der Mosel (2011)):

Theorem 3.2. *Let $\mathcal{C}_A(M_g)$ be the class of all closed, compact and connected Lipschitz surfaces $\Sigma^2 \subset \mathbb{R}^3$ without boundary, ambiently isotopic to a fixed reference manifold $M_g \subset \mathbb{R}^n$ of genus g with the uniform area bound $\mathscr{H}^2(\Sigma) \le A$. Then, for any $p > 2(2+2) = 8$, there is an $\mathscr{M}_p$-minimizing surface $\Sigma_g \in \mathcal{C}_A(M_g)$, i.e., $\inf_{\mathcal{C}_A(M_g)} \mathscr{M}_p = \mathscr{M}_p(\Sigma_g)$. Moreover, there are only finitely many isotopy types of surfaces that can be represented under a given $\mathscr{M}_p$-energy level.*

The last statement should be compared to finiteness theorems in smooth Riemannian geometry by Anderson & Cheeger (1991), but in the present setting we are below the C^2-category.

The general strategy to prove Theorem 3.1 is, albeit much more technical, the same as for curves. One tries to estimate the beta numbers, say in the two-dimensional case ($k = 2$),

$$\beta_\Sigma(x,d) := \inf\Big\{ \sup_{y\in\Sigma\cap B_d(x)} \frac{\operatorname{dist}(y,P)}{d} \; : \; P \text{ is a 2-dimensional affine plane through } x \Big\}. \qquad (12)$$

for $x \in \Sigma$ and $d > 0$, to show local flatness in a first step, which leads to an initial (non-optimal) oscillation estimate for tangent planes. A second technical step, as for curves, improves the Hölder exponent via slicing and iteration.

To obtain in the first step such beta number estimates under a suitable balance condition one uses elementary geometry in a contradiction argument: assuming a fourth surface point $\xi \in \Sigma \cap B_d(x)$ too far away from a flat slab around the plane $\langle x, y, z\rangle$ spanned by $x, y, z \in \Sigma$, one can deduce a lower bound on the integrand $\mathcal{K}(T)$ for that particular tetrahedron $T = (\xi, x, y, z)$. But here comes the catch: to derive a contradiction one needs similar bounds on sufficiently many tetrahedra T' close to T. But are there "sufficiently many" such tetrahedra with vertices in small balls around the four vertices of T? What was automatic for curves (cf. (6)) is not at all true for arbitrary surfaces. Even smooth surfaces $\Sigma \subset \mathbb{R}^3$ may exhibit very slender fingers so that there might be points $x \in \Sigma$ such that the surface area $\mathscr{H}^2(\Sigma\cap B_r(x))$ is not all comparable to r^2; see Figure 3 again. So, apart from the technical complications in higher dimensions, one needs an additional step to prove *uniform Ahlfors regularity* for finite energy surfaces: *there is a universal constant $R = R(E, p)$ depending only on p and the energy level E such that*

$$\mathscr{H}^2(\Sigma \cap B_r(x)) \ge \frac{\pi}{2} r^2 \qquad \textit{for all } \; r \le R. \qquad (13)$$

The idea to prove this can be described as follows. Assume, that the tangent plane $T_x\Sigma$ of Σ at $x \in \Sigma$ exists. Then one can imagine this point x as a "seed" from which we can grow double cones $C(x)$ with opening angle, say, $\pi/4$ and cone axis perpendicular to $T_x\Sigma$, until we hit another part of the

[4] This result extends to higher dimensions and co-dimensions due to the uniform estimates on the graph patches; see Kolasiński *et al.* (2013*b*).

surface after some positive distance, the "stopping distance" $d(x)$ depending on the starting point x. (Notice that one half of the cone is in the interior, the other in the exterior of Σ, which, as a closed surface in $\mathbb{R}^3$, separates space into a bounded interior and an unbounded exterior. So eventually, this increasing cone must hit the compact surface Σ at some surface point $\xi \in \Sigma \setminus \{x\}$.) As long as there is no hit one has large orthogonal projections of Σ onto the tangent plane $T_x\Sigma$, which in turn gives the (not yet uniform) lower Ahlfors estimate

$$\mathscr{H}^2(\Sigma \cap B_r(x)) \geq \frac{\pi}{2} r^2 \quad \text{for all } r \leq d(x). \tag{14}$$

If the first hit yields a surface point $\xi \in \Sigma$ fairly close to the cone axis, then it is easy to find two other points $y, z \in \Sigma$ to generate a "voluminous" tetrahedron $T = (\xi, x, y, z)$ whose smallest height (of one point over the plane spanned by the remaining three points) is comparable to the diameter of T. Indeed, any segment parallel to the cone axis connecting the two components of $C(x) \setminus \{x\}$ must hit the (connected) surface Σ since they connect the surface's interior to its exterior. Just choose two such appropriately spaced segments so that their intersection points y, z with Σ together with x and ξ span a voluminous tetrahedron; see Figure 7.

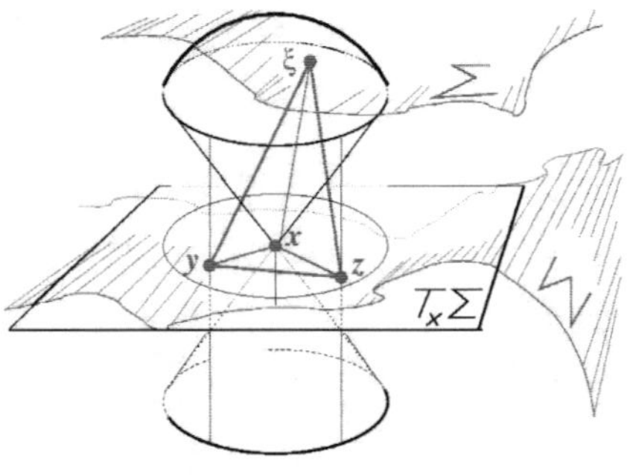

Figure 7. Voluminous tetrahedra.

Such a voluminous tetrahedron T has potential to contribute to the energy. Indeed, perturbing all vertices just a little leads to similar estimates for the integrand $\mathcal{K}(T')$ on neighbouring tetrahedra T'. And if $\operatorname{diam} T$ (and hence $\operatorname{diam} T'$ for all such T' near T) is small, then one expects a high contribution to the energy. This observation can be used for surfaces Σ with $\mathscr{M}_p(\Sigma) \leq E < \infty$ to actually bound the stopping distances $d(x)$ from below: there is a constant $R_0(E,p) > 0$ depending only on E and p, such that $d(x) \geq R_0(E,p)$ for all $x \in \Sigma$ such that $T_x\Sigma$ exists, which implies the desired uniform Ahlfors regularity (13).

The only question that remains is: does the growing cone $C(x)$ always hit Σ in a point $\xi \in \Sigma$ close to the cone axis? The answer is "no", but in that case a subtle algorithm can unfold: the growing cone can be turned suitably depending on the geometric situation, and one can grow further by doubling the cone's size, possibly ignoring some parts closer to x, but always keeping a connected set with one portion in the interior and the other in the exterior of Σ. One continues that algorithm by growing and turning (possibly neglecting some subsets closer to x). After a finite number of steps one has to hit the compact surface Σ in a suitable point to find the voluminous tetrahedron; otherwise Σ would be unbounded. The quantitative reasoning to make this algorithm work is quite involved, but the outcome is valid for arbitrary surfaces and may be of independent interest (Strzelecki & von der Mosel (2011)):

Theorem 3.3. *For any closed, compact, connected Lipschitz surface Σ and every point $x \in \Sigma$ where the tangent plane $T_x\Sigma$ exists, one finds a positive stopping distance $d(x)$ and a voluminous tetrahedron $T = (\xi, x, y, z)$. In addition, the lower Ahlfors estimate* (14) *holds at such x.*

After this informal description it becomes clear that one does not need Lipschitz surfaces (guaranteeing tangent planes at almost every point). One simply needs a dense set of starting points $x \in \Sigma$, and the corresponding planes $P(x)$ orthogonal to which one can start growing cones for some positive distance. The second ingredient is topology: in co-dimension one, i.e., for hypersurfaces Σ, this is granted by a bounded interior and an unbounded exterior of Σ. In higher co-dimensions one has to add suitable linking conditions to define the class of admissible surfaces; see Strzelecki & von der Mosel (2013); Kolasiński *et al.* (2013*a*); Kolasiński (2011).

Acknowledgments

We would like to thank the organizers of the Workshop "Quantised Flux in Tightly Knotted and Linked Systems" at the Isaac Newton Institute for Mathematical Sciences at Cambridge University, UK, in December 2012. This research was partially supported by Deutsche Forschungsgemeinschaft (project no. Mo 966/4-1) and the Polish National Science Centre (project no. 2012/07/B/ST1/03366).

References

ANDERSON, M. T. & CHEEGER, J. 1991 Diffeomorphism finiteness for manifolds with Ricci curvature and $L^{n/2}$-norm of curvature bounded. *Geom. Funct. Anal.* **1** (3), 231–252.

BANAVAR, J. R., GONZALEZ, O., MADDOCKS, J. H. & MARITAN, A. 2003 Self-interactions of strands and sheets. *J. Statist. Phys.* **110** (1-2), 35–50.

BLATT, S. 2013 A note on integral Menger curvature. *Mathematische Nachrichten* **286** (2-3).

BLATT, S. & KOLASIŃSKI, S. 2012 Sharp boundedness and regularizing effects of the integral Menger curvature for submanifolds. *Adv. Math.* **230** (3), 839–852.

BUCK, G. & SIMON, J. 1997 Energy and length of knots. In *Lectures at KNOTS '96 (Tokyo), Ser. Knots Everything*, vol. 15, pp. 219–234. World Sci. Publ., River Edge, NJ.

CANTARELLA, J., ELLIS, J., FU, J. H. & MASTIN, M. 2012 Symmetric criticality for tight knots. ArXiv:1208.3879v3 [math.DG].

CANTARELLA, J., FU, J. H. G., KUSNER, R. & SULLIVAN, J. M. 2011 Ropelength criticality. ArXiv:1102.3234v2 [math.DG].

CANTARELLA, J., FU, J. H. G., KUSNER, R., SULLIVAN, J. M. & WRINKLE, N. C. 2006 Criticality for the Gehring link problem. *Geom. Topol.* **10**, 2055–2116 (electronic).

CANTARELLA, J., KUSNER, R. B. & SULLIVAN, J. M. 2002 On the minimum ropelength of knots and links. *Invent. Math.* **150** (2), 257–286.

DAVID, G. 1999 Analytic capacity, Calderón-Zygmund operators, and rectifiability. *Publ. Mat.* **43** (1), 3–25.

FREEDMAN, M. H., HE, Z.-X. & WANG, Z. 1994 Möbius energy of knots and unknots. *Ann. of Math. (2)* **139** (1), 1–50.

GERLACH, H. & VON DER MOSEL, H. 2011*a* On sphere-filling ropes. *Amer. Math. Monthly* **118** (10), 863–876.

GERLACH, H. & VON DER MOSEL, H. 2011*b* What are the longest ropes on the unit sphere? *Arch. Ration. Mech. Anal.* **201** (1), 303–342.

GONZALEZ, O. & DE LA LLAVE, R. 2003 Existence of ideal knots. *J. Knot Theory Ramifications* **12** (1), 123–133.

GONZALEZ, O. & MADDOCKS, J. H. 1999 Global curvature, thickness, and the ideal shapes of knots. *Proc. Natl. Acad. Sci. USA* **96** (9), 4769–4773 (electronic).

GONZALEZ, O., MADDOCKS, J. H., SCHURICHT, F. & VON DER MOSEL, H. 2002 Global curvature and self-contact of nonlinearly elastic curves and rods. *Calc. Var. Partial Differential Equations* **14** (1), 29–68.

JONES, P. W. 1990 Rectifiable sets and the traveling salesman problem. *Invent. Math.* **102** (1), 1–15.

KOLASIŃSKI, S. 2011 Integral Menger curvature for sets of arbitrary dimension and codimension. PhD thesis, Institute of Mathematics, University of Warsaw. ArXiv:1011.2008v4.

KOLASIŃSKI, S. 2012 Geometric Sobolev-like embedding using high-dimensional Menger-like curvature. ArXiv:1205.4112v1; Trans. Amer. Math. Soc., accepted for publication.

KOLASIŃSKI, S., STRZELECKI, P. & VON DER MOSEL, H. 2013*a* Characterizing $W^{2,p}$-submanifolds by p-integrability of global curvatures. *Geom. Funct. Anal.* **23** (3), 937–984.

KOLASIŃSKI, S., STRZELECKI, P. & VON DER MOSEL, H. 2013*b* Compactness and finiteness results for sets with equibounded integral Menger curvature. In preparation.

KOLASIŃSKI, S. & SZUMAŃSKA, M. 2013 Minimal Hölder regularity implying finiteness of integral Menger curvature. *Manuscripta Math.* **141** (1), 125147.

LÉGER, J.-C. 1999 Menger curvature and rectifiability. *Ann. of Math. (2)* **149** (3), 831–869.

MATTILA, P. 2004 Search for geometric criteria for removable sets of bounded analytic functions. *Cubo* **6** (4), 113–132.

MENGER, K. 1930 Untersuchungen über allgemeine Metrik. *Math. Ann.* **103** (1), 466–501.

O'HARA, J. 1991 Energy of a knot. *Topology* **30** (2), 241–247.

O'HARA, J. 1992*a* Energy functionals of knots. In *Topology Hawaii (Honolulu, HI, 1990)*, pp. 201–214. World Sci. Publ., River Edge, NJ.

O'HARA, J. 1992*b* Family of energy functionals of knots. *Topology Appl.* **48** (2), 147–161.

SCHOLTES, S. 2011 For which positive p is the integral Menger curvature $\mathcal{M}_p$ finite for all simple polygons? ArXiv:1202.0504v1.

SCHURICHT, F. & VON DER MOSEL, H. 2003 Euler-Lagrange equations for nonlinearly elastic rods with self-contact. *Arch. Ration. Mech. Anal.* **168** (1), 35–82.

SCHURICHT, F. & VON DER MOSEL, H. 2004 Characterization of ideal knots. *Calc. Var. Partial Differential Equations* **19** (3), 281–305.

STRZELECKI, P. & VON DER MOSEL, H. 2005 On a mathematical model for thick surfaces. In *Physical and numerical models in knot theory*, *Ser. Knots Everything*, vol. 36, pp. 547–564. World Sci. Publ., Singapore.

STRZELECKI, P. & VON DER MOSEL, H. 2006 Global curvature for surfaces and area minimization under a thickness constraint. *Calc. Var. Partial Differential Equations* **25** (4), 431–467.

STRZELECKI, P. & VON DER MOSEL, H. 2011 Integral Menger curvature for surfaces. *Adv. Math.* **226**, 2233–2304.

STRZELECKI, P. & VON DER MOSEL, H. 2013 Tangent-point repulsive potentials for a class of non-smooth m-dimensional sets in $\mathbb{R}^n$. Part I: Smoothing and self- avoidance effects. *J. Geometric Analysis* **23** (3), 1085–1139.

STRZELECKI, P., SZUMAŃSKA, M. & VON DER MOSEL, H. 2010 Regularizing and self-avoidance effects of integral Menger curvature. *Ann. Sc. Norm. Super. Pisa Cl. Sci. (5)* **9** (1), 145–187.

STRZELECKI, P., SZUMAŃSKA, M. & VON DER MOSEL, H. 2013 On some knot energies involving Menger curvature. *Topology Applications* **160**, 1507–1529.

SULLIVAN, J. M. 2002 Approximating ropelength by energy functions. In *Physical knots: knotting, linking, and folding geometric objects in* $\mathbb{R}^3$ *(Las Vegas, NV, 2001)*, *Contemp. Math.*, vol. 304, pp. 181–186. Amer. Math. Soc., Providence, RI.

SZUMAŃSKA, M. 2009 Integral versions of Menger curvature: smoothing potentials for rectifiable curves. (In Polish). PhD thesis, Institute of Mathematics, University of Warsaw.

TOLSA, X. 2006 Analytic capacity, rectifiability, and the Cauchy integral. In *International Congress of Mathematicians. Vol. II*, pp. 1505–1527. Eur. Math. Soc., Zürich.

Journal of Physics: Conference Series **544** (2014) 012019 doi:10.1088/1742-6596/544/1/012019

Möbius invariant energy of tori of revolution

H. Funaba & J. O'Hara

Department of Mathematics and Information Sciences, Tokyo Metropolitan University, 1-1 Minami-Ohsawa, Hachiouji-Shi, Tokyo 192-0397 Japan

E-mail: ohara@tmu.ac.jp

Abstract. We show that among tori of revolution the Clifford torus gives the minimum value of the Möbius invariant surface energy defined by Auckly and Sadun.

1. Introduction and main result

Recently Fernando C. Marques and André Neves (Marques & Neves) proved the Willmore conjecture, namely, they showed the following. Let Σ be an immersed torus in $\mathbb{R}^3$. Let κ_1 and κ_2 be principal curvatures. The Willmore functional is given by

$$\mathcal{W}(\Sigma) = \int_\Sigma \left(\frac{\kappa_1+\kappa_2}{2}\right)^2 d\Sigma = \int_\Sigma \left(\frac{\kappa_1-\kappa_2}{2}\right)^2 d\Sigma,$$

where the second equation is the consequence of the Gauss-Bonnet theorem. It is known to be invariant under Möbius transformations of $\mathbb{R}^3$. Then the Willmore conjecture, now the theorem of Marques and Neves, asserts that $\mathcal{W}(\Sigma) \geq 2\pi^2$ and that the equality holds if and only if Σ is a torus of revolution whose generating circle has radius 1 and center at distance $\sqrt{2}$ from the axis of revolution up to a Möbius transformation, in other words, if and only if Σ is the image of a stereographic projection of the Clifford torus

$$\left\{(z_1, z_2) \in \mathbb{C}\times\mathbb{C} \,\middle|\, |z_1| = |z_2| = 1/\sqrt{2}\right\} \subset S^3 = \left\{(z_1, z_2) \in \mathbb{C}\times\mathbb{C} \,\middle|\, |z_1|^2 + |z_2|^2 = 1\right\}.$$

In this paper, we give another characterization of the Clifford torus using the surface energy introduced by David Auckly and Lorenzo Sadun ((1)), which is also invariant under Möbius transformations. To be precise, we have not yet succeeded in proving that the Clifford torus gives the minimum energy among all the immersed tori[1]. We only show that it gives the minimum energy among one-parameter family of tori of revolution. Since the energy we use is conformally invariant, it follows that the Clifford torus gives the minimum energy among Dupin cyclides. As the surface energy that we use in this paper is generelization of knot energy, we start with the review of it.

Energy of knots was introduced in (5) motivated to give a functional on the space of knots that can produce a representative configuration of a knot for each knot type as an embedding

[1] To show it, it suffice to show that the Clifford torus gives the minimum energy among all the embedded tori, since the energy blows up if a torus has a double point

Journal of Physics: Conference Series **544** (2014) 012019 doi:10.1088/1742-6596/544/1/012019

that minimizes the energy in the knot type. Let K be a knot and x be a point on it. Define

$$V(x;K) = \lim_{\varepsilon\to 0}\left(\int_{K\setminus B_\varepsilon(x)} \frac{dy}{|x-y|^2} - \frac{2}{\varepsilon}\right), \quad \text{and} \quad E(K) = \int_K V(x;K)dx, \tag{1}$$

where $B_\varepsilon(x)$ is a ball with center x and radius ε. Let us call a process as in the definition of $V(x;K)$ in (1) the *renormalization* in this paper. In general, when we are interested in a diverging integral, we first restrict the integration to the complement of an ε-neighbourhood of the set where the integrand blows up, then expand the result in a Laurent series in ε, and finally take the constant term. In the case of a knot, the integrand of V in (1) blows up at the one-point set $\{x\}$. There are two ways to define an ε-neighbourhood of it, according to the choice of the distance between a pair of points on the knot; either the chord length as in (1) or the arc-length along the knot as in (5). Both types of the renormalization give the same result ((6)).

The energy $E(K)$ in (1) was proved to be invariant under Möbius transformations by Freedman, He, and Wang ((2)), which is the reason why it is sometimes called the *Möbius energy* of knots.

After this energy was found, it has been generalized to functionals that can measure geometric complexity of knots, surfaces, and in general, submanifolds ((1), (3), et al.). Among several ways of generalization to surface energy, in this paper we study the one by Auckly and Sadun that uses a similar renormalization process as in (1).

Let S be an embedded surface in $\mathbb{R}^3$ without boundary and x be a point in S. Define the *renormalized r^{-4}-potential* V and the *renormalized r^{-4}-potential energy* E by

$$\begin{aligned} V(x;S) &= \lim_{\varepsilon\to 0}\left(\int_{S\setminus B_\varepsilon(x)} \frac{d^2y}{|x-y|^4} - \frac{\pi}{\varepsilon^2} + \frac{\pi\Delta(x)}{16}\log\left(\Delta(x)\varepsilon^2\right) + \frac{\pi K(x)}{4}\right), \\ E(S) &= \int_S V(x;S)d^2x, \end{aligned} \tag{2}$$

where d^2y and d^2x mean the volume element of S, $\Delta(x)$ is given by $\Delta(x) = (\kappa_1(x) - \kappa_2(x))^2$, and $K(x)$ is the Gauss curvature; $K = \kappa_1\kappa_2$. This energy $E(S)$ was proved to be invariant under Möbius transformations in (1). It blows up as S degenerates to an immersed surface with double points.

As was pointed out in (1), the choice of the log-term in (2) is not the unique reasonable one. The reason why there is a factor $\Delta(x)$ in the log-term is to make the resulting energy scale invariant, but $c\Delta(x)$ $(c \neq 0)$ also has the same effect. Thus there is ambiguity in the definition of the renormalized potential.

In (1), Auckly and Sadun has computed the energy of spheres and planes and the potentials V of an infinitely long straight cylinder and a surface called dimple. In this article we compute the energies of one-parameter family of tori of revolution.

Theorem: *Let T_R be a torus of revolution whose generating circle has radius 1 and center at distance R $(R > 1)$ from the axis of revolution. Then the renormalized potential energy is given by*

$$E(T_R) = \frac{\pi^3}{2\sqrt{R^2-1}}\left(R^2\left(3\log 2 - 1\right) + 2 - \frac{2}{R^2}\right).$$

Corollary: *Among tori of revolution, a stereographic projection of the Clifford torus gives the minimum energy.*

Proof of Corollary: Since

$$\frac{d}{dR}E(T_R) = \frac{\pi^3(R^2-2)\left((R^2-2)^2 + 3\,R^4\log(4/e)\right)}{4R^3\,(R^2-1)^{3/2}},$$

the energy takes the minimum value $\pi^3(6\log 2 - 1)/2$ when $R=\sqrt{2}$. □

Problem: (1) Does $T_{\sqrt{2}}$ give the minimum energy among all the embedded tori in $\mathbb{R}^3$, hence among all the immersed tori in $\mathbb{R}^3$?

(2) When we change the power of $|x-y|$ in the denominator in (2) from 4 to any number λ, we obtain a new potential energy $E_{r^{-\lambda}}$ after suitable renormalization. It is no longer scale invariant when $\lambda \neq 4$. What is $R = R(\lambda)$ that makes T_R give the minimum energy $E_{r^{-\lambda}}$ after rescaling to have area 1?

2. Computation of the energy of a torus

Proof of Theorem: Let T be a torus of revolution parametrized by

$$p(u,v) = ((R+\cos u)\cos v, (R+\cos u)\sin v, \sin u).$$

Let $x = p(\alpha,0) = (R+\cos\alpha, 0, \sin\alpha)$ be a point on T. First we fix α and compute the renormalized r^{-4}-potential of T at x. This is the main part of the paper. Some of the complicated computation to obtain expansion in seiries in ε has been done with the help of Maple.

Let Dist $=$ Dist(u,v) be the distance between x and a point $y = p(u,v)$:

$$\begin{aligned}\mathrm{Dist}^2 &= |x-y|^2 = |p(u,v)-p(\alpha,0)|^2\\ &= ((R+\cos\alpha)-(R+\cos u)\cos v)^2 + (R+\cos u)^2\sin^2 v + (\sin\alpha - \sin u)^2\\ &= 2R^2 + 2 + 2R(\cos\alpha + \cos u) - 2\sin\alpha\sin u - 2\,(R+\cos\alpha)\,(R+\cos u)\cos v.\end{aligned}$$

To compute the integral $\int_{T\setminus B_\varepsilon(x)} |x-y|^{-4}d^2y$, we will divide the domain of the integration into four parts and use several kinds of changes of variables. We may assume without loss of generality that $u\in[\alpha-\pi,\alpha+\pi]$ and $v\in[-\pi,\pi]$. Put

$$\theta = \frac{\pi}{2} - \frac{u-\alpha}{2},\quad t = 2\cos\theta = 2\sin\frac{u-\alpha}{2}.$$

Then, as $(u-\alpha)/2 \in [-\pi/2, \pi/2]$ we have $\theta\in[0,\pi]$ and $\cos(u-\alpha)/2 \geq 0$. Therefore,

$$\cos u = \frac{2-t^2}{2}\cos\alpha - \frac{t\sqrt{4-t^2}}{2}\sin\alpha,\quad \sin u = \frac{t\sqrt{4-t^2}}{2}\cos\alpha + \frac{2-t^2}{2}\sin\alpha,$$

which implies that the distance can be expressed as

$$\begin{aligned}\mathrm{Dist}^2 &= t^2 + \left[4(R+\cos\alpha)^2 - 2(R+\cos\alpha)\cos\alpha\cdot t^2 - 2(R+\cos\alpha)\sin\alpha\cdot t\sqrt{4-t^2}\right]\sin^2\frac{v}{2}\\ &= 4\left[\cos^2\theta + (R+\cos\alpha)\left(R+\cos\alpha - 2\cos\alpha\cos^2\theta - 2\sin\alpha\sin\theta\cos\theta\right)\sin^2\frac{v}{2}\right]\\ &= 4\left[\cos^2\theta + (R+\cos\alpha)\,(R-\cos(\alpha-2\theta))\sin^2\frac{v}{2}\right].\end{aligned}$$

Journal of Physics: Conference Series **544** (2014) 012019 doi:10.1088/1742-6596/544/1/012019

Define $b = b(\theta)$ and $c = c(t)$ by

$$\begin{aligned} c(t) &= 2(R+\cos\alpha)\left[2(R+\cos\alpha) - \cos\alpha\cdot t^2 - \sin\alpha\cdot t\sqrt{4-t^2}\right] \\ b &= \frac{c}{4} = (R+\cos\alpha)\,(R-\cos(\alpha-2\theta)). \end{aligned} \tag{3}$$

Then

$$\mathrm{Dist}^2 = t^2 + c(t)\sin^2\frac{v}{2} = 4\left(\cos^2\theta + b\sin^2\frac{v}{2}\right).$$

Put

$$V(\varepsilon,x) = \iint_{\mathrm{Dist}\geq\varepsilon} \frac{d^2y}{|x-y|^2}.$$

Since the area element of T is given by $d^2y = (R+\cos u)\,dudv$, $V(\varepsilon,x)$ is given by

$$\iint_{\mathrm{Dist}\geq\varepsilon} \frac{(R+\cos u)}{\left(2R^2+2+2R\,(\cos\alpha+\cos u) - 2\sin\alpha\sin u - 2\,(R+\cos\alpha)\,(R+\cos u)\cos v\right)^2}\,dudv$$

$$= \iint_{\mathrm{Dist}\geq\varepsilon} \frac{R+\left(\frac{2-t^2}{2}\cos\alpha - \frac{t\sqrt{4-t^2}}{2}\sin\alpha\right)}{\left(t^2+c(t)\sin^2\frac{v}{2}\right)^2}\,\frac{2dt}{\sqrt{4-t^2}}\,dv \tag{4}$$

$$= \frac{1}{8}\iint_{\mathrm{Dist}\geq\varepsilon} \frac{(R-\cos(\alpha-2\theta))}{\left(\cos^2\theta + b\sin^2\frac{v}{2}\right)^2}\,d\theta dv. \tag{5}$$

The integration by dv an be executed as

$$\begin{aligned} &\int \frac{dv}{(A^2+B\sin^2\frac{v}{2})^2} \\ &= \frac{2A^2+B}{A^3(A^2+B)^{\frac{3}{2}}}\tan^{-1}\left(\frac{\sqrt{A^2+B}}{A}\tan\frac{v}{2}\right) + \frac{B\tan\frac{v}{2}}{A^2(A^2+B)\left((A^2+B)\tan^2\frac{v}{2}+A^2\right)}. \end{aligned} \tag{6}$$

We devide the domain of the integration $T\setminus B_\varepsilon(x)$ into eight parts as is illustrated in Figure 1. Using symmetry, we have only to compute integrals over the following four regions

$$\begin{aligned} &\left\{p(u,v)\,|\,2\sin^{-1}(\varepsilon/2) \leq u-\alpha \leq \pi, 0\leq v\leq\pi\right\}, \\ &\left\{p(u,v)\,|\,0 \leq u-\alpha \leq \sin^{-1}(\varepsilon/2), 0\leq v\leq\pi, |p(u,v)-x|\geq\varepsilon\right\}, \\ &\left\{p(u,v)\,|\,-\pi \leq u-\alpha \leq -2\sin^{-1}(\varepsilon/2), 0\leq v\leq\pi\right\}, \\ &\left\{p(u,v)\,|\,-2\sin^{-1}(\varepsilon/2) \leq u-\alpha \leq 0, 0\leq v\leq\pi, |p(u,v)-x|\geq\varepsilon\right\}, \end{aligned}$$

which we denote by I_1, I_2, I_3, and I_4 respectively. We remark that the regions given by the first and the third lines do not have intersection with $B_\varepsilon(x)$.

Let us first compute I_2 and I_4 using variables (t,v). Since $0 \leq u-\alpha \leq 2\sin^{-1}(\varepsilon/2)$ we have

$$0 \leq t = 2\sin((u-\alpha)/2) \leq \varepsilon,$$

and the condition $|p(u,v)-x| \geq \varepsilon$ implies

$$\sin^2\frac{v}{2} = \frac{\mathrm{Dist}^2-t^2}{c(t)} \geq \frac{\varepsilon^2-t^2}{c(t)}, \text{ i.e., } v \geq 2\sin^{-1}\left(\sqrt{\frac{\varepsilon^2-t^2}{c(t)}}\right).$$

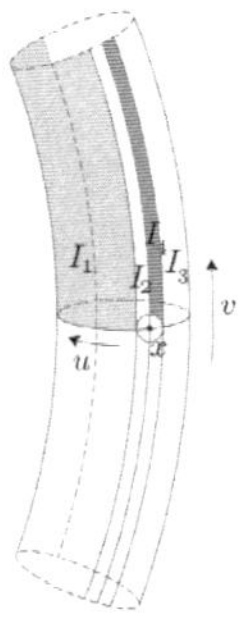

Figure 1. Parts of the domains of integration of I_1 (blue or light grey) and I_4 (red or dark grey) near the point x

Therefore, I_2 is given by

$$I_2 = \int_0^{\varepsilon} \int_{2\sin^{-1}\left(\sqrt{\frac{\varepsilon^2-t^2}{c(t)}}\right)}^{\pi} \frac{2R + \cos\alpha \cdot (2-t^2) - \sin\alpha \cdot t\sqrt{4-t^2}}{(t^2 + c(t)\sin^2\frac{v}{2})^2} dv \, \frac{dt}{\sqrt{4-t^2}} \, .$$

By the formula (6) we get

$$\begin{aligned} I_2 = & \int_0^{\varepsilon} \left\{ \frac{2R + \cos\alpha \cdot (2-t^2) - \sin\alpha \cdot t\sqrt{4-t^2}}{\sqrt{4-t^2}} \right. \\ & \times \left[\frac{2t^2 + c(t)}{t^3(t^2+c(t))^{\frac{3}{2}}} \tan^{-1}\left(\frac{\sqrt{t^2+c(t)}}{t} \tan\frac{v}{2} \right) \right. \\ & \left.\left. + \frac{c(t)\tan\frac{v}{2}}{t^2(t^2+c(t))\left((t^2+c(t))\tan^2\frac{v}{2} + t^2\right)} \right]_{v=2\sin^{-1}\left(\sqrt{\frac{\varepsilon^2-t^2}{c(t)}}\right)}^{\pi} \right\} dt \\ = & \int_0^{\varepsilon} \frac{2R + \cos\alpha \cdot (2-t^2) - \sin\alpha \cdot t\sqrt{4-t^2}}{\sqrt{4-t^2}} \\ & \times \left\{ \frac{2t^2 + c(t)}{t^3(t^2+c(t))^{\frac{3}{2}}} \left(\frac{\pi}{2} - \tan^{-1}\left(\frac{\sqrt{\varepsilon^2 - t^2}}{t} \sqrt{\frac{c(t)+t^2}{c(t)+t^2-\varepsilon^2}} \right) \right) - \frac{\sqrt{\varepsilon^2-t^2}\sqrt{c(t)+t^2-\varepsilon^2}}{\varepsilon^2 t^2 (t^2+c(t))} \right\} dt \end{aligned}$$

Putting $\tau = \sin^{-1}(t/\varepsilon)$, i.e., $t = \varepsilon\sin\tau$, we have

$$\begin{aligned} I_2 = & \int_0^{\frac{\pi}{2}} \frac{2R + \cos\alpha \cdot (2 - \varepsilon^2\sin^2\tau) - \varepsilon\sin\alpha\sin\tau\sqrt{4-\varepsilon^2\sin^2\tau}}{\sqrt{4-\varepsilon^2\sin^2\tau}} \\ & \times \left\{ \frac{\cos\tau\left(2\varepsilon^2\sin^2\tau + c(\varepsilon\sin\tau)\right)}{\varepsilon^2\sin^3\tau(\varepsilon^2\sin^2\tau + c(\varepsilon\sin\tau))^{\frac{3}{2}}} \left(\frac{\pi}{2} - \tan^{-1}\left(\cot\tau\sqrt{\frac{c(\varepsilon\sin\tau)+\varepsilon^2\sin^2\tau}{c(\varepsilon\sin\tau)+\varepsilon^2\sin^2\tau-\varepsilon^2}} \right) \right) \right. \\ & \left. - \frac{\cot^2\tau\sqrt{c(\varepsilon\sin\tau)+\varepsilon^2\sin^2\tau-\varepsilon^2}}{\varepsilon^2(\varepsilon^2\sin^2\tau + c(\varepsilon\sin\tau))} \right\} d\tau. \end{aligned}$$

Journal of Physics: Conference Series **544** (2014) 012019 doi:10.1088/1742-6596/544/1/012019

Expanding the integrand in a series in ε with the help of Maple we obtain

$$\begin{aligned}I_2 =& \frac{1}{2\varepsilon^2}\int_0^{\frac{\pi}{2}}\left(\tau\frac{\cos\tau}{\sin^3\tau}-\frac{\cos^2\tau}{\sin^2\tau}\right)d\tau - \frac{\sin\alpha}{4(R+\cos\alpha)\varepsilon}\int_0^{\frac{\pi}{2}}\left(\tau\frac{\cos\tau}{\sin^2\tau}-\frac{\cos^2\tau}{\sin\tau}\right)d\tau \\ &+\frac{R^2}{16(R+\cos\alpha)^2}\int_0^{\frac{\pi}{2}}\left(\tau\frac{\cos\tau}{\sin\tau}-\cos^2\tau\right)d\tau + \frac{1}{8(R+\cos\alpha)^2}\int_0^{\frac{\pi}{2}}\cos^2\tau d\tau + O(\varepsilon).\end{aligned} \tag{7}$$

Direct integration shows

$$\begin{aligned}&\int_0^{\frac{\pi}{2}}\left(\tau\frac{\cos\tau}{\sin^3\tau}-\frac{\cos^2\tau}{\sin^2\tau}\right)d\tau = \frac{1}{2}\left[\frac{\tau+\sin\tau\cos\tau-2\tau\cos^2\tau}{\sin^2\tau}\right]_0^{\frac{\pi}{2}} = \frac{\pi}{4}, \\ &\int_0^{\frac{\pi}{2}}\left(\tau\frac{\cos\tau}{\sin^2\tau}-\frac{\cos^2\tau}{\sin\tau}\right)d\tau = \left[-\cos\tau-\frac{\tau}{\sin\tau}\right]_0^{\frac{\pi}{2}} = \frac{4-\pi}{2}.\end{aligned} \tag{8}$$

On the other hand, $\int_0^{\frac{\pi}{2}}\left(\tau\cot\tau-\cos^2\tau\right)d\tau$ can be computed as follows.

Put

$$I = \int_0^{\frac{\pi}{2}}\log(\sin\tau)d\tau.$$

Then

$$I = \int_0^{\frac{\pi}{2}}\log\left(\sin\left(\frac{\pi}{2}-\tau'\right)\right)d\tau' = \int_0^{\frac{\pi}{2}}\log(\cos\tau)d\tau.$$

Adding the two above, we get

$$\begin{aligned}2I &= \int_0^{\frac{\pi}{2}}\log(\sin\tau\cos\tau)d\tau \\ &= \int_0^{\frac{\pi}{2}}\log(\sin 2\tau)d\tau - \frac{\pi}{2}\log 2 \\ &= \frac{1}{2}\int_0^{\pi}\log(\sin\tau')d\tau' - \frac{\pi}{2}\log 2 \\ &= \frac{1}{2}\left(\int_0^{\frac{\pi}{2}}\log(\sin\tau)d\tau + \int_{\frac{\pi}{2}}^{\pi}\log(\sin\tau)d\tau\right) - \frac{\pi}{2}\log 2 \\ &= I - \frac{\pi}{2}\log 2,\end{aligned}$$

which implies

$$I = \int_0^{\frac{\pi}{2}}\log(\sin\tau)d\tau = -\frac{\pi}{2}\log 2.$$

Now we have

$$\begin{aligned}\int_0^{\frac{\pi}{2}}\left(\tau\frac{\cos\tau}{\sin\tau}-\cos^2\tau\right)d\tau &= \Big[\tau\log(\sin\tau)\Big]_0^{\frac{\pi}{2}} - \int_0^{\frac{\pi}{2}}\log(\sin\tau)d\tau - \frac{\pi}{4} \\ &= \frac{\pi}{4}(2\log 2 - 1).\end{aligned} \tag{9}$$

Substituting (8) and (9) to (7) we obtain

$$I_2 = \frac{\pi}{8\varepsilon^2} + \frac{(\pi-4)\sin\alpha}{8(R+\cos\alpha)\varepsilon} + \frac{\pi R^2}{64(R+\cos\alpha)^2}(2\log 2 - 1) + \frac{\pi}{32(R+\cos\alpha)^2} + O(\varepsilon).$$

Journal of Physics: Conference Series **544** (2014) 012019 doi:10.1088/1742-6596/544/1/012019

The integral I_4 can be obtained from I_2 by changing α to $-\alpha$:

$$I_4 = \int_{-\varepsilon}^{0} \int_{2\sin^{-1}\left(\sqrt{\frac{\varepsilon^2 - t^2}{c(t)}}\right)}^{\pi} \frac{2R + \cos\alpha\cdot(2-t^2) - \sin\alpha\cdot t\sqrt{4-t^2}}{(t^2 + c(t)\sin^2\frac{v}{2})^2} \frac{dt}{\sqrt{4-t^2}}\, dv$$
$$= \frac{\pi}{8\varepsilon^2} - \frac{(\pi-4)\sin\alpha}{8(R+\cos\alpha)\varepsilon} + \frac{\pi R^2}{64(R+\cos\alpha)^2}(2\log 2 - 1) + \frac{\pi}{32(R+\cos\alpha)^2} + O(\varepsilon).$$

Let us next compute I_1 and I_3. Since $u - \alpha \in [2\sin^{-1}(\varepsilon/2), \pi]$ $\theta = \pi/2 - (u-\alpha)/2 \in [0, \cos^{-1}(\varepsilon/2)]$. The integral I_1 is given by

$$I_1 = \frac{1}{8}\int_0^{\cos^{-1}\frac{\varepsilon}{2}} \int_0^{\pi} \frac{R - \cos(\alpha - 2\theta)}{(\cos^2\theta + b\sin^2\frac{v}{2})^2}\, dv d\theta,$$

where $b = (R+\cos\alpha)\,(R - \cos(\alpha - 2\theta))$ as was given by (3). Put $a = \cos\theta$. Then (6) shows

$$\int_0^{\pi} \frac{dv}{(a^2 + b\sin^2\frac{v}{2})^2} = \frac{2a^2+b}{a^3(a^2+b)^{\frac{3}{2}}}\cdot\frac{\pi}{2}.$$

Since

$$R - \cos(\alpha - 2\theta) = R + \cos\alpha - 2\cos\alpha\cos^2\theta - 2\sin\alpha\sin\theta\cos\theta,$$

we have

$$I_1 = \frac{\pi}{16}\int_0^{\cos^{-1}\frac{\varepsilon}{2}} (R - \cos(\alpha - 2\theta))\frac{2a^2+b}{a^3(a^2+b)^{\frac{3}{2}}} d\theta$$
$$= \frac{\pi}{16}\int_0^{\cos^{-1}\frac{\varepsilon}{2}} \left[(R+\cos\alpha) - 2\cos\alpha\cos^2\theta - 2\sin\alpha\sin\theta\cos\theta\right]\frac{(a^2+b)+a^2}{a^3(a^2+b)^{\frac{3}{2}}}\, d\theta$$
$$= \frac{\pi(R+\cos\alpha)}{16} I_{11} - \frac{\pi\cos\alpha}{8} I_{12} - \frac{\pi\sin\alpha}{8} I_{13} + \frac{\pi(R+\cos\alpha)}{16} I_{14} - \frac{\pi\cos\alpha}{8} I_{15} - \frac{\pi\sin\alpha}{8} I_{16},$$

where

$$I_{11} = \int_0^{\cos^{-1}\frac{\varepsilon}{2}} \frac{d\theta}{a^3\sqrt{a^2+b}},\quad I_{12} = \int_0^{\cos^{-1}\frac{\varepsilon}{2}} \frac{\cos^2\theta\, d\theta}{a^3\sqrt{a^2+b}},\quad I_{13} = \int_0^{\cos^{-1}\frac{\varepsilon}{2}} \frac{\sin\theta\cos\theta\, d\theta}{a^3\sqrt{a^2+b}},$$
$$I_{14} = \int_0^{\cos^{-1}\frac{\varepsilon}{2}} \frac{d\theta}{a(a^2+b)^{\frac{3}{2}}},\quad I_{15} = \int_0^{\cos^{-1}\frac{\varepsilon}{2}} \frac{\cos^2\theta\, d\theta}{a(a^2+b)^{\frac{3}{2}}},\quad I_{16} = \int_0^{\cos^{-1}\frac{\varepsilon}{2}} \frac{\sin\theta\cos\theta\, d\theta}{a(a^2+b)^{\frac{3}{2}}}.$$

These integrals can be computed using

$$a^2 + b = \cos^2\theta\left(R^2 + ((R+\cos\alpha)\tan\theta - \sin\alpha)^2\right),\quad \tan\left(\cos^{-1}\left(\frac{\varepsilon}{2}\right)\right) = \frac{\sqrt{4-\varepsilon^2}}{\varepsilon}.$$

Journal of Physics: Conference Series **544** (2014) 012019 doi:10.1088/1742-6596/544/1/012019

We have used Maple to expand them in series in ε (the last step of each computation of I_{1j}).

$$
\begin{aligned}
I_{15} &= \int_0^{\cos^{-1}\frac{\varepsilon}{2}} \frac{1}{\cos^2\theta\left(R^2+((R+\cos\alpha)\tan\theta-\sin\alpha)^2\right)^{\frac{3}{2}}}\,d\theta \\
&= \left[\frac{(R+\cos\alpha)\tan\theta-\sin\alpha}{R^2(R+\cos\alpha)\sqrt{R^2+((R+\cos\alpha)\tan\theta-\sin\alpha)^2}}\right]_0^{\cos^{-1}\frac{\varepsilon}{2}} \\
&= \frac{(R+\cos\alpha)\sqrt{4-\varepsilon^2}-(\sin\alpha)\varepsilon}{R^2(R+\cos\alpha)\sqrt{R^2\varepsilon^2+((R+\cos\alpha)\sqrt{4-\varepsilon^2}-(\sin\alpha)\varepsilon)^2}}+\frac{\sin\alpha}{R^2(R+\cos\alpha)\sqrt{R^2+\sin^2\alpha}} \\
&= \frac{1}{R^2(R+\cos\alpha)}+\frac{\sin\alpha}{R^2(R+\cos\alpha)\sqrt{R^2+\sin^2\alpha}}+O(\varepsilon),
\end{aligned}
$$

$$
\begin{aligned}
I_{16} &= \int_0^{\cos^{-1}\frac{\varepsilon}{2}} \frac{\sin\theta}{\cos^3\theta\left(R^2+((R+\cos\alpha)\tan\theta-\sin\alpha)^2\right)^{\frac{3}{2}}}\,d\theta \\
&= \left[\frac{\sin\alpha\,((R+\cos\alpha)\tan\theta-\sin\alpha)-R^2}{R^2(R+\cos\alpha)^2\sqrt{R^2+((R+\cos\alpha)\tan\theta-\sin\alpha)^2}}\right]_0^{\cos^{-1}\frac{\varepsilon}{2}} \\
&= \frac{\sin\alpha\left((R+\cos\alpha)\sqrt{4-\varepsilon^2}-(\sin\alpha)\varepsilon\right)-R^2\varepsilon}{R^2(R+\cos\alpha)^2\sqrt{R^2\varepsilon^2+\left((R+\cos\alpha)\sqrt{4-\varepsilon^2}-(\sin\alpha)\varepsilon\right)^2}}+\frac{\sqrt{R^2+\sin^2\alpha}}{R^2(R+\cos\alpha)^2} \\
&= \frac{\sin\alpha}{R^2(R+\cos\alpha)^2}+\frac{\sqrt{R^2+\sin^2\alpha}}{R^2(R+\cos\alpha)^2}+O(\varepsilon),
\end{aligned}
$$

$$
\begin{aligned}
I_{12} &= \int_0^{\cos^{-1}\frac{\varepsilon}{2}} \frac{1}{\cos^2\theta\sqrt{R^2+((R+\cos\alpha)\tan\theta-\sin\alpha)^2}}\,d\theta \\
&= \left[\frac{1}{R+\cos\alpha}\sinh^{-1}\left(\frac{(R+\cos\alpha)\tan\theta-\sin\alpha}{R}\right)\right]_0^{\cos^{-1}\frac{\varepsilon}{2}} \\
&= \frac{1}{R+\cos\alpha}\left\{\sinh^{-1}\left(\frac{(R+\cos\alpha)\sqrt{4-\varepsilon^2}-(\sin\alpha)\varepsilon}{R\varepsilon}\right)+\sinh^{-1}\left(\frac{\sin\alpha}{R}\right)\right\} \\
&= -\frac{1}{2(R+\cos\alpha)}\log\left(\frac{R^2\varepsilon^2}{16(R+\cos\alpha)^2}\right)+\frac{1}{R+\cos\alpha}\sinh^{-1}\left(\frac{\sin\alpha}{R}\right)+O(\varepsilon),
\end{aligned}
$$

$$\begin{aligned}
I_{13} &= \int_0^{\cos^{-1}\frac{\varepsilon}{2}} \frac{\sin\theta}{\cos^3\theta\sqrt{R^2+((R+\cos\alpha)\tan\theta-\sin\alpha)^2}}\,d\theta \\
&= \frac{\sin\alpha}{R+\cos\alpha}I_{12} + \frac{1}{(R+\cos\alpha)^2}\left[\sqrt{R^2+((R+\cos\alpha)\tan\theta-\sin\alpha)^2}\right]_0^{\cos^{-1}\frac{\varepsilon}{2}} \\
&= \frac{\sin\alpha}{R+\cos\alpha}I_{12} + \frac{\sqrt{R^2\varepsilon^2+\left((R+\cos\alpha)\sqrt{4-\varepsilon^2}-(\sin\alpha)\varepsilon\right)^2}}{(R+\cos\alpha)^2\varepsilon} - \frac{\sqrt{R^2+\sin^2\alpha}}{(R+\cos\alpha)^2} \\
&= \frac{\sin\alpha}{R+\cos\alpha}I_{12} + \left\{\frac{2}{(R+\cos\alpha)\varepsilon} - \frac{\sin\alpha}{(R+\cos\alpha)^2} + O(\varepsilon)\right\} - \frac{\sqrt{R^2+\sin^2\alpha}}{(R+\cos\alpha)^2},
\end{aligned}$$

$$\begin{aligned}
I_{14} &= \int_0^{\cos^{-1}\frac{\varepsilon}{2}} \frac{1}{\cos^4\theta\left(R^2+((R+\cos\alpha)\tan\theta-\sin\alpha)^2\right)^{\frac{3}{2}}}\,d\theta \\
&= \frac{1}{(R+\cos\alpha)^2}I_{12} - \frac{(1-2R\cos\alpha-2\cos^2\alpha)}{(R+\cos\alpha)^2}I_{15} + \frac{2\sin\alpha}{R+\cos\alpha}I_{16} \\
&= \frac{1}{(R+\cos\alpha)^2}I_{12} + \frac{1+2R\cos\alpha}{R^2(R+\cos\alpha)^3} + \frac{(1+2R^2)\sin\alpha+2R^2\sin\alpha\cos\alpha}{R^2(R+\cos\alpha)^3\sqrt{R^2+\sin^2\alpha}} + O(\varepsilon),
\end{aligned}$$

and

$$\begin{aligned}
I_{11} &= \int_0^{\cos^{-1}\frac{\varepsilon}{2}} \frac{1}{\cos^4\theta\sqrt{R^2+((R+\cos\alpha)\tan\theta-\sin\alpha)^2}}\,d\theta \\
&= \frac{R^2-1+4R\cos\alpha+3\cos^2\alpha}{2(R+\cos\alpha)^2}I_{12} + \frac{3\sin\alpha}{2(R+\cos\alpha)}I_{13} \\
&\quad + \left[\frac{\sin\theta\sqrt{R^2+((R+\cos\alpha)\tan\theta-\sin\alpha)^2}}{2(R+\cos\alpha)^2\cos\theta}\right]_0^{\cos^{-1}\frac{\varepsilon}{2}} \\
&= \frac{R^2-1+4R\cos\alpha+3\cos^2\alpha}{2(R+\cos\alpha)^2}I_{12} + \frac{3\sin\alpha}{2(R+\cos\alpha)}I_{13} \\
&\quad + \frac{\sqrt{4-\varepsilon^2}\sqrt{R^2\varepsilon^2+\left((R+\cos\alpha)\sqrt{4-\varepsilon^2}-(\sin\alpha)\varepsilon\right)^2}}{2(R+\cos\alpha)^2\varepsilon^2} \\
&= \frac{R^2-1+4R\cos\alpha+3\cos^2\alpha}{2(R+\cos\alpha)^2}I_{12} + \frac{3\sin\alpha}{2(R+\cos\alpha)}I_{13} \\
&\quad + \left\{\frac{2}{(R+\cos\alpha)\varepsilon^2} - \frac{\sin\alpha}{(R+\cos\alpha)^2\varepsilon} - \frac{R^2+4R\cos\alpha+2\cos^2\alpha}{4(R+\cos\alpha)^3} + O(\varepsilon)\right\}.
\end{aligned}$$

Journal of Physics: Conference Series **544** (2014) 012019 doi:10.1088/1742-6596/544/1/012019

Therefore

$$\begin{aligned}
I_1 =& \frac{\pi(R+\cos\alpha)}{16} I_{11} - \frac{\pi\cos\alpha}{8} I_{12} - \frac{\pi\sin\alpha}{8} I_{13} + \frac{\pi(R+\cos\alpha)}{16} I_{14} - \frac{\pi\cos\alpha}{8} I_{15} - \frac{\pi\sin\alpha}{8} I_{16} \\
=& \frac{\pi R^2}{32(R+\cos\alpha)} I_{12} + \frac{\pi}{8\varepsilon^2} - \frac{\sin\alpha}{8(R+\cos\alpha)\varepsilon} - \frac{\pi(R^2+4R\cos\alpha+2\cos^2\alpha)}{64(R+\cos\alpha)^2} - \frac{\pi}{16R^2(R+\cos\alpha)^2} \\
&+ \frac{\pi\sin^2\alpha}{32(R+\cos\alpha)^2} - \frac{\pi\sin\alpha}{16R^2(R+\cos\alpha)^2\sqrt{R^2+\sin^2\alpha}} + \frac{\pi\sin\alpha\sqrt{R^2+\sin^2\alpha}}{32(R+\cos\alpha)^2} + O(\varepsilon) \\
=& -\frac{\pi R^2}{64(R+\cos\alpha)^2}\log\left(\frac{R^2\varepsilon^2}{16(R+\cos\alpha)^2}\right) + \frac{\pi R^2}{32(R+\cos\alpha)^2}\sinh^{-1}\left(\frac{\sin\alpha}{R}\right) \\
&+ \frac{\pi}{8\varepsilon^2} - \frac{\sin\alpha}{8(R+\cos\alpha)\varepsilon} - \frac{\pi(R^2+4R\cos\alpha+2\cos^2\alpha)}{64(R+\cos\alpha)^2} - \frac{\pi}{16R^2(R+\cos\alpha)^2} + \frac{\pi\sin^2\alpha}{32(R+\cos\alpha)^2} \\
&- \frac{\pi\sin\alpha}{16R^2(R+\cos\alpha)^2\sqrt{R^2+\sin^2\alpha}} + \frac{\pi\sin\alpha\sqrt{R^2+\sin^2\alpha}}{32(R+\cos\alpha)^2} + O(\varepsilon).
\end{aligned}$$

The integral I_3 can be obtained from I_1 by changing α to $-\alpha$:

$$\begin{aligned}
I_3 =& \frac{1}{8}\int_0^{\cos^{-1}\frac{\varepsilon}{2}}\int_0^{\frac{\pi}{2}} \frac{R-\cos(-\alpha-2\theta)}{(\cos^2\theta+(R+\cos\alpha)(R-\cos(-\alpha-2\theta))\sin^2\frac{v}{2})^2}\,dvd\theta \\
=& -\frac{\pi R^2}{64(R+\cos\alpha)^2}\log\left(\frac{R^2\varepsilon^2}{16(R+\cos\alpha)^2}\right) - \frac{\pi R^2}{32(R+\cos\alpha)^2}\sinh^{-1}\left(\frac{\sin\alpha}{R}\right) \\
&+ \frac{\pi}{8\varepsilon^2} + \frac{\sin\alpha}{8(R+\cos\alpha)\varepsilon} - \frac{\pi(R^2+4R\cos\alpha+2\cos^2\alpha)}{64(R+\cos\alpha)^2} - \frac{\pi}{16R^2(R+\cos\alpha)^2} + \frac{\pi\sin^2\alpha}{32(R+\cos\alpha)^2} \\
&+ \frac{\pi\sin\alpha}{16R^2(R+\cos\alpha)^2\sqrt{R^2+\sin^2\alpha}} - \frac{\pi\sin\alpha\sqrt{R^2+\sin^2\alpha}}{32(R+\cos\alpha)^2} + O(\varepsilon).
\end{aligned}$$

By putting all the formulae together we obtain

$$\begin{aligned}
V(\varepsilon,x) =& 2(I_1+I_2+I_3+I_4) \\
=& \frac{\pi}{\varepsilon^2} - \frac{\pi R^2}{16(R+\cos\alpha)^2}\log\left(\frac{R^2\varepsilon^2}{(R+\cos\alpha)^2}\right) + \frac{\pi R^2}{8(R+\cos\alpha)^2}3\log 2 \\
&- \frac{\pi}{8} - \frac{\pi}{4R^2(R+\cos\alpha)^2} + \frac{\pi(1+\sin^2\alpha)}{8(R+\cos\alpha)^2}.
\end{aligned}$$

As the Gauss curvature and $\Delta = (\kappa_1-\kappa_2)^2$ at the point $x = p(\alpha, 0)$ is given by

$$K(x) = \frac{\cos\alpha}{(R+\cos\alpha)}, \quad \Delta(x) = \frac{R^2}{(R+\cos\alpha)^2},$$

the renormalized potential is given by

$$\begin{aligned}
V(x;T) =& \lim_{\varepsilon\to 0}\left(V(\varepsilon,x) - \frac{\pi}{\varepsilon^2} + \frac{\pi\Delta(x)}{16}\log\left(\Delta(x)\varepsilon^2\right) + \frac{\pi K(x)}{4}\right) \\
=& \frac{\pi R^2}{8(R+\cos\alpha)^2}3\log 2 - \frac{\pi}{8} - \frac{\pi}{4R^2(R+\cos\alpha)^2} + \frac{\pi(1+\sin^2\alpha)}{8(R+\cos\alpha)^2} + \frac{\pi\cos\alpha}{4(R+\cos\alpha)}.
\end{aligned}$$

Finally, it implies that the renormalized r^{-4}-potential energy of the torus T is given by

$$\begin{aligned}
E(T) &= \int_T V(x;T)\,d^2x \\
&= 2\pi \int_0^{2\pi} \left(\frac{\pi R^2}{8(R+\cos\alpha)^2} 3\log 2 - \frac{\pi}{8} - \frac{\pi}{4R^2(R+\cos\alpha)^2} + \frac{\pi(1+\sin^2\alpha)}{8(R+\cos\alpha)^2} \right. \\
&\qquad \left. + \frac{\pi\cos\alpha}{4(R+\cos\alpha)} \right)(R+\cos\alpha)\,d\alpha \\
&= \frac{\pi^2}{2} \int_0^{\pi} \left(\left(R^2(3\log 2 - 1) + 2 - \frac{2}{R^2} \right) \frac{1}{R+\cos\alpha} \right) d\alpha \\
&= \frac{\pi^2}{2} \left[\left(R^2(3\log 2 - 1) + 2 - \frac{2}{R^2} \right) \frac{2}{\sqrt{R^2-1}} \tan^{-1}\left(\sqrt{\frac{R-1}{R+1}} \tan\frac{\alpha}{2} \right) \right]_0^{\pi} \\
&= \frac{\pi^3}{2\sqrt{R^2-1}} \left(R^2(3\log 2 - 1) + 2 - \frac{2}{R^2} \right).
\end{aligned}$$

□

3. Application as another motivation

First we state our framework ((O'Hara & Solanes)). Let M be an m-dimensional compact orientable submanifold of $\mathbb{R}^n$ and λ be a real number. Put

$$\Delta_\varepsilon = \{(x,y) \in \mathbb{R}^n \times \mathbb{R}^n \mid |x-y| < \varepsilon\}.$$

It is interesting to see the asymptotics of

$$E_{r^\lambda}(\varepsilon, M) = \iint_{M\times M\setminus\Delta_\varepsilon} |x-y|^\lambda \, d^m x d^m y,$$

where $d^m x$ and $d^m y$ denote the standard Lebesgue measure of M. To be precise, we expand the above in a series in ε and study the coefficients. For example, if K is a knot in $\mathbb{R}^3$ and $\lambda = -2$ then

$$E_{r^{-2}}(\varepsilon, K) = \frac{2L(K)}{\varepsilon} + E(K) + O(\varepsilon),$$

where $L(K)$ is the length of the knot and $E(K)$ denotes the knot energy given by (1), and if S is a closed surface in $\mathbb{R}^3$ and $\lambda = -4$ then

$$E_{r^{-4}}(\varepsilon, S) = \frac{\pi}{\varepsilon^2} A(S) - \frac{\pi}{8} \log\varepsilon \int_S \Delta(x)\,d^2x + E(S) - \frac{\pi}{16}\int_S \Delta(x)\log\Delta(x)\,d^2x - \frac{\pi^2}{2}\chi(S),$$

where $\chi(S)$ is the Euler characteristic of S, and if Ω is a 2-dimensional compact submanifold of $\mathbb{R}^2$ then

$$E_{r^{-4}}(\varepsilon, \Omega) = \frac{\pi}{\varepsilon^2} A(\Omega) - \frac{2}{\varepsilon} L(\partial\Omega) + E_{OS}(\Omega) - \frac{\pi^2}{4}\chi(\Omega) + O(\varepsilon),$$

where $A(\Omega)$ is the area of Ω and E_{OS} is the energy defined in (O'Hara & Solanes). We remark that E_{OS} is also invariant under Möbius transformations. We conjecture that a similar formula holds for compact bodies in $\mathbb{R}^3$.

Let us focus on the constant term of the series of $E_{r^\lambda}(\varepsilon, M)$, which, after some modification if necessary, we call the *renormalized r^λ-potential energy* of M, denoted by $E_{r^\lambda}(M)$.

Now we can define functionals for knots as follows. Let $N_\varepsilon(K)$ be an ε-tubular neighbourhood of K. Expand $E_{r^\lambda}(\partial N_\varepsilon(K))$ and $E_{r^{\lambda'}}(N_\varepsilon(K))$ in series of ε. We conjecture that functionals that can capture global properties of knots appear as a coefficient of ε^2-term of $E_{r^\lambda}(\partial N_\varepsilon(K))$ and as a coefficient of ε^4-term of $E_{r^{\lambda'}}(N_\varepsilon(K))$.

When $K_\circ$ is a round circle with radius 1, our main theorem implies

$$\begin{aligned} E_{r^{-4}}(\partial N_\varepsilon(K_\circ)) &= \frac{\pi^3\varepsilon}{2\sqrt{1-\varepsilon^2}}\left(\frac{3\log 2-1}{\varepsilon^2}+2-2\varepsilon^2\right) \\ &= \frac{\pi^3(3\log 2-1)}{2\varepsilon}+\frac{3\pi^3\,(\log 2+1)}{4}\varepsilon+\frac{\pi^3\,(9\log 2-11)}{16}\varepsilon^3+O(\varepsilon^5). \end{aligned}$$

Therefore, the functional thus obtained from the ε^2-term of $E_{r^{-4}}(\partial N_\varepsilon(K))$ vanishes for round circles.

References

[1] Auckly, D. & Sadun, L. 1997 *A family of Möbius invariant 2-knot energies*, Geometric Topology (Proceedings of the 1993 Georgia International Topology Conference). AMS/IP Studies in Adv. Math., W. H. Kazez ed. Amer. Math. Soc. and International Press, Cambridge, MA. 235–258.

[2] Freedman, M.H., He, Z-X., & Wang, Z. 1994 *Möbius energy of knots and unknots.* Ann. of Math. **139**, 1 – 50.

[3] Kusner, R. & Sullivan, J.M. 1997 *Möbius energies for knots and links, surfaces and submanifolds,* Geometric Topology (Proceedings of the 1993 Georgia International Topology Conference). AMS/IP Studies in Adv. Math., W. H. Kazez ed. Amer. Math. Soc. and International Press, Cambridge, MA. 570 – 604.

[Marques & Neves] Marques, F.C. & Neves, A. 2014 *Min-Max theory and the Willmore conjecture*, Ann. of Math. **179** 683-782.

[5] O'Hara, J. 1991 *Energy of a knot*, Topology **30** no. 2, 241–247.

[6] O'Hara, J. 2003 *Energy of knots and conformal geometry,* Series on Knots and Everything Vol. **33**, World Scientific, Singapore, xiv + 288 pages.

[O'Hara & Solanes] O'Hara, J. & Solanes, G. *Möbius invariant energies and average linking with circles*, to appear in Tohoku Math. J.

[O'Hara & Solanes] O'Hara, J. & Solanes, G. *Renormalized potential energies and their asymptotics*, in preparation.

How nice are critical knots?

Regularity theory for knot energies

Simon Blatt[1], Philipp Reiter[2]

[1] Workgroup Applied Analysis, Karlsruhe Institute of Technology, Kaiserstraße 89-93, 76133 Karlsruhe, Germany

[2] Fakultät für Mathematik der Universität Duisburg-Essen, Forsthausweg 2, 47057 Duisburg, Germany

E-mail: `simon.blatt@kit.edu`, `philipp.reiter@uni-due.de`

Abstract. In this note we report on some recent developments in *geometric knot theory* which aims at finding links between geometric properties of a given knotted curve and its knot type. The central object of this field are so-called *knot energies* which are defined on closed embedded curves.

First we present three important examples of two-parameter knot energy families, namely O'Hara's energies, the (generalized) integral Menger curvature, and the (generalized) tangent-point energies.

Subsequently we outline the main steps that lead to C^∞-regularity of stationary points—especially minimizers—in the non-degenerate sub-critical range of parameters.

Particular attention is devoted to the appearing parallels between these energies which, surprisingly at first glance, are quite similar from an analyst's perspective.

1. Introduction

Is the "cable spaghetti" on the floor really knotted or is it enough to pull on both ends of the wire to completely unfold it?

This is the basic question that led to the formation of *geometric knot theory*. During the past twenty five years, this new subfield emerged from the search of particularly "nice" shapes of knots (whatever that means), in contrast to classical knot theory [Burde & Zieschang, 2003] making use of fine analytic tools rather than methods involving algebraic topology. The general aim is investigating geometric properties of a given knotted curve in order to gain information on its knot type. In a broader sense, it is part of *geometric curvature energies* which include geometric integrals measuring smoothness and bending for objects that *a priori* do not have to be smooth, also covering the higher-dimensional case.

The first steps into this new area were paved by Fukuhara [1988] who considered knotted polygons. He thought of a new approach to the fundamental question of classical knot theory—to decide whether two given knots belong to the same knot class. His idea was to design a sort of simplification process for complicate knots which of course has to ensure that the simplified knot still belongs to the same knot class. It should be easier to answer that question for the simplified versions than for the original ones.

So Fukuhara defined an "energy" that basically models *self-avoidance*, i.e. the functional blows up on embedded curves converging to a curve with a self-intersection. Following the negative gradient flow of this functional should simultaneously "untangle" the curve and prevent it from leaving the ambient knot class.

Such a functional can in fact be seen as a measure of "entangledness" and consequently one is interested in finding minimizers within a given knot class, hoping to find the desired "optimal shape", having strands particularly wide apart. Therefore it is a crucial task to design and investigate self-avoiding functionals that are bounded below, which are referred to as *knot energies* according to O'Hara [2003, Def. 1.1].

Knot energies can help to model repulsive forces of fibres, whenever self-interaction of strands should be avoided. In fact, there is indication for DNA molecules seeking to attain a minimum state of a suitable energy [Moffatt, 1996]. There have been several attempts to employ self-avoiding energies for mathematical models in microbiology, see e.g. Banavar et al. [2003] and references therein for links to polymer science and protein science. Attraction phenomena may also be modeled by a corresponding positive gradient flow [Alt *et al.*, 2009].

O'Hara's energies

Later on, Fukuhara's ideas were adapted by O'Hara [1992; 1994] who defined a two-parameter family of knot functionals on smooth curves. Motivated by the Coulomb potential of an equidistributed charge on a smooth curve, he set

$$E^{\alpha,p}(\gamma) := \int_{\mathbb{R}/\mathbb{Z}} \int_{-1/2}^{1/2} \left(\frac{1}{|\gamma(u+w) - \gamma(u)|^{\alpha}} - \frac{1}{D_{\gamma}(u+w,u)^{\alpha}} \right)^{p} |\gamma'(u+w)|\,|\gamma'(u)| \,\mathrm{d}w\,\mathrm{d}u. \quad (1)$$

Here $\alpha, p > 0$, and $\gamma \in C^{0,1}(\mathbb{R}/\mathbb{Z}, \mathbb{R}^n)$. The quantity $D_{\gamma}(u+w,u)$ measures the intrinsic distance between $\gamma(u+w)$ and $\gamma(u)$ on the curve γ.

The mathematical idea behind this definition is quite easy to understand. To obtain a self-avoiding functional, one intends to penalize pairs of points $(\gamma(u), \gamma(v))$ having a small Euclidean distance. This motivates the first term $|\gamma(u+w) - \gamma(u)|^{-1}$ which has to be taken to an adequate power to produce a singularity. On the other hand, this *Ansatz* does not take into account that neighboring points $\gamma(u), \gamma(u+\varepsilon)$ naturally have a small Euclidean distance. Therefore we have to add some sort of regularization: substracting the intrinsic distance $D_{\gamma}(u,v)$ taken to the same negative power as the Euclidean distance, the singularities stemming from neighboring points are cancelled while those for distant points are not essentially affected. Averaging over all pairs of points is achieved by the integration process, and the factors $|\gamma'(u+w)|\,|\gamma'(u)|$ guarantee invariance under reparametrization.

Be aware that not all functionals of the family (1) are in fact knot energies. Some are far too singular, taking the value $+\infty$ only, while others are not strong enough to ensure self-avoidance. In fact, in order to gain knot energies one has to satisfy $\alpha p \geq 2$ and $(\alpha - 2)p < 1$.

Particular attention was devoted to $E^{2,1}$ which O'Hara introduced in [1991]. Later on, Freedman, He, and Wang [1994] coined the name *Möbius energy* due to their seminal discovery that it is in fact invariant under Möbius transformations. The immediate consequence that circles are unique minimizers among all closed curves could be generalized by analytical means to the entire family (1) by Abrams, Cantarella, Fu, Ghomi, and Howard [2003].

In the case of the Möbius energy, Freedman, He, and Wang [1994] could establish the existence of minimizers within prime knot classes only—Kusner and Sullivan [1998] even conjecture that this

Journal of Physics: Conference Series **544** (2014) 012020 doi:10.1088/1742-6596/544/1/012020

statement does not hold for composite knots—whereas for $\alpha p > 2$ the existence of minimizers within any knot class has already been proven by O'Hara [1994].

We briefly mention some further contributions regarding the Möbius energy. Kim and Kusner [1993] constructed extremizing torus knots that are explicitly computable using the calculus of residues, Kusner and Sullivan [1998] performed numerical experiments, Rawdon and co-authors [2006; 2010] established error estimates.

The subsequent article by He on the Euler-Lagrange equation and heat flow for the Möbius energy [2000] was the starting point both for investigating the corresponding gradient flow [Blatt, 2010, 2012*b*] and the regularity of stationary points [Reiter, 2010, 2012], both for the one-parameter sub-family $E^{\alpha,1}$, $\alpha \in [2,3)$. After a first attempt [Blatt & Reiter, 2008], the identification of the energy spaces [Blatt, 2012*a*] led to a significant improvement of the latter result [Blatt & Reiter, 2013*a*; Blatt *et al.*, 2012].

Integral Menger curvature

In the sequel, many other knot energies came up. One important example is the reciprocal of *thickness* which has been investigated by Litherland, Simon, Durumeric, and Rawdon [1999] for C^2-curves. Diao, Ernst, and Janse van Rensburg [1999] considered a slightly different notion on C^1-curves. See Ashton et al. [2011] for further references.

Gonzalez and Maddocks [1999] provided an alternative characterization—not requiring any initial regularity—by the minimum value of the circumradius function $R(\gamma(s), \gamma(t), \gamma(u))$ over all triplets of points of γ. Here

$$R(x,y,z) := \frac{|y-z|\,|y-x|\,|z-x|}{|(y-x)\wedge(z-x)|} = \frac{|y-z|}{\sin\angle(y-x,z-x)}, \qquad x,y,z \in \mathbb{R}^d, \tag{2}$$

denotes the radius of the circle passing through the three points x, y, and z. The reciprocal of R is called *Menger curvature*, crediting Menger's work [1930]. Minimizers of $\gamma \mapsto \sup_{s\neq t\neq u\neq s} R(\gamma(s), \gamma(t), \gamma(u))^{-1}$ (with fixed length, in a prescribed isotopy class) are referred to as *tight knots* (or *ideal knots*).

Existence of tight knots has been proven in [Gonzalez *et al.*, 2002*a*; Cantarella *et al.*, 2002; Gonzalez & de la Llave, 2003], for regularity questions we refer to [Schuricht & von der Mosel, 2003, 2004; Cantarella *et al.*, 2011]. Despite of the lack of an explicit analytical characterization of the shape of a (non-trivial) tight knot, there are several contributions to discretization and numerical visualization, see [Ashton *et al.*, 2011; Cantarella *et al.*, 2005; Carlen & Gerlach, 2012; Carlen *et al.*, 2005; Gerlach, 2010; Gonzalez *et al.*, 2002*b*; Smutny, 2004]. An interesting packing problem, namely *maximizing* length for prescribed thickness on the two-dimensional sphere $\mathbb{S}^2$, has been tackled by Gerlach and von der Mosel [2011*a*; 2011*b*].

The difficulties regarding regularity of tight knots is essentially due to taking infima which is a non-smooth operation. It is reasonable to replace the infima by integration which leads to the functional

$$\mathscr{M}_p(\gamma) := \iiint_{(\mathbb{R}/\mathbb{Z})^3} \frac{\mathrm{d}s\,\mathrm{d}t\,\mathrm{d}u}{R(s,t,u)^p}, \tag{3}$$

that appears as $U_{p,3}[\cdot]$ in [Gonzalez & Maddocks, 1999] amongst several intermediate versions. The element $\mathscr{M}_2$ is referred to as *total Menger curvature*. Léger [1999] could show that one-dimensional Borel sets with finite total Menger curvature are 1-rectifiable, for $\mathscr{M}_p$, $p \neq 2$, see Lin and Mattila [2001]. Mel'nikov and Verdera [1995; 1995; 2001] provided a new proof of the L^2-boundedness of the Cauchy integral operator on Lipschitz graphs exploiting an identity involving the Menger curvature. A more general setting on metric spaces is discussed in Hahlomaa [2008].

Journal of Physics: Conference Series **544** (2014) 012020 doi:10.1088/1742-6596/544/1/012020

A characterization of the Cauchy integral in terms of analytic capacity and uniform rectifiability is given in Mattila et al. [1996]. For the relation to Painlevé's problem we refer to Dudziak [2010] and references therein.

The functional (3) has been thoroughly investigated by Strzelecki, Szumańska and von der Mosel [2010; 2012] in the case $p > 3$ which we will presume throughout this paragraph. Their main result is a geometric Morrey-Sobolev embedding theorem [2010, Thm 1.2], stating that the arc-length reparametrization of any finite-energy curve being a local homeomorphism is $C^{1,1-3/p}$ and its image is C^1-diffeomorphic to the circle. It is a consequence of a more general geometric Morrey space embedding theorem [2010, Thm 1.3]. The former result can be strengthened as follows: any arc-length parametrized curve being a local homeomorphism has finite energy if and only if it is an $W^{2-2/p,p}$-embedding [Blatt, 2013]. The functional $\mathscr{M}_p$ is self-repulsive, i.e. a knot energy, and possesses minimizers within any knot class [Strzelecki *et al.*, 2012, Cor. 2.3].

Hermes [2012] computed the first variation and performed several numerical experiments on the gradient flow. The regularizing behaviour of this energy extends to general measurable sets $X \subset \mathbb{R}^n$ (instead of a curve γ), see Scholtes [2012] and references therein. For higher-dimensional objects one first has to find an appropriate notion for R. We refer to [Lerman & Whitehouse, 2011, 2009] (for $\mathscr{M}_2$), [Strzelecki & von der Mosel, 2005, 2006, 2011*a*] and [Blatt & Kolasiński, 2011; Kolasiński, 2012*a*,*b*; Kolasiński *et al.*, 2012; Kolasiński & Szumańska, 2011].

Introducing powers in the definition of the circumcircle

$$R^{(p,q)}(x,y,z) := \frac{(|y-z|\,|y-x|\,|z-x|)^p}{|(y-x)\wedge(z-x)|^q} = \frac{|y-z|^p\,|y-x|^{p-q}\,|z-x|^{p-q}}{\sin\angle\,(y-x,z-x)^q} \tag{2'}$$

gives rise to a full two-parameter family of *generalized integral Menger curvature* functionals [Blatt & Reiter, 2013*b*]

$$\operatorname{intM}^{(p,q)} := \iiint_{\mathbb{R}/\mathbb{Z}\times[-\frac{1}{2},\frac{1}{2}]^2} \frac{|\gamma'(u)|\,|\gamma'(u+v)|\,|\gamma'(u+w)|}{R^{(p,q)}(\gamma(u),\gamma(u+v),\gamma(u+w))}\,\mathrm{d}w\,\mathrm{d}v\,\mathrm{d}u, \qquad p,q>0. \tag{3'}$$

Of course, $\mathscr{M}_p = 2^p\operatorname{intM}^{(p,p)}$. As before, $\gamma \in C^{0,1}(\mathbb{R}/\mathbb{Z},\mathbb{R}^n)$. The elements of this family are knot energies for certain ranges of parameters only. More precisely, they are self-avoiding if and only if $p \geq \frac{2}{3}q+1$. On the other hand, they are non-singular iff $p < q + \frac{2}{3}$.

In contrast to classical energies $\mathscr{M}_p$, the integrand of the generalized energies lacks an appealing geometric interpretation. But we will see that they have nicer analytic properties, basically due to the fact that their first variation leads to non-degenerate elliptic operator if $q = 2$.

Tangent-point energies

Another variant of these three-point circle-based functionals, already appearing as $U_{p,2}[\mathcal{C}]$ in the article by Gonzalez and Maddocks [1999], are the *tangent-point energies*

$$\mathscr{E}_q(\Gamma) := \iint_{[0,L]^2} \frac{\mathrm{d}s\,\mathrm{d}t}{r_\Gamma(s,t)^q}, \qquad q \geq 2, \tag{4}$$

where $r_\Gamma(s,t) := \lim_{u\to t} R(\gamma(s),\gamma(t),\gamma(u))$ denotes the radius of the smallest circle tangent to $\gamma(t)$ and passing through $\gamma(s)$. It amounts to half the ratio of the squared distances $|\Gamma(s)-\Gamma(t)|^2$ to the distance of the current tangent line $\ell(t) = \Gamma(t) + \mathbb{R}\Gamma'(t)$ from $\Gamma(s)$, i.e.

$$r_\Gamma(s,t) = \frac{|\Gamma(s)-\Gamma(t)|^2}{2\operatorname{dist}(\Gamma(s),\ell(t))}. \tag{5}$$

Journal of Physics: Conference Series **544** (2014) 012020 doi:10.1088/1742-6596/544/1/012020

The first main result by Strzelecki and von der Mosel on the tangent-point energies defined in (4) is that the image of Γ is a one-dimensional topological manifold if $\mathscr{E}_q(\Gamma)$, $q \geq 2$, is finite [Strzelecki & von der Mosel, 2010, Thms. 1.1 and 1.4]. Restricting to $q > 2$ for the remainder of this paragraph, the manifold is of class $C^{1,1-2/q}$ [2010, Thm. 1.3]. If the Hausdorff distance of two given curves is bounded by some threshold only depending on the respective tangent-point energy values, they are ambient isotopic [2010, Thm. 1.2]. Furthermore, $\mathscr{E}_q$ is in fact a knot energy [2010, Prop. 5.1], which improves an earlier result requiring higher regularity by Sullivan [2002], who used this functional to approach *ropelength*. Furthermore, there are minimizers within every knot class [Strzelecki *et al.*, 2012, Cor. 2.3].

The above-mentioned result can be refined as follows, providing an explicit characterization of finite-energy curves: the image of Γ is an embedded manifold of class $W^{2-1/q,q} \subset C^{1,1-2/q}$ if and only if $\mathscr{E}_q(\Gamma)$ is finite [Blatt, 2011]. Results for higher-dimensional analoga to $\mathscr{E}_q$ can be found in [Strzelecki & von der Mosel, 2011*b*; Blatt, 2011; Kolasiński *et al.*, 2012].

Starting point for our investigation in [Blatt & Reiter, 2012] is again decoupling of powers in the nominator and denominator of r_Γ. Using $\operatorname{dist}(\gamma(u+w), \ell(u)) = \left|P^\perp_{\gamma'(u)}\left(\gamma(u+w) - \gamma(u)\right)\right|$, where

$$P_{\gamma'(u)}a := \left\langle a, \frac{\gamma'(u)}{|\gamma'(u)|}\right\rangle \frac{\gamma'(u)}{|\gamma'(u)|}, \qquad P^\perp_{\gamma'(u)}a := a - P_{\gamma'(u)}a \qquad \text{for } a \in \mathbb{R}^n \tag{6}$$

denote the projection onto the tangential and normal part along γ respectively, we define

$$\tilde{r}_\Gamma^{(p,q)}(t,s) := \frac{|\Gamma(s) - \Gamma(t)|^p}{\operatorname{dist}(\ell(t), \Gamma(s))^q} \qquad \text{for } p, q \geq 1 \tag{5'}$$

which leads to the *generalized tangent-point energies*

$$\mathrm{TP}^{(p,q)}(\gamma) = \int_{\mathbb{R}/\mathbb{Z}} \int_{-1/2}^{1/2} \frac{\left|P^\perp_{\gamma'(u)}\left(\gamma(u+w) - \gamma(u)\right)\right|^q}{|\gamma(u+w) - \gamma(u)|^p} |\gamma'(u+w)|\,|\gamma'(u)|\, \mathrm{d}w\, \mathrm{d}u \tag{4'}$$

where $p \in [q+2, 2q+1)$. Note that for $\gamma \in C^{1,1}$, by

$$P^\perp_{\gamma'(u)}\left(\gamma(u+w) - \gamma(u)\right) = P^\perp_{\gamma'(u)}\left(\gamma(u+w) - \gamma(u) - w\gamma'(u)\right),$$

the integrand in (4′) behaves like $\mathcal{O}\left(|w|^{2q-p}\right)$ as $w \to 0$.

Considering only continuously differentiable curves, our method of proof [Blatt & Reiter, 2012]—based on techniques developed in [Blatt, 2011]—works entirely without using the techniques by von der Mosel and Strzelecki [2010] which are quite involved and technical. In fact, we developed a completely independent and fast approach towards existence and regularity of minimizers completely avoiding decay estimates of Jones' beta numbers that play a fundamental rôle in [Strzelecki & von der Mosel, 2010].

In fact, there are many desirable properties of a knot energy one could ask for. Does it penalize pulling-tight (which does not lead to a self-intersection) of small knotted arcs? Are there only finitely many distinct knot types under each energy level? Does the knot energy detect the unknot? Is the round circle the unique minimizer among all curves? A recent account that comments on these questions for several energy functionals including those presented in this text, providing proofs and references, is given by Strzelecki, Szumańska and von der Mosel [2012].

In the following we will concentrate on the analysis of stationary points. Of course, by identifying energy spaces as in the following section we obtain *a priori* estimates which are useful for proving the above-mentioned properties.

2. How fractional Sobolev spaces come into play

The first important step in the investigation of these functionals is to identify the energy spaces, i.e. the subset of embedded $C^{0,1}$-curves on which the respective energy takes finite values. In order to avoid strange parametrizations, we restrict ourselves to arc-length parametrizations.

As to O'Hara's energies, the integrand then reads (for $p = 1$)

$$\frac{1}{|\gamma(u+w)-\gamma(u)|^\alpha} - \frac{1}{|w|^\alpha} = \left|\frac{\gamma(u+w)-\gamma(u)}{w}\right|^{-\alpha} \frac{\left(1-\left|\frac{\gamma(u+w)-\gamma(u)}{w}\right|^\alpha\right)}{|w|^\alpha}. \tag{7}$$

The main idea is now that, using arc-length parametrization and $|a-b|^2 = 2 - 2\langle a, b\rangle$ for $|a| = |b| = 1$ as well as $1 - x^\alpha \approx 1 - x^2$ for $0 < c \le x \le 1$,

$$\begin{aligned} 1-\left|\frac{\gamma(u+w)-\gamma(u)}{w}\right|^\alpha &\approx 1-\left|\frac{\gamma(u+w)-\gamma(u)}{w}\right|^2 \\ &= 1 - \iint_{[0,1]^2} \langle \gamma'(u+\vartheta_1 w), \gamma'(u+\vartheta_2 w)\rangle \,\mathrm{d}\vartheta_1\,\mathrm{d}\vartheta_2 \\ &= \tfrac{1}{2}\iint_{[0,1]^2} |\gamma'(u+\vartheta_1 w)-\gamma'(u+\vartheta_2 w)|^2 \,\mathrm{d}\vartheta_1\,\mathrm{d}\vartheta_2. \end{aligned}$$

Furthermore, $(u, w) \mapsto \left|\frac{\gamma(u+w)-\gamma(u)}{w}\right|$ is continuous, taking values in some compactum in $(0, 1]$ on $\mathbb{R}/\mathbb{Z} \times [-\frac{1}{2}, \frac{1}{2}]$. By means of the functions $g_\alpha : x \mapsto x^{-\alpha}$ and $h_\alpha : x \mapsto \frac{1-x^\alpha}{1-x^2}$, which are analytic and non-negative on $(0, 1]$ provided $a \ge 2$, we may express the right-hand side of (7) by

$$\tfrac{1}{2} g_\alpha\left(\left|\tfrac{\gamma(u+w)-\gamma(u)}{w}\right|\right) h_\alpha\left(\left|\tfrac{\gamma(u+w)-\gamma(u)}{w}\right|\right) \iint_{[0,1]^2} \frac{|\gamma'(u+\vartheta_1 w)-\gamma'(u+\vartheta_2 w)|^2}{|w|^\alpha} \,\mathrm{d}\vartheta_1\,\mathrm{d}\vartheta_2.$$

For arbitrary $p > 0$ this leads us to

$$\left(\frac{1}{|\gamma(u+w)-\gamma(u)|^\alpha} - \frac{1}{|w|^\alpha}\right)^p \approx \frac{|\gamma'(u+\vartheta_1 w)-\gamma'(u+\vartheta_2 w)|^{2p}}{|w|^{\alpha p}}.$$

Integrating over $\mathbb{R}/\mathbb{Z} \times [-\frac{1}{2}, \frac{1}{2}]$ this resembles (the $2p$-th power of) the *Sobolev-Slobodeckiĭ seminorm*

$$[f]_{W^{s,\varrho}} := \left(\int_{\mathbb{R}/\mathbb{Z}} \int_{-1/2}^{1/2} \frac{|f(u+w)-f(u)|^\varrho}{|w|^{1+\varrho s}} \,\mathrm{d}w\,\mathrm{d}u\right)^{1/\varrho} \tag{8}$$

where $f = \gamma'$, $s = \frac{\alpha p - 1}{2p}$, and $\varrho = 2p$. Full details of this argument can be found in [Blatt, 2012*a*]. In other words, an embedded arc-length parametrized curve γ belongs to $W^{1+\alpha/2-1/(2p),2p}$ if and only if its energy $E^{\alpha,p}(\gamma)$ is finite. Here, $W^{k+s,\varrho}(\mathbb{R}/\mathbb{Z},\mathbb{R}^n) := \left\{f \in W^{k,\varrho}(\mathbb{R}/\mathbb{Z},\mathbb{R}^n) \,\middle|\, \|f\|_{W^{k+s,\varrho}} < \infty\right\}$ is equipped with the norm $\|f\|_{W^{k+s,\varrho}} := \|f\|_{W^{k,\varrho}} + \left[f^{(k)}\right]_{W^{s,\varrho}}$ for $k \in \mathbb{N} \cup \{0\}$, $s \in (0, 1)$, and $\varrho \in [1, \infty)$, whereas $W^{k,\varrho}(\mathbb{R}/\mathbb{Z},\mathbb{R}^n)$ denotes the usual Sobolev space. One of the most important features is the (compact) embedding

$$W^{k+s,\varrho}(\mathbb{R}/\mathbb{Z},\mathbb{R}^n) \hookrightarrow C^{k,s-1/\varrho}(\mathbb{R}/\mathbb{Z},\mathbb{R}^n) \tag{9}$$

in the *sub-critical** range $s \in (\varrho^{-1}, 1)$.

* The range of parameters that lead to a *sub*-critical Euler-Lagrange equation is referred to as *sub*-critical in this text. It is also characterized by the existence of the embedding (9).—In contrast, the corresponding range is called *super*-critical in [Strzelecki *et al.*, 2012] as it lies above the respective critical value for which the corresponding energy is scale-invariant.

Journal of Physics: Conference Series **544** (2014) 012020 doi:10.1088/1742-6596/544/1/012020

For the tangent-point energies, we may proceed in a similar way, making use of $\gamma'' \perp \gamma'$. The integrand reads

$$\frac{\left|P^{\perp}_{\gamma'(u)}\left(\gamma(u+w)-\gamma(u)\right)\right|^q}{|\gamma(u+w)-\gamma(u)|^p} \approx \frac{\left|P^{\perp}_{\gamma'(u)}\int_0^1 \gamma'(u+\vartheta w)\,\mathrm{d}\vartheta\right|^q}{|w|^{p-q}}$$
$$= \frac{\left|\int_0^1 P^{\perp}_{\gamma'(u)}\left(\gamma'(u+\vartheta w)-\gamma'(u)\right)\mathrm{d}\vartheta\right|^q}{|w|^{p-q}} \approx \int_0^1 \frac{|\gamma'(u+\vartheta w)-\gamma'(u)|^q}{|w|^{p-q}}\,\mathrm{d}\vartheta,$$

so $\mathrm{TP}^{(p,q)}(\gamma) < \infty \iff \gamma \in W^{(p-1)/q,q}$ for arc-length parametrized γ. Of course, this is merely a hand-waving argument that illustrates the idea, see [Blatt, 2011; Blatt & Reiter, 2012] for details.

The situation is a little bit more delicate for the integral Menger curvature as we have to deal with three integrations, cf. [Blatt, 2013; Blatt & Reiter, 2013*b*]. We skip this part for sake of simplicity.

3. Towards regularity theory

The next step is to compute the derivative. Formally this is a straight-forward calculation, but one has to take into account that this involves interchanging differentiation and integration. Especially in the case of O'Hara's energies this is a quite subtle process which can be tackled by a careful analysis of approximating energy functionals, see [Blatt & Reiter, 2013*a*]. In this particular situation, the first variation turns out to be a principal value integral of type $\lim_{\varepsilon\searrow 0}\iint_{|w|\geq\varepsilon}\cdots$.

The situation is quite different for the integral Menger curvature which has been pointed out by Hermes [2012]. Here the integrand of the first variation is L^1. The same applies to the tangent-point energies, cf. [Blatt & Reiter, 2012]. This might in fact be seen as an advantage of Menger curvature based knot energies over O'Hara's energies.

For the majority of parameters, the Euler-Lagrange operators seem to be somehow "degenerate". Therefore our regularity result applies to the non-degenerate sub-critical case, namely $E^{\alpha,1}$, $\alpha \in (2,3)$, for O'Hara's energies, $\mathrm{intM}^{(p,2)}$, $p \in (\frac{7}{3},\frac{8}{3})$, for the integral Menger curvature, and $\mathrm{TP}^{(p,2)}$, $p \in (4,5)$, for the tangent-point energies. As these energies are not scaling invariant, we have to fix length.

THEOREM (Stationary points are smooth [Blatt & Reiter, 2013*a*, 2012, 2013*b*]). *Any stationary point of* $E^{\alpha,1}$, $\mathrm{intM}^{(p,2)}$, *and* $\mathrm{TP}^{(p,2)}$ *in the non-degenerate sub-critical case with respect to fixed length, being injective and parametrized by arc-length, is* C^∞*-smooth.*

Surprisingly, up to some technical difficulties which have to be dealt with separately, the respective proofs roughly follow the same lines—yet another indication that the three families of energies are not too different from the perspective of an analyst.

Let us briefly outline the common structure of the proofs with some details for the case of O'Hara's energies. First, one has to identify the highest-order term in the Euler-Lagrange equation and show that it is in fact an elliptic operator. Now the remainder consists of lower-order terms that can be brought into a common form which allows to deal with them simultaneously.

For O'Hara's knot energies $E^{\alpha,1}$, $\alpha \in (2,3)$, these two steps show that a stationary curve $\gamma \in C^{0,1}(\mathbb{R}/\mathbb{Z},\mathbb{R}^n)$ of finite energy parametrized by arc-length satisfies the Euler-Lagrange equation

$$Q^{(\alpha)}(\gamma; h) = R^{(\alpha)}(\gamma; h) \tag{10}$$

for all $h \in C^\infty(\mathbb{R}/\mathbb{Z}, \mathbb{R}^n)$ where $Q^{(\alpha)}$ is a symmetric bilinear operator given by

$$Q^{(\alpha)}(\gamma; h) = \sum_{k\in\mathbb{Z}} q_k |k|^{1+\alpha} \hat{\gamma}(k)\overline{\hat{h}(k)} \tag{11}$$

for some sequence $(q_k)_{k\in\mathbb{Z}} \in \ell^\infty(\mathbb{R})$ with $\lim_{|k|\to\infty} q_k \to q_\infty > 0$. Here $\hat{f}(k)$ denotes the k-th Fourier coefficient of f. Note that

$$f \in W^{s,2} \iff \sum_{k\in\mathbb{Z}} |k|^{2s} \left|\hat{f}_k\right|^2 < \infty. \tag{12}$$

If now $\gamma \in W^{\frac{\alpha+1}{2}+\sigma,2}$ for some $\sigma \geq 0$, the remainder term $R^{(\alpha)}(\gamma; \cdot)$ defines a bounded (conjugate-) linear operator on $W^{\frac{3}{2}-\tilde{\sigma}}$ for all $\tilde{\sigma} < \sigma$; so

$$R^{(\alpha)}(\gamma; h) = \sum_{k\in\mathbb{Z}} r_k |k|^{\frac{3}{2}-\tilde{\sigma}} \overline{\hat{h}(k)}$$

for some sequence $(r_k)_{k\in\mathbb{Z}} \in \ell^2(\mathbb{C})$. Plugging this into (10), using (11), and testing with basis functions $h(u) = e^{2\pi i k u}$, we arrive at

$$q_k |k|^{1+\alpha} \hat{\gamma}(k) = r_k |k|^{\frac{3}{2}-\tilde{\sigma}} \qquad \text{for all } k \in \mathbb{Z},$$

thus

$$\left(q_k |k|^{(1+\alpha)-\frac{3}{2}+\tilde{\sigma}} \hat{\gamma}(k)\right)_{l\in\mathbb{Z}} \in \ell^2(\mathbb{C}).$$

Since $q_k \to q_\infty > 0$ as $|k| \to \infty$, Equation (12) implies

$$\gamma \in W^{1+\alpha-\frac{3}{2}+\tilde{\sigma},2} = W^{\frac{\alpha+1}{2}+\sigma+\left(\frac{\alpha-2}{2}+\tilde{\sigma}-\sigma\right),2}$$

for all $\tilde{\sigma} < \sigma$. As $\alpha > 2$, we can finally iterate the argument to show that critical points are smooth and thus prove the Theorem.

Although our method fundamentally relies on the sub-critical range on which we can make use of the embedding (9), the statement of the theorem above should also hold for the non-degenerate *critical case.* For the Möbius-energy $E^{2,1}$, we could establish a corresponding result [Blatt *et al.*, 2012] that—using sophisticated methods from harmonic analysis—improves previous results by Freedman, He, Wang [1994] and He [2000]. In light of the technical difficulties arising here we expect these situation for $\mathrm{intM}^{(7/3,2)}$ and $\mathrm{TP}^{(4,2)}$ to be much more involved.

Acknowledgments

The second author was supported by DFG Transregional Collaborative Research Centre SFB TR 71.

References

Abrams, A., Cantarella, J., Fu, J. H. G., Ghomi, M. & Howard, R. 2003 Circles minimize most knot energies. *Topology* **42** (2), 381–394.

Alt, W., Felix, D., Reiter, Ph. & von der Mosel, H. 2009 Energetics and dynamics of global integrals modeling interaction between stiff filaments. *J. Math. Biol.* **59** (3), 377–414.

Ashton, T., Cantarella, J., Piatek, M. & Rawdon, E. J. 2011 Knot tightening by constrained gradient descent. *Exp. Math.* **20** (1), 57–90.

Banavar, J. R., Gonzalez, O., Maddocks, J. H. & Maritan, A. 2003 Self-interactions of strands and sheets. *J. Statist. Phys.* **110** (1-2), 35–50.

Blatt, S. 2010 The gradient flow of O'Hara's knot energies. Preprint.

Blatt, S. 2011 The energy spaces of the tangent point energies. Preprint.

Blatt, S. 2012*a* Boundedness and regularizing effects of O'Hara's knot energies. *J. Knot Theory Ramifications* **21** (1), 1250010, 9.

Blatt, S. 2012*b* The gradient flow of the Möbius energy near local minimizers. *Calc. Var. Partial Differential Equations* **43** (3-4), 403–439.

Blatt, S. 2013 A note on integral Menger curvature for curves. *Math. Nachr.* **286** (2-3), 149–159.

Blatt, S. & Kolasiński, S. 2011 Sharp Boundedness and Regularizing effects of the integral Menger curvature for submanifolds. *ArXiv e-prints* .

Blatt, S. & Reiter, Ph. 2008 Does finite knot energy lead to differentiability? *J. Knot Theory Ramifications* **17** (10), 1281 – 1310.

Blatt, S. & Reiter, Ph. 2012 Regularity theory for tangent-point energies: The non-degenerate sub-critical case. *ArXiv e-prints* .

Blatt, S. & Reiter, Ph. 2013*a* Stationary points of O'Hara's knot energies. *Manuscripta Mathematica* **140**, 29–50.

Blatt, S. & Reiter, Ph. 2013*b* Towards a regularity theory for integral Menger curvature. *ArXiv e-prints* .

Blatt, S., Reiter, Ph. & Schikorra, A. 2012 Hard analysis meets critical knots (Stationary points of the Möbius energy are smooth). *ArXiv e-prints* .

Burde, G. & Zieschang, H. 2003 **Knots**, 2nd edn. *de Gruyter Studies in Mathematics*, vol. 5. Walter de Gruyter & Co., Berlin.

Cantarella, J., Fu, J. H. G., Kusner, R. & Sullivan, J. M. 2011 Ropelength Criticality. *ArXiv e-prints* .

Cantarella, J., Kusner, R. B. & Sullivan, J. M. 2002 On the minimum ropelength of knots and links. *Invent. Math.* **150** (2), 257–286.

Cantarella, J., Piatek, M. & Rawdon, E. 2005 Visualizing the tightening of knots. In *VIS '05: Proceedings of the conference on Visualization '05*, pp. 575–582. IEEE Computer Society, Washington, DC, USA.

Carlen, M. & Gerlach, H. 2012 Fourier approximation of symmetric ideal knots. *Journal of Knot Theory and Its Ramifications* **21** (05), 1250057.

Carlen, M., Laurie, B., Maddocks, J. H. & Smutny, J. 2005 Biarcs, global radius of curvature, and the computation of ideal knot shapes. In *Physical and numerical models in knot theory*, *Ser. Knots Everything*, vol. 36, pp. 75–108. World Sci. Publ., Singapore.

Diao, Y., Ernst, C. & Janse van Rensburg, E. J. 1999 Thicknesses of knots. *Math. Proc. Cambridge Philos. Soc.* **126** (2), 293–310.

Dudziak, J. J. 2010 **Vitushkin's conjecture for removable sets**. Springer, New York.

Freedman, M. H., He, Z.-X. & Wang, Z. 1994 Möbius energy of knots and unknots. *Ann. of Math. (2)* **139** (1), 1–50.

Fukuhara, S. 1988 Energy of a knot. In *A fête of topology*, pp. 443–451. Academic Press, Boston, MA.

Gerlach, H. 2010 Ideal knots and other packing problems of tubes. PhD thesis, EPF Lausanne.

Gerlach, H. & von der Mosel, H. 2011*a* On sphere-filling ropes. *Amer. Math. Monthly* **118** (10), 863–876.

Gerlach, H. & von der Mosel, H. 2011*b* What are the longest ropes on the unit sphere? *Archive for Rational Mechanics and Analysis* **201**, 303–342.

Gonzalez, O. & de la Llave, R. 2003 Existence of ideal knots. *J. Knot Theory Ramifications* **12** (1), 123–133.

Gonzalez, O. & Maddocks, J. H. 1999 Global curvature, thickness, and the ideal shapes of knots. *Proc. Natl. Acad. Sci. USA* **96** (9), 4769–4773 (electronic).

Gonzalez, O., Maddocks, J. H., Schuricht, F. & von der Mosel, H. 2002*a* Global curvature and self-contact of nonlinearly elastic curves and rods. *Calc. Var. Partial Differential Equations* **14** (1), 29–68.

Gonzalez, O., Maddocks, J. H. & Smutny, J. 2002*b* Curves, circles, and spheres. In *Physical knots: knotting, linking, and folding geometric objects in* $\mathbb{R}^3$ *(Las Vegas, NV, 2001)*, *Contemp. Math.*, vol. 304, pp. 195–215. Amer. Math. Soc., Providence, RI.

Hahlomaa, I. 2008 Menger curvature and rectifiability in metric spaces. *Adv. Math.* **219** (6), 1894–1915.

He, Z.-X. 2000 The Euler-Lagrange equation and heat flow for the Möbius energy. *Comm. Pure Appl. Math.* **53** (4), 399–431.

Hermes, T. 2012 Analysis of the first variation and a numerical gradient flow for integral Menger curvature. PhD thesis, RWTH Aachen. `http://darwin.bth.rwth-aachen.de/opus3/volltexte/2012/4186`.

Kim, D. & Kusner, R. 1993 Torus knots extremizing the Möbius energy. *Experiment. Math.* **2** (1), 1–9.

KOLASIŃSKI, S. 2012*a* Geometric Sobolev-like embedding using high-dimensional Menger-like curvature. *ArXiv e-prints* .

KOLASIŃSKI, S. 2012*b* Geometric Sobolev-like embedding using high-dimensional Menger-like curvature. *ArXiv e-prints* .

KOLASIŃSKI, S., STRZELECKI, P. & VON DER MOSEL, H. 2012 Characterizing $W^{2,p}$ submanifolds by p-integrability of global curvatures. *ArXiv e-prints* .

KOLASIŃSKI, S. & SZUMAŃSKA, M. 2011 Minimal Hölder regularity implying finiteness of integral Menger curvature. *ArXiv e-prints* .

KUSNER, R. B. & SULLIVAN, J. M. 1998 Möbius-invariant knot energies. In *Ideal knots*, *Ser. Knots Everything*, vol. 19, pp. 315–352. World Sci. Publishing, River Edge, NJ.

LÉGER, J. C. 1999 Menger curvature and rectifiability. *Ann. of Math. (2)* **149** (3), 831–869.

LERMAN, G. & WHITEHOUSE, J. T. 2009 High-dimensional Menger-type curvatures. II. d-separation and a menagerie of curvatures. *Constr. Approx.* **30** (3), 325–360.

LERMAN, G. & WHITEHOUSE, J. T. 2011 High-dimensional Menger-type curvatures. Part I: Geometric multipoles and multiscale inequalities. *Rev. Mat. Iberoam.* **27** (2), 493–555.

LIN, Y. & MATTILA, P. 2001 Menger curvature and C^1 regularity of fractals. *Proc. Amer. Math. Soc.* **129** (6), 1755–1762 (electronic).

LITHERLAND, R. A., SIMON, J., DURUMERIC, O. & RAWDON, E. 1999 Thickness of knots. *Topology Appl.* **91** (3), 233–244.

MATTILA, P., MELNIKOV, M. S. & VERDERA, J. 1996 The Cauchy integral, analytic capacity, and uniform rectifiability. *Ann. of Math. (2)* **144** (1), 127–136.

MEL′NIKOV, M. S. 1995 Analytic capacity: a discrete approach and the curvature of measure. *Mat. Sb.* **186** (6), 57–76.

MEL′NIKOV, M. S. & VERDERA, J. 1995 A geometric proof of the L^2 boundedness of the Cauchy integral on Lipschitz graphs. *Internat. Math. Res. Notices* (7), 325–331.

MENGER, K. 1930 Untersuchungen über allgemeine Metrik. *Math. Ann.* **103** (1), 466–501.

MOFFATT, H. K. 1996 Pulling the knot tight. *Nature* **384**, 114.

O'HARA, J. 1991 Energy of a knot. *Topology* **30** (2), 241–247.

O'HARA, J. 1992 Family of energy functionals of knots. *Topology Appl.* **48** (2), 147–161.

O'HARA, J. 1994 Energy functionals of knots. II. *Topology Appl.* **56** (1), 45–61.

O'HARA, J. 2003 **Energy of knots and conformal geometry**. *Series on Knots and Everything*, vol. 33. World Scientific Publishing Co. Inc., River Edge, NJ.

RAWDON, E. J. & SIMON, J. K. 2006 Polygonal approximation and energy of smooth knots. *J. Knot Theory Ramifications* **15** (4), 429–451.

RAWDON, E. J. & WORTHINGTON, J. 2010 Error analysis of the minimum distance energy of a polygonal knot and the Möbius energy of an approximating curve. *J. Knot Theory Ramifications* **19** (8), 975–1000.

REITER, PH. 2010 Regularity theory for the Möbius energy. *Commun. Pure Appl. Anal.* **9** (5), 1463–1471.

REITER, PH. 2012 Repulsive knot energies and pseudodifferential calculus for O'Hara's knot energy family $E^{(\alpha)}$, $\alpha \in [2,3)$. *Math. Nachr.* **285** (7), 889–913.

SCHOLTES, S. 2012 Tangency properties of sets with finite geometric curvature energies. *Fund. Math.* **218**, 165–191.

SCHURICHT, F. & VON DER MOSEL, H. 2003 Euler-Lagrange equations for nonlinearly elastic rods with self-contact. *Arch. Ration. Mech. Anal.* **168** (1), 35–82.

SCHURICHT, F. & VON DER MOSEL, H. 2004 Characterization of ideal knots. *Calc. Var. Partial Differential Equations* **19** (3), 281–305.

SMUTNY, J. 2004 Global radii of curvature, and the biarc approximation of space curves: In pursuit of ideal knot shapes. PhD thesis, EPF Lausanne.

STRZELECKI, P. & VON DER MOSEL, H. 2005 On a mathematical model for thick surfaces. In *Physical and numerical models in knot theory*, *Ser. Knots Everything*, vol. 36, pp. 547–564. World Sci. Publ., Singapore.

STRZELECKI, P. & VON DER MOSEL, H. 2006 Global curvature for surfaces and area minimization under a thickness constraint. *Calc. Var. Partial Differential Equations* **25** (4), 431–467.

STRZELECKI, P. & VON DER MOSEL, H. 2011*a* Integral Menger curvature for surfaces. *Advances in Mathematics* **226** (3), 2233–2304.

STRZELECKI, P. & VON DER MOSEL, H. 2011*b* Tangent-point repulsive potentials for a class of non-smooth m-dimensional sets in $\mathbb{R}^n$. Part I: Smoothing and self-avoidance effects. *Journal of Geometric Analysis* pp. 1–55.

STRZELECKI, P., SZUMAŃSKA, M. & VON DER MOSEL, H. 2010 Regularizing and self-avoidance effects of integral Menger curvature. *Ann. Sc. Norm. Super. Pisa Cl. Sci. (5)* **IX** (1), 145–187.

STRZELECKI, P., SZUMAŃSKA, M. & VON DER MOSEL, H. 2012 On some knot energies involving Menger curvature. *ArXiv e-prints* .

STRZELECKI, P. & VON DER MOSEL, H. 2010 Tangent-point self-avoidance energies for curves. *ArXiv e-prints* Published in *Journal of Knot Theory and Its Ramifications* 21(05):1250044, 2012.

SULLIVAN, J. M. 2002 Approximating ropelength by energy functions. In *Physical knots: knotting, linking, and folding geometric objects in $\mathbb{R}^3$ (Las Vegas, NV, 2001)*, *Contemp. Math.*, vol. 304, pp. 181–186. Amer. Math. Soc., Providence, RI.

VERDERA, J. 2001 L^2 boundedness of the Cauchy integral and Menger curvature. In *Harmonic analysis and boundary value problems (Fayetteville, AR, 2000)*, *Contemp. Math.*, vol. 277, pp. 139–158. Amer. Math. Soc., Providence, RI.

Journal of Physics: Conference Series **544** (2014) 012021 doi:10.1088/1742-6596/544/1/012021

Topological stirring of two-dimensional atomic Bose-Einstein condensates

A. C. White[1,2], **N. P. Proukakis**[1] **& C. F. Barenghi**[1]

[1] Joint Quantum Centre (JQC), Durham-Newcastle, School of Mathematics and Statistics, Newcastle University, Newcastle upon Tyne, NE1 7RU, United Kingdom.

[2] School of Mathematical Sciences, University of Nottingham, University Park, Nottingham, NG7 2RD, United Kingdom.

E-mail: ang.c.white@gmail.com

Abstract. We stir vortices into a trapped quasi two-dimensional atomic Bose-Einstein condensate by moving three laser stirrers. We apply stirring protocols introduced by Boyland *et al.* (2000), that efficiently build in topological chaos in classical fluids and are classified as Pseudo-Anosov stirring protocols. These are compared to their inefficient mixing counterparts, finite-order stirring protocols. We investigate if inefficient stirring protocols result in a more clustered distribution of vortices. The efficiency with which vortices are 'mixed' or distributed in a condensate is important for investigating dynamics of continuously forced quantum turbulence and the existence of the inverse cascade in turbulent two-dimensional superfluids.

1. Introduction

Jupiter's Great Red Spot is a beautiful demonstration of a large spot of vorticity, one of the defining features of two dimensional (2D) classical turbulence, arising from the inverse cascade process (Sommeria *et al.* (1988); Marcus (1988)). This process describes the flow of energy from small scales, given by the scale at which energy is injected into the system, to larger scales, manifesting in the clustering of vortices of like-winding. The concept of an inverse energy cascade process was introduced by Onsager (1949), who showed that clusters of like-signed point vortices in a large Onsager gas of many point-vortices possess negative temperature. Quantum vortices in superfluid He can be realistically modelled as point vortices or vortex lines, due their small vortex core size ($\sim$ 1nm), quantised vorticity and large ratio of inter-vortex spacing to vortex core radius ($\sim 10^5 - 10^6$). At large scales greater than the average inter-vortex spacing, three dimensional turbulent quantum fluids are also known to exhibit the same classical Kolmogorov energy spectrum with scaling $k^{-5/3}$ (where k is the wavenumber) as classical fluids (see Nore *et al.* (1997); Maurer & Tabeling (1998); Stalp *et al.* (1999); Araki *et al.* (2002); Kobayashi & Tsubota (2005); Salort *et al.* (2010); Baggaley & Barenghi (2011) and Baggaley *et al.* (2012)). For these reasons, one might expect the inverse cascade process to occur in other two dimensional quantum fluids, such as Bose-Einstein condensates (BECs), where vortices also have quantised circulation. While at large scales turbulent quantum fluids are known to have similarities to turbulent classical fluids for three dimensional systems, for two dimensional systems such a crossover is yet to be established. There have been some numerical simulations working towards this goal in trapped and homogeneous condensates, however, to date there is no conclusive demonstration of the inverse cascade process for harmonically trapped atomic BECs (Parker

& Adams (2005); Horng *et al.* (2009); Numasato *et al.* (2010); Numasato & Tsubota (2010); White *et al.* (2012); Bradley & Anderson (2012); Reeves *et al.* (2012*a*,*b*)). Unlike in superfluid He, atomic BECs of low dimensions can be routinely created and manipulated experimentally and so they provide a promising system for such investigations. One important component of any two dimensional turbulent set-up is the introduction of vortices into the condensate in a controlled manner, such that their distribution is well-mixed. Finding such a well-mixed vortex distribution will be the focus of this work.

Recently White *et al.* (2012) showed that the trajectory a single laser stirrer takes through the condensate determines if the resulting vortices are clustered or more randomly distributed. We found that a single laser moving on a circular trajectory creates an initial cluster of like-signed vortices. While the extent of clustering was found to increase with stirring, it did not persist and decayed after the laser stirrer was turned off. This clustering of vortices manifests due to the trajectory of the laser stirrer, and not as a result of an inverse cascade process. Indeed it was shown that altering the trajectory of the laser stirrer resulted in less-extensive clustering and vortices could even be created in more randomly distributed configurations. In this work we employ three laser stirrers and measure the distribution and clustering of vortices by applying statistical measures of clustering previously developed in White *et al.* (2012). In particular, we stir the condensate in a manner that is known to efficiently mix classical fluids, known as pseudo-Anosov stirring protocols. These are contrasted to finite-order stirring protocols which have been shown to be inefficient mixers for classical fluids.

In this paper we proceed in the following way. Firstly we give an overview of topological stirring protocols and then outline the model applied to simulate vortex dynamics in atomic BECs. Finally we introduce statistical measures of clustering that are applied to give insight into how two quite different stirring protocols influence vortex dynamics in a surprisingly similar way. While this is unexpected from the perspective of classical fluids, it is a remarkable demonstration of the potential flow of quantum fluids.

2. Topological stirring protocols

Topological concepts were introduced to address the question of how to most efficiently mix classical fluids. Boyland *et al.* (2000) determined that for flows with the topology of certain braids, given only the topology of the flow, it is possible to deduce the material stretch rate. For stokes flow (characterised by Reynolds number, $Re \ll 1$), chaotic advection of fluid particles is built into the flow faster by stirring schemes that build in topological chaos. Such protocols are classified as pseudo-Anosov stirring protocols and it is established that stirring a fluid in such a way results in exponential stretching of material lines. Stirring protocols that do not have the correct topology to mix effectively are known as finite-order stirring protocols (Boyland *et al.* (2000); Finn *et al.* (2003); Gouillart *et al.* (2006)). In classical fluids, finite-order stirring protocols have been shown to stretch material lines linearly and therefore do not mix as efficiently as their pseudo-Anosov counterparts.

The trajectory of the stirring lasers can be represented by braid-diagrams in space-time, with braid letters σ_i denoting the exchange of stirrers labelled i and $i+1$ in a clockwise direction. This can be depicted in a braid diagram by the strand corresponding to the stirrer labelled $i+1$ passing over the strand showing the ith stirrer. Similarly, an anti-clockwise path of stirrers through the condensate is denoted by braid letters σ_{-i} which represents stirrers $i+1$ and i exchanging position in an anti-clockwise direction. This interchange is shown on a braid-diagram by the strand depicting the $i+1$ stirrer passing under the ith strand. Figure 1 (A) illustrates the braid diagram corresponding to the exchange of stirrers following a protocol that can be classified as pseudo-Anosov.

Each stirring operation, corresponding to a braid letter, can be represented by a braid matrix. For the stirring operations corresponding to the σ_1, σ_{-1} and σ_2 braid letters, which we apply in this work, the corresponding braid matrices take the form (Boyland *et al.* (2000); Finn *et al.* (2003))

$$\Sigma_1 = \begin{bmatrix} 1 & -1 \\ 0 & 1 \end{bmatrix}, \quad \Sigma_{-1} = \begin{bmatrix} 1 & 1 \\ 0 & 1 \end{bmatrix}, \quad \Sigma_2 = \begin{bmatrix} 1 & 0 \\ 1 & 1 \end{bmatrix}. \tag{1}$$

The time-like sequence of braid letters composes a braid word, where the rightmost letter gives the first stirring operation and the leftmost letter denotes the most recent stirring operation. We concentrate on two stirring protocols, which we label A and B that are represented by the braid words

$$A = (\sigma_1\sigma_2)^n \quad \text{and} \quad B = (\sigma_{-1}\sigma_2)^n \ , \tag{2}$$

here n denotes the number of repetitions of braid letter operations. Similarly, the corresponding action of the braid matrices takes the form

$$A = [\Sigma_1\Sigma_2]^n \quad \text{and} \quad B = [\Sigma_{-1}\Sigma_2]^n \ . \tag{3}$$

Stirring protocol A is a finite-order protocol, while protocol B is a pseudo-Anosov stirring protocol which builds topological chaos into the flow. The topology of the fluid flow is completely characterised by the braid word, regardless of the nature of the fluid flow, i.e. if it is compressible, incompressible, viscous or inviscid. For three stirrers, the magnitude of the largest eigenvalue of the associated braid matrix, called the spectral radius, gives a rate at which at least one material line is stretched. Finn *et al.* (2003) computed the stretching in time of a finite material line element and showed that the chaotic region of the fluid flow is of the same extent as the region of fluid the stirrers are moved through.

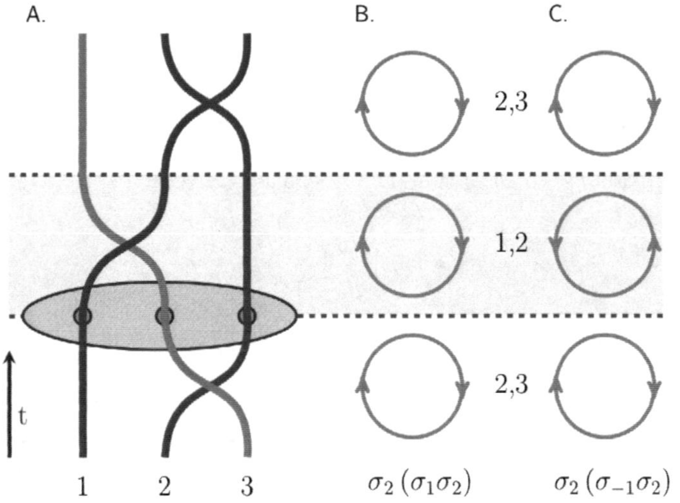

Figure 1. (A) Braid representation in space and time depicting the exchange of stirrers for a pseudo-Anosov stirring protocol. Three stirring operations are shown, corresponding to the exchange of stirrers $\sigma_2 = (2,3)$, then $\sigma_{-1} = (1,2)$ and again stirrers $\sigma_2 = (2,3)$. The rotation stirrers trace out are depicted in (B) for a finite-order stirring protocol and (C) for the pseudo-Anosov stirring protocol depicted in (A).

In the remainder of this paper we present numerical simulations of three laser stirrers in an atomic BEC tracing out the finite-order and pseudo-Anosov stirring schemes A and B, presented above. Note the stirrers trace out a figure eight path, depicted in figure 2. We choose the velocity of the laser stirrers to be larger than the critical velocity for vortex nucleation in order to determine if a pseudo-Anosov stirring protocol (B) also results in a more random distribution of vortices than a finite-order stirring scheme (A).

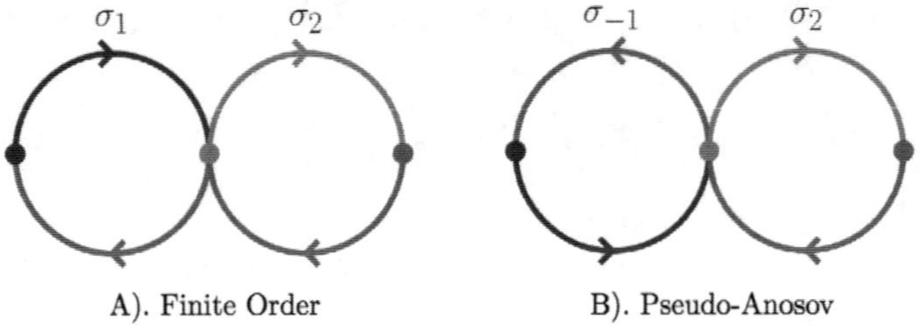

Figure 2. Stirrer paths for first two stirring operations of A) finite-order stirring protocol and B) pseudo-Anosov stirring protocol.

3. Modelling vortex dynamics

The second quantised hamiltonian describing a trapped, weakly-interacting atomic Bose-Einstein condensate takes the form

$$H = \int d\tilde{\mathbf{r}} \left[-\hat{\Psi}^{\dagger}(\tilde{\mathbf{r}}) \frac{\hbar^2}{2m} \nabla^2 \hat{\Psi}(\tilde{\mathbf{r}}) + (V(\tilde{\mathbf{r}}) + V_P) \hat{\Psi}^{\dagger}(\tilde{\mathbf{r}}) \hat{\Psi}(\tilde{\mathbf{r}}) + \frac{U_0}{2} \hat{\Psi}^{\dagger}(\tilde{\mathbf{r}}) \hat{\Psi}^{\dagger}(\tilde{\mathbf{r}}) \hat{\Psi}(\tilde{\mathbf{r}}) \hat{\Psi}(\tilde{\mathbf{r}}) \right]. \quad (4)$$

Here $\hat{\Psi}(\tilde{\mathbf{r}})^{\dagger}$ and $\hat{\Psi}(\tilde{\mathbf{r}})$ are the creation and annihilation operators for bosons of mass m, trapped in a potential $V(\tilde{\mathbf{r}})$. The laser stirrers are described by V_P. The boson-boson interaction is given by a contact potential $U_0 = 4\pi\hbar^2 a/m$, where a is the s-wave scattering length. The evolution of the field operator can be obtained from the Heisenberg equations of motion. At low temperatures, we can describe the condensed state by a classical mean field $\Psi = \langle\hat{\Psi}\rangle$ and excitations $\langle\phi\rangle$, decomposing the boson operators as $\hat{\Psi} = \Psi + \langle\phi\rangle$. The dynamics of the mean-field condensate wavefunction are then given by the nonlinear Schrödinger equation

$$i\hbar \frac{\partial \Psi(\tilde{\mathbf{r}})}{\partial \tilde{t}} = -\frac{\hbar^2}{2m} \nabla^2 \Psi(\tilde{\mathbf{r}}) + (V(\tilde{\mathbf{r}}) + V_P) \Psi(\tilde{\mathbf{r}}) + U_0 |\Psi(\tilde{\mathbf{r}})|^2 \Psi(\tilde{\mathbf{r}}). \quad (5)$$

We consider a condensate trapped such that the axial trapping frequency is much greater than the radial trapping frequency, $\omega_z \gg \omega_r$. In this case the condensate is quasi-two dimensional. For our simulations we model the dimensionless form of the 2D non-linear Schrödinger equation,

$$i\frac{\partial \psi}{\partial t} = -\frac{1}{2}\nabla^2\psi + \frac{r^2}{2}\psi + V_P\psi + \kappa_{2d}|\psi|^2\psi. \quad (6)$$

Here the 2D interaction strength takes the form $\kappa_{2d} = 2\sqrt{2\pi}aN/a_z$, where N is the total number of atoms in the condensate. The 2D condensate wavefunction is written as $\psi = a_r\Psi/\sqrt{N}$ and we have imposed the normalisation $\int d\mathbf{x}|\psi|^2 = 1$. The radial (axial) trapping frequency, ω_r (ω_z), determines the radial (axial) harmonic oscillator length $a_r = \sqrt{\hbar/(m\omega_r)}$ ($a_z = \sqrt{\hbar/(m\omega_z)}$). We choose the scaling for lengths and time to be $\tilde{r}/a_r = r$ and $\tilde{t}\omega_r = t$, where $\tilde{r}$ and $\tilde{t}$ are dimensional with units of meters and seconds respectively. We solve Eq.(6) pseudo-spectrally, applying an adaptive 4-5th order Runge-Kutta method in time (Dennis *et al.* (2013)). We select a stirring strength $V_0 \sim 2.6\mu$, where μ is the chemical potential of the BEC. We choose $\kappa = 10400$, which taking ^{23}Na atoms corresponds to parameters of $\omega_z = 2\pi \times 50$Hz, $\omega_r = 2\pi \times 5$Hz, $N = 2.2 \times 10^6$ and $a = 2.75$nm. Simulations are run on a 512^2 grid of spatial extent -20 to 20. Snap-shots of time-evolution of the condensate density profile is depicted in figure 3.

3.1. Decomposition of the Kinetic Energy contribution

For a time-independent potential, the total energy of the system is conserved within the nonlinear Schrödinger equation. Writing the mean-field condensate wavefunction in terms of the condensate density $n = |\psi|^2$, and phase θ, as $\psi = \sqrt{n}\exp[i\theta]$, the kinetic energy component

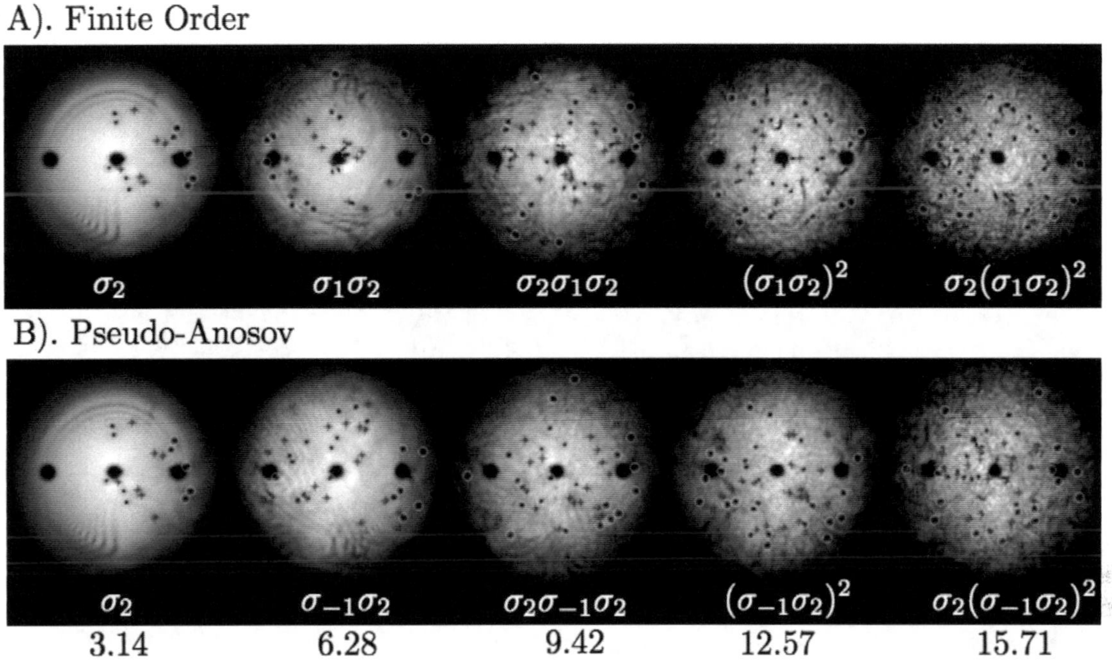

Figure 3. Vortex Dynamics: Condensate density profiles after simulation times given along the bottom of the figure. Magenta + and cyan − symbols denote vortices with positive and negative winding respectively. Row A.) finite-order stirring protocol at times corresponding to braid-words σ_2, $\sigma_1\sigma_2$, $\sigma_2\sigma_1\sigma_2$, $(\sigma_1\sigma_2)^2$ and $\sigma_2\,(\sigma_1\sigma_2)^2$. Row B.) corresponds to pseudo-Anosov stirring protocols at identical times, corresponding to braid words σ_2, $\sigma_{-1}\sigma_2$, $\sigma_2\sigma_{-1}\sigma_2$, $(\sigma_{-1}\sigma_2)^2$ and $\sigma_2\,(\sigma_{-1}\sigma_2)^2$.

of the total energy can be decomposed into contributions from the quantum pressure and kinetic energy density (Nore *et al.* (1997))

$$E_{\mathsf{KE}} = \int \mathrm{d}\mathbf{x} \left(\frac{1}{2}|\nabla\psi|^2\right) = \int \mathrm{d}\mathbf{x} \left(\frac{1}{2}|\nabla\sqrt{n}|^2 + \frac{1}{2}|\sqrt{n}\mathbf{v}|^2\right) . \tag{7}$$

Here we have defined the velocity field $\mathbf{v} = \nabla\theta$. The kinetic energy density contribution can be further decomposed into a compressible component for which the density weighted velocity field $\boldsymbol{\Upsilon} = \sqrt{n}\mathbf{v}$, satisfies $\nabla \times \boldsymbol{\Upsilon}^c = 0$ and an incompressible, divergence free component $\nabla \cdot \boldsymbol{\Upsilon}^i = 0$, such that $\boldsymbol{\Upsilon} = \boldsymbol{\Upsilon}^i + \boldsymbol{\Upsilon}^c$. The total compressible energy is a useful measure of the total sound in the system and the total incompressible kinetic energy gives the total kinetic energy of vortices

$$E^{i,c}_{\mathsf{KE}} = \frac{1}{2}\int \mathrm{d}\mathbf{x} \left|\boldsymbol{\Upsilon}^{i,c}\right|^2 . \tag{8}$$

One can subsequently define compressible and incompressible kinetic energy spectrums, however due to the small size of trapped atomic condensates, only a short range of length scales is accessible in these systems, in comparison to other 2D turbulent systems. Consequently, evaluating scaling exponents from such spectrums is of limited use and the derived scaling laws should be interpreted with care. We refrain from such an analysis here.

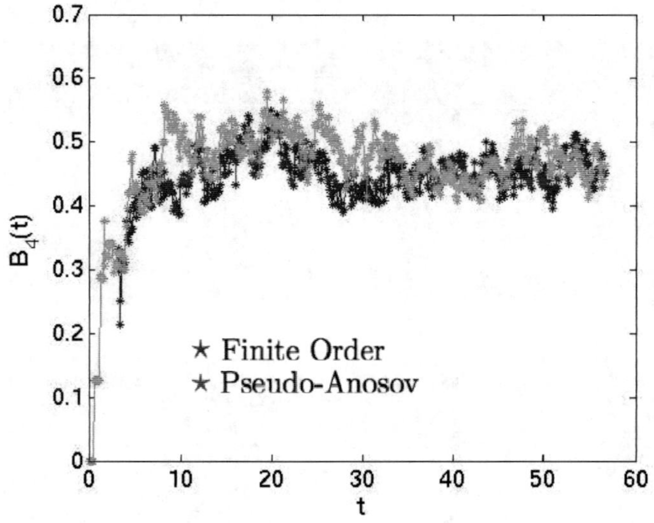

Figure 4. Evolution of vortex distribution statistics given by fourth nearest neighbor vortices for finite-order (blue $*$) and pseudo-Anosov (green $*$) stirring schemes.

3.2. Statistical measures of clustering

Motivated by the seminal work of Ripley (1976, 1977), White *et al.* (2012) developed statistical measures of clustering that are appropriate for measuring if vortices are clustered or more randomly distributed in atomic Bose-Einstein condensates. These measures have the advantage of being experimentally accessible as they rely only on knowledge of the position and winding of vortices, taking into account both the sign and distance between a reference vortex i and its neighbouring vortices j, up to its Jth nearest-neighbor,

$$B_J(t) = \frac{1}{N}\sum_{i=1}^{N}\sum_{j=1}^{J}\frac{b_{ij}(t)}{J}\,. \tag{9}$$

N denotes all vortices that are at a distance $d_i \geq R_E$ from the condensate edge. If vortex i and its jth nearest neighbor are of opposite sign, $b_{ij} = 0$. If vortex i and its jth nearest neighbor are of the same sign and separation distance between them, $d_{ij} \leq R_c$, then $b_{ij} = 1$. If vortex i and vortex j are separated by a distance larger than R_c, ($d_{ij} \geq R_c$), then $b_{ij} = 0$. In this work we take the distance $R_c = R_E$ to be greater than the average inter-vortex separation distance and of the order of the largest cluster size. These constraints are included in the statistical measure of vortex distribution in Eq.(9), to account for edge effects that arise as we are dealing with a small system of finite size. The parameter limiting separation distance between vortices i and j gives a means to investigate the variation of clustering over spatial regions. We take vortex distribution to be clustered for values of $B_J(t) > 0.5$, while for more random vortex distributions $B_J(t) < 0.5$. Related measures of clustering have been recently applied to look at two dimensional turbulence in Bradley & Anderson (2012) and Reeves *et al.* (2012*b*).

4. Discussion and Conclusions

From the point of view of classical fluid dynamics, visualising and tracking the evolution of fluid material lines through the use of tracer particles such as dye, are standard and straightforward techniques that enable one to distinguish between underlying chaotic and regular dynamics in the fluid flow. For atomic Bose-Einstein condensates, the best method for flow visualisation is not so obvious, however some information about the condensate flow can be provided with current experimental techniques. The positions of vortices can be visualised through standard absorption imaging techniques and the condensate phase profile has the potential to be measured

Journal of Physics: Conference Series **544** (2014) 012021 doi:10.1088/1742-6596/544/1/012021

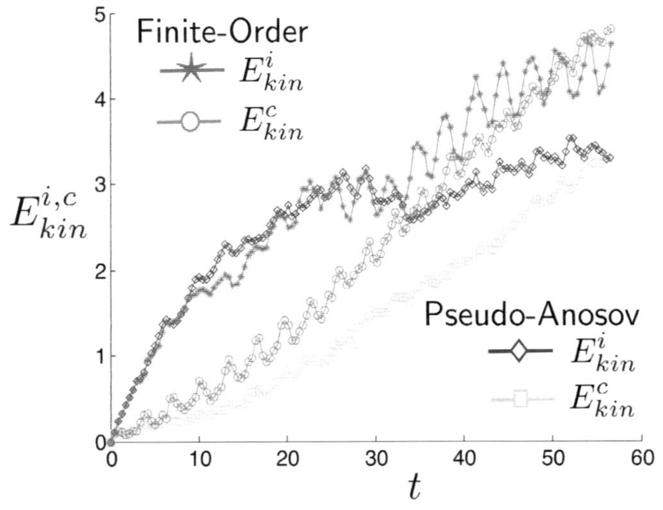

Figure 5. Incompressible and compressible kinetic energy distribution for finite-order stirring given by red stars and magenta circles respectively, and pseudo-Anosov stirring, given by blue diamonds and cyan squares respectively.

by applying interferometric measurements giving information about vortex winding. For these reasons we employ the statistical measure of clustering introduced above to analyse condensate flow.

We employ Eq.(9) up to fourth-nearest neighbor vortices and track the corresponding statistical measure of clustering, which is plotted in figure 4. Evidently, neither the finite-order nor the pseudo-Anosov stirring protocol that we have chosen results in a vortex distribution that is significantly clustered. After a few stirring operations, the vortices appear randomly distributed, independent of stirring protocol. This result is an indication of the potential nature of quantum fluids. One might expect underlying chaotic dynamics would result in a vortex distribution that is less clustered than underlying regular dynamics. Although material line elements are expected to be exponentially stretched for pseudo-Anosov stirring in contrast to the linear stretching of material line elements in the finite-order case, this does not appear to influence the vortex distribution. In other words, the vortex distribution is independent of if topological chaos is built into the flow by the method of stirring.

Another explanation for our observations may also lie in the speed at which the condensate is stirred. From a classical fluid perspective, when a container of classical fluid, such as water is stirred with very fast moving obstacles, the fluid will become turbulent and well-mixed, independent of the motion of the stirring obstacles provided their paths traverse a significant extent of the container surface. In this limit linear or exponential stretching of material line elements in the fluid may be present, but not visible. For a quantum fluid we can define limits on the speed at which the condensate is stirred. The lower limit on stirring speed is given by the condensate critical velocity, below which an obstacle moving through the condensate will not nucleate vortices. The upper limit is set by the speed at which fragmentation of the condensate occurs (Parker & Adams (2005)). In our simulations the condensate is stirred at speeds faster than the critical velocity for vortex nucleation, but not so fast that we enter the regime of condensate fragmentation. Although the speed of stirring is chosen to be fast enough to nucleate vortices, this does not necessarily imply we are in the quantum analogue of the classical scenario described above, where linear or exponential stretching of material lines may be present but not visible due to the very fast stirring speeds. Defining the analogue of such a limit is an interesting question we leave for future work.

From the analysis of the incompressible and compressible kinetic energy distribution (see figure 5), it is found that pseudo-Anosov stirring results in less sound being generated in

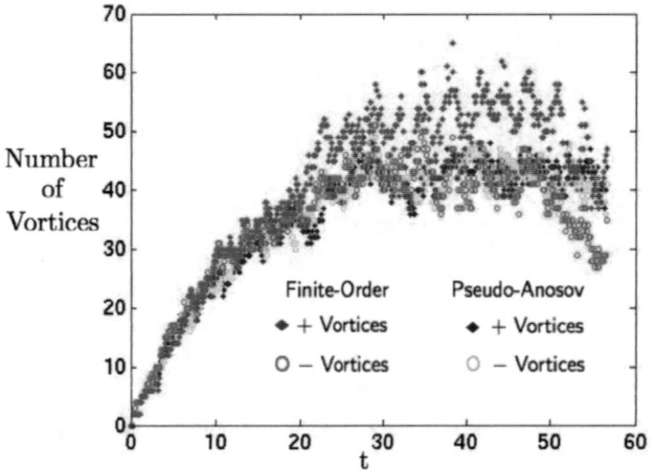

Figure 6. Evolution of number of vortices with positive and negative winding given by red diamonds and magenta circles respectively for finite-order stirring and blue diamonds and blue circles respectively for pseudo-Anosov stirring.

comparison to finite-order stirring protocols. The finite-order stirring scheme creates more sound in the condensate as result of the abrupt change in stirring direction during sequential stirring operations in this scheme. Although stirrers in both schemes follow trajectories that trace out a figure of eight shape, such abrupt changes in stirring direction do not occur in the pseudo-Anosov stirring scheme employed. Although at times $t < 30$ the total incompressible kinetic energy is initially similar for both stirring protocols, at later times more incompressible kinetic energy is generated for finite-order stirring. This difference is also reflected in the total vortex number for both stirring protocols (refer to figure 6). Finite-order stirring generates a larger imbalance in the total number of positive and negative vortices than pseudo-Anosov stirring. The oscillations in the vortex number and kinetic energy distribution for finite order stirring reflect the oscillations in the total condensate volume induced by this stirring protocol.

In conclusion, we have applied both pseudo-Anosov and finite-order stirring protocols to introduce vortices in a quasi-two dimensional Bose-Einstein condensate. Efficient mixing, as described by exponential stretching of material lines, could be thought to also translate into efficient vortex mixing. However, we observe that for a trapped inhomogeneous Bose-Einstein condensate, there is no appreciable increase in 'mixing' of vortices, as determined by the random, or clustered nature of the vortex distribution. That is, we do not find any significant correlation between if topological chaos is being introduced into the flow and how random the resulting vortex distribution is. In future work, we set up a stirring scheme that induces topological chaos in three dimensional quantum fluid flow and also analyse the onset of chaos for both finite-order and pseudo-Anosov stirring schemes.

Acknowledgments

We thank the anonymous reviewer for their insightful comments and especially for raising the alternative explanation for our observations we have discussed. This work was supported by the EPSRC. ACW acknowledges funding from EPSRC grant No. EP/H027777/1 and CFB and NPP acknowledge funding from EPSRC grant No. EP/I019413/1.

References

Araki, T., Tsubota, M. & Nemirovskii, S. K. 2002 Energy spectrum of superfluid turbulence with no normal–fluid component. *Phys. Rev. Lett.* **89**, 145301.

Baggaley, A. W. & Barenghi, C. F. 2011 Quantum turbulent velocity statistics and quasiclassical limit. *Phys. Rev. E* **84**, 067301.

BAGGALEY, A. W., LAURIE, J. & BARENGHI, C. F. 2012 Vortex-density fluctuations, energy spectra, and vortical regions in superfluid turbulence. *Phys. Rev. Lett.* **109**, 205304.

BOYLAND, P. L., AREF, H. & STREMLER, M. A. 2000 Topological fluid mechanics of stirring. *J. Fluid Mech.* **403**, 277.

BRADLEY, A. S. & ANDERSON, B. P. 2012 Energy Spectra of Vortex Distributions in Two-Dimensional Quantum Turbulence. *Phys. Rev. X* **2**, 041001.

DENNIS, G. R., HOPE, J. J. & JOHNSSON, M. T. 2013 Xmds2: Fast, scalable simulation of coupled stochastic partial differential equations. *Computer Physics Communications* **184** (1), 201 – 208.

FINN, M. D., COX, S. M. & BYRNE, H. M. 2003 Topological chaos in inviscid and viscous mixers. *J. Fluid Mech.* **493**, 345.

GOUILLART, E., THIFFEAULT, J.-L. & FINN, M. D. 2006 Topological mixing with ghost rods. *Phys. Rev. E* **73**, 036311.

HORNG, T.-L., HSUEH, C.-H., SU, S.-W., KAO, Y.-M. & GOU, S.-C. 2009 Two-dimensional quantum turbulence in a nonuniform Bose-Einstein condensate. *Phys. Rev. A* **80**, 023618.

KOBAYASHI, M. & TSUBOTA, M. 2005 Kolmogorov Spectrum of Superfluid Turbulence: Numerical Analysis of the Gross-Pitaevskii Equation with a Small-Scale Dissipation. *Phys. Rev. Lett.* **94**, 065302.

MARCUS, P. S. 1988 Numerical simulation of Jupiter's great red spot. *Nature* **331**, 693.

MAURER, J. & TABELING, P. 1998 Local investigation of superfluid turbulence. *Europhys. Lett.* **43**, 29.

NORE, C., ABID, M. & BRACHET, M. E. 1997 Kolmogorov Turbulence in Low-Temperature Superflows. *Phys. Rev. Lett.* **78**, 3896.

NUMASATO, R. & TSUBOTA, M. 2010 Possibility of inverse energy cascade in two-dimensional quantum turbulence. *J. Low Temp. Phys.* **158**, 415.

NUMASATO, R., TSUBOTA, M. & L'VOV, V. S. 2010 Direct energy cascade in two-dimensional compressible quantum turbulence. *Phys. Rev. A* **81**, 063630.

ONSAGER, L. 1949 *Nuovo Cimento (suppl.)* **6**, 279.

PARKER, N. G. & ADAMS, C. S. 2005 Emergence and decay of turbulence in stirred atomic Bose-Einstein condensates. *Phys. Rev. Lett.* **95**, 145301.

REEVES, M. T., ANDERSON, B. P. & BRADLEY, A. S. 2012*a* Classical and quantum regimes of two-dimensional turbulence in trapped Bose-Einstein condensates. *Phys. Rev. A* **86**, 053621.

REEVES, M. T., BILLAM, T. B., ANDERSON, B. P. & BRADLEY, A. S. 2012*b* Inverse Energy Cascade in Forced 2D Quantum Turbulence ArXiv:1209.5824 [cond-mat.quant-gas].

RIPLEY, B. D. 1976 The second-order analysis of stationary point processes. *Biophysical Journal* **13**, 255.

RIPLEY, B. D. 1977 Modelling spatial patterns. *J. Royal. Stat. Soc. B* **39**, 172.

SALORT, J., BAUDET, C., CASTAING, B., CHABAUD, B., DAVIAUD, F., DEDELOT, T., DIRIBARNE, P., DUBRULLE, B., GAGNE, Y., GAUTHIER, F., GIRARD, A., HÉBRAL, B., ROUSSET, B., THIBAULT, P. & ROCHE, P.-E. 2010 Turbulent velocity spectra in superfluid flows. *Phys. Fluids* **22**, 125102.

SOMMERIA, J., MEYERS, S. D. & SWINNEY, H. L. 1988 Laboratory simulation of Jupiter's great red spot. *Nature* **331**, 689.

STALP, S. R., SKRBEK, L. & DONNELLY, R. J. 1999 Decay of Grid Turbulence in a Finite Channel. *Phys. Rev. Lett.* **82**, 4831.

WHITE, A. C., BARENGHI, C. F. & PROUKAKIS, N. P. 2012 Creation and characterization of vortex clusters in atomic Bose-Einstein condensates. *Phys. Rev. A.* **86**, 103635.

Torus quantum vortex knots in the Gross-Pitaevskii model for Bose-Einstein condensates

D. Proment[1], M. Onorato[2,3] & C.F. Barenghi[4]

[1] School of Mathematics, University of East Anglia, Norwich Research Park, Norwich, NR4 7RJ, United Kingdom
[2] Dipartimento di Fisica, Università degli Studi di Torino, Via Pietro Giuria 1, 10125 Torino, Italy
[3] INFN, Sezione di Torino, Via Pietro Giuria 1, 10125 Torino, Italy
[4] School of Mathematics and Statistics, Newcastle University, Newcastle upon Tyne, NE1 7RU, United Kingdom

E-mail: davideproment@gmail.com

Abstract. We examine the static and dynamic properties of quantum knots in a Bose-Einstein condensate. In particular, we consider the Gross-Pitaevskii model and revise a technique to construct *ab initio* the condensate wave-function of a generic torus knot. After analysing its excitation energy, we study its dynamics, relating the topological parameter to its translational velocity and characteristic size. We also investigate the breaking mechanisms of non shape-preserving torus knots, confirming an evidence of universal decaying behaviour previously observed.

1. Introduction

A Bose-Einstein condensate (BEC) is a peculiar state of matter that manifests macroscopically the effects of quantum mechanics. Essentially, when a system of weakly interacting bosons is cooled to very low temperatures, the great majority of them occupies the single particle *ground state*, i.e. the lowest energy quantum mechanical level. Then, an *order parameter* complex field ψ, usually called condensate wave-function, can be used to describe the collective behaviour of all bosons. Predicted by Bose and Einstein in the 1920s, BECs were first created experimentally by cooling specially confined alkali atom gases in 1995 [Anderson *et al.* (1995); Davis *et al.* (1995)]. Nowadays, a condensate can be obtained using also many other systems of interacting bosons like magnons [Demokritov *et al.* (2006)], exciton polaritons [Kasprzak *et al.* (2006)], and photons [Klaers *et al.* (2010)].

Among many models used to describe the dynamics of BECs, without any doubt the most simple one is the Gross-Pitaevskii equation (GPE) [Pitaevskii & Stringari (2003)]

$$i\hbar\frac{\partial\psi(\mathbf{r},t)}{\partial t} = -\frac{\hbar^2}{2m}\nabla^2\psi(\mathbf{r},t) + \frac{4\pi\hbar^2 a_s}{m}|\psi(\mathbf{r},t)|^2\psi(\mathbf{r},t) + V_{ext}(\mathbf{r},t)\,\psi(\mathbf{r},t)\,, \tag{1}$$

where ψ denotes the condensate wave-function in space and time, m and a_s are the mass and the s-wave scattering length of the boson respectively, V_{ext} is an external potential acting on the condensate, and $\hbar$ is the reduced Planck constant. This nonlinear partial differential equation, also known in other research fields as the nonlinear Schrödinger equation, mimics well the BEC dynamics in the case of a very dilute gas of bosons in the limit of zero temperature, that is to say when almost all bosons are in the ground state. It is well known that the GPE is not integrable in more than one spatial dimension and so no generic analytical solutions may be provided.

For simplicity we will consider in the following no external potential acting on the system, i.e. $V(\mathbf{r},t)_{ext} \equiv 0$, and the same non-dimensional version of GPE chosen in Berloff (2004). By rescaling the wave-function using the uniform density at infinity as $\psi \to \sqrt{\rho_\infty}\,\psi$, time as $t \to m/(8\pi\hbar\, a_s\, \rho_\infty)\, t$, and lengths as $\mathbf{r} \to \xi\, \mathbf{r}$, equation (1) results in

$$i\,\frac{\partial\psi(\mathbf{r},t)}{\partial t} + \frac{1}{2}\nabla^2\psi(\mathbf{r},t) - \frac{1}{2}|\psi(\mathbf{r},t)|^2\psi(\mathbf{r},t) = 0\,, \quad \text{with} \quad \xi = \sqrt{\frac{1}{8\pi\, a_s\, \rho_\infty}}\,. \tag{2}$$

We underline that ξ, usually called the *healing length* or *coherence length*, is the only characteristic length of the system. It turns out to be only a function of the boson scattering length and the uniform density at infinity of the BEC, or equivalently the total number of bosons per the total volume when only few localised fluctuations on a uniform solution are present.

It is straightforward to prove that a BEC modelled by the GPE is also a superfluid, that is to say it is a fluid having no viscosity. Indeed, one may use the Madelung transformation $\psi(\mathbf{r},t) = \sqrt{\rho(\mathbf{r},t)}\,\exp i\theta(\mathbf{r},t)$ to rewrite the complex order parameter wave-function in terms of two real fields $\rho(\mathbf{r},t)$ and $\theta(\mathbf{r},t)$. Substituting this into equation (2), splitting real and imaginary parts, and defining the vector field $\mathbf{v}(\mathbf{r},t) = \nabla\theta(\mathbf{r},t)$, one obtains the following system of equations:

$$\begin{aligned} &\frac{\partial\rho}{\partial t} + \nabla\cdot(\rho\,\mathbf{v}) = 0\,, \\ &\frac{\partial\mathbf{v}}{\partial t} + (\mathbf{v}\cdot\nabla)\,\mathbf{v} = -\frac{\nabla\rho}{2} + \nabla\left(\frac{\nabla^2\sqrt{\rho}}{2\sqrt{\rho}}\right)\,. \end{aligned} \tag{3}$$

These are nothing but the continuity and linear momentum conservation equations for a barotropic, compressible, inviscid, and irrotational fluid where the two fields $\rho(\mathbf{r},t)$ and $\mathbf{v}(\mathbf{r},t)$ are simply the fluid density and velocity. The last term in the latter equation is called the *quantum stress tensor*: it becomes important at scales of the order of the healing length and is the key difference with the classical Euler equation.

Even though the superfluid BEC is irrotational, vortical structures called quantized vortices can appear as the wave-function ψ is still single-valued if the phase field θ varies by a multiple of 2π. Indeed, the circulation $\mathcal{C}$ around the closed curve γ results in

$$\mathcal{C} = \oint_\gamma \mathbf{v}\cdot d\mathbf{l} = \oint_\gamma \nabla\theta\cdot d\mathbf{l} = \Delta\theta = \pm n\, 2\pi\,, \quad \text{with} \quad n \in \mathcal{N}\,. \tag{4}$$

If $n \neq 0$, the circulation takes non-zero quantized values $\pm n\,\kappa$, being $\kappa = 2\pi$ the quantum of circulation, and for Stokes' theorem the domain inside γ is no more simply connected. In two-dimensional slices of the three-dimensional space these quantized vortices take the form of zeros of the density field (the density goes regularly to zero in the vicinity of the defect) where the phase twists by $\pm\kappa$. In three dimensions the quantized vortices form closed rings or lines that end at the system boundaries. The vortex lines interact with each other similarly to steady current-carrying wires, may form helicoidal Kelvin waves and can eventually reconnect rearranging their topology.

Vortex lines have been widely observed in BECs forming arrays under rotation [Matthews *et al.* (1999); Madison *et al.* (2000); Abo-Shaeer *et al.* (2001) or interacting and decaying in quasi two-dimensional geometries [Weiler *et al.* (2008); Neely *et al.* (2010)]. Vortex ring observation has been on the contrary quite elusive so far [Anderson *et al.* (2001)] as the visualisation techniques used in BECs do not allow for a clear measure of the density and velocity fields and in most cases are destructive measurements. Quantum vortices have also been experimentally observed in superlfuid Helium II [Donnelly (1991)] and even reconnection events have been directly detected using hydrogen particle tracers [Bewley *et al.* (2008)].

More topologically complex quantum vortices like knots have not yet been detected in a superfluid. Several theoretical and numerical studies have been carried out to understand the stability and dynamics of knots using the Biot-Savart model and its local induction approximation (LIA) [Donnelly (1991)]. Only recently we have been able to create an *ab initio* wave-function describing the simpler vortex knots in the GPE [Proment *et al.* (2012)]. This provided a basis for studying the shape preserving properties and dynamics of topological structures in BECs. Very similar results regarding trefoil knot dynamics and breaking mechanisms were observed experimentally in classical fluid where a specifically-designed airfoil was created to produce the vortex knot [Kleckner & Irvine (2013)].

In this contribution we extend our previous work on knots in the GPE, analysing the energy of simpler torus knots, reporting new results on more topologically complicated knots and presenting open questions and future perspectives in this field.

2. The generic torus vortex knot $T_{p,q}$ wave-function

In Proment *et al.* (2012) we proposed a technique to generate the initial wave-function of a trefoil and its dual knot based on the fact that several knots may be built as closed lines on a torus. Indeed, by twisting a curve p and q times around the toroidal and poloidal circles of the torus before tying it the line, one can build topologically non-trivial curves defined as $\mathcal{T}_{p,q}$. The easier example is the trefoil knot which is the $\mathcal{T}_{2,3}$ knot and its dual, the $\mathcal{T}_{3,2}$ knot.

We may ask "how is it possible to create a quantum vortex knot $\mathcal{T}_{p,q}$ on a torus axisymmetric with respect to the z axis set at the origin of a cartesian three-dimensional space?". The main idea is to notice that each plane sOz perpendicular to the z axis of the torus identifies a particular slice of the three-dimensional space where $2q$ two-dimensional quantum vortices are present. In particular, q vortices lie on the circumference of radius R_1 centered at $(-R_0, 0)$ and the other q vortices lie on the circumference of radius R_1 centered at $(R_0, 0)$, given R_0 and R_1 the toroidal and poloidal radii of the torus respectively.

In Berloff (2004) a Padé approximation for a two-dimensional wave-function $\Psi_{2D}(s-s_0, z-z_0)$ describing a single quantum vortex centered at (s_0, z_0) is given; moreover, to create a quantum vortex with opposite circulation it is sufficient to apply the complex conjugate operator $(\cdot)^*$ to the wave-function. To the first approximation, one can consider the two-dimensional wave-function describing $2q$ quantum vortices just as a superposition of the single vortex solutions, that is to say the multiplication of the single vortex wave-functions Ψ_{2D}. Thus, one can come back to the three-dimensional space knowing the map $s(x, y)$ and controlling the position of the vortices in each slice sOz so that the three-dimensional quantum vortex line tie into the desired $\mathcal{T}_{p,q}$ knot. More specifically, the three-dimensional wave-function is given by

$$\begin{aligned}\psi_{p,q}(x,y,z) = &\prod_{i=1}^{q} \Psi_{2D}\left\{ s(x,y) - R_0 - R_1\cos\left[\alpha(x,y) + i\frac{2\pi\, p}{q}\right], z - R_1 \sin\left[\alpha(x,y) + i\frac{2\pi\, p}{q}\right]\right\} \\ \times &\prod_{i=1}^{q} \Psi^*_{2D}\left\{ s(x,y) + R_0 + R_1\cos\left[\alpha(x,y) + i\frac{2\pi\, p}{q}\right], z - R_1 \sin\left[\alpha(x,y) + i\frac{2\pi\, p}{q}\right]\right\},\end{aligned} \tag{5}$$

where

$$s(x,y) = \mathrm{sgn}(x)\sqrt{x^2+y^2} \quad \text{and} \quad \alpha(x,y) = \frac{q\,\mathrm{atan2}(x,y)}{p}. \tag{6}$$

Journal of Physics: Conference Series **544** (2014) 012022 doi:10.1088/1742-6596/544/1/012022

Here sgn(·) is the sign function and atan2(·, ·) is the arctangent function with two argument that gather the information on the signs of the inputs in order to return the appropriate quadrant of the computed angle.

Quantum vortices are easily detectable as points where the density field $\rho(x, y, z)$ vanishes. In a system where the mean spatial density over the entire volume is $\bar{\rho}$ it is usually sufficient to plot the iso-surfaces corresponding to the threshold $\rho_{th} = 0.2\,\bar{\rho}$ to show vortex defects and to filter out eventual sound density waves. By using equation (5) and setting $R_0 = 20$ and $R_1 = 4$ we can plot the iso-surfaces at ρ_{th} for various knots, as shown in Figure 1, and use their wave-function as initial condition to study their dynamics.

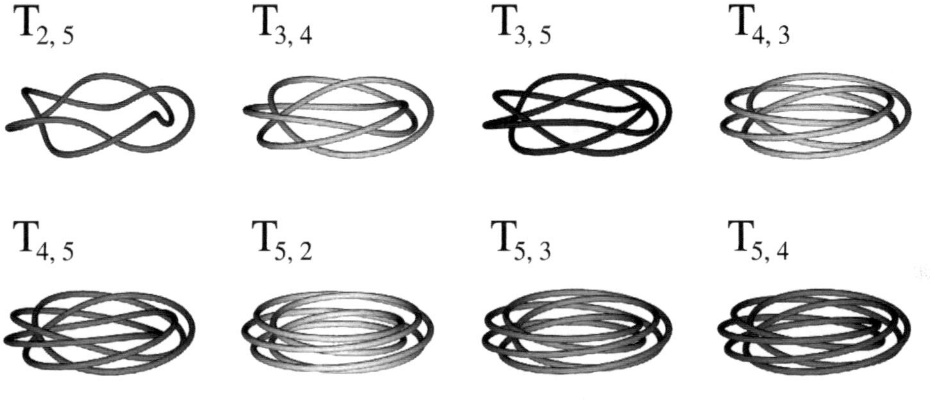

Figure 1. Iso-surface density plots at $\rho_{th} = 0.2\,\bar{\rho}$ showing the initial configuration of different quantum knots $\mathcal{T}_{p,q}$ based on the wave-function (5) having $R_0 = 20$ and $R_1 = 4$ (lengths are expressed in healing length units).

3. Vortex knot energy

The above structures modeled by GPE represent collective excitations of the superfluid. The non-dimensional GPE (2) has two conserved quantities which are the total number of bosons N and the total energy E, resulting in the following integrals over the space volume V:

$$N = \int_V |\psi(\mathbf{r}, t)|^2 dV \quad \text{and} \quad E = \int_V \left(\frac{1}{2}|\nabla\psi(\mathbf{r}, t)|^2 + \frac{1}{4}|\psi(\mathbf{r}, t)|^4 \right) dV\,, \tag{7}$$

where $dV = dxdydz$ and $\mathbf{r} = (x, y, z)$. Set the mean density $\bar{\rho} = N/V$, one can identify the energy corresponding to an excitation given its wave-function. It is well-know that in the absence of an external potential the ground state wave-function for the GPE (2) is the uniform density solution $\psi_{gs}(x, y, z) = \sqrt{\bar{\rho}}\exp\left[-i(\bar{\rho}/2\,t + \alpha_0)\right]$, where $\alpha_0 \in [0, 2\pi)$ is an overall phase that depends on the initial conditions. Thus, the excitation energy of a particular structure with respect to the ground state is simply

$$\begin{aligned} E_{exc} &= \int_V \left(\frac{1}{2}|\nabla\psi(\mathbf{r}, t)|^2 + \frac{1}{4}|\psi(\mathbf{r}, t)|^4 \right) dV - \int_V \left(\frac{1}{2}|\nabla\psi_{gs}(\mathbf{r}, t)|^2 + \frac{1}{4}|\psi_{gs}(\mathbf{r}, t)|^4 \right) dV \\ &= \int_V \left(\frac{1}{2}|\nabla\psi(\mathbf{r}, t)|^2 + \frac{1}{4}|\psi(\mathbf{r}, t)|^4 \right) dV - \frac{\bar{\rho}^2\,V}{4}\,. \end{aligned} \tag{8}$$

One may then ask what is the excitation energy of a generic $\mathcal{T}_{p,q}$ torus knot. We have addressed this problem numerically by considering a trefoil $\mathcal{T}_{2,3}$ knot and its dual $\mathcal{T}_{3,2}$ knot positioned at the centre of a finite cubic volume of side $L = 441\,\Delta x$ (the numerical algorithms use a uniform grid spacing $\Delta x = \xi/2$ and 441 points are selected to increase efficiency of the cosine transforms). Results obtained by considering R_0 and R_1 as parameters are presented in Figures 2 and 3 corresponding to the wave-function (5) of the $\mathcal{T}_{2,3}$ and $\mathcal{T}_{3,2}$ torus knots respectively. We

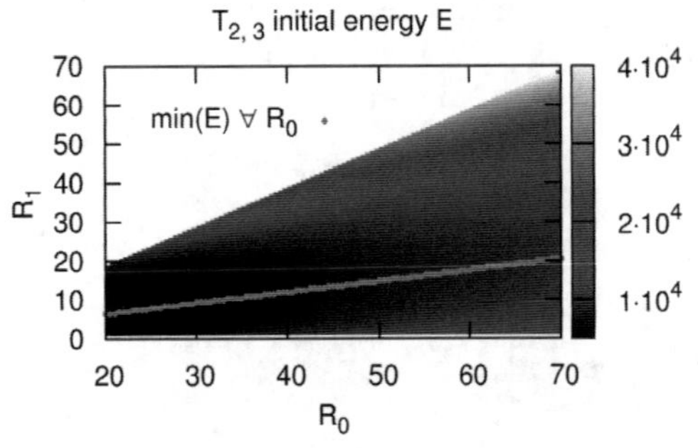

Figure 2. Computed excitation energy (8) for the quantum vortex knot $\mathcal{T}_{2,3}$ as a function of the torus radii R_0 and R_1. The red points are the energy minima evaluated by fixing the toroidal radius R_0 (lengths are expressed in healing length units).

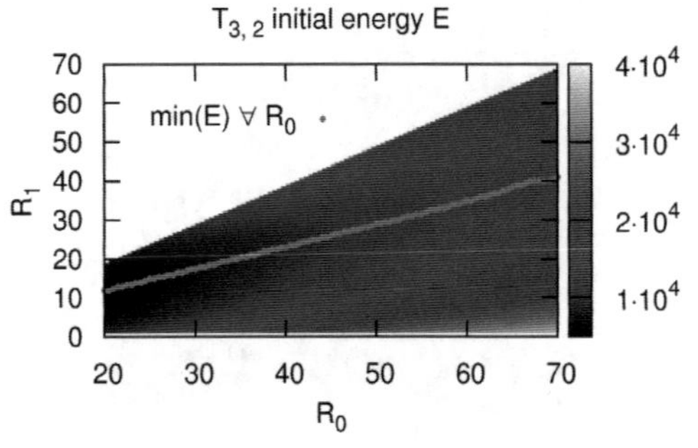

Figure 3. Computed excitation energy (8) for the quantum vortex knot $\mathcal{T}_{3,2}$ as a function of the torus radii R_0 and R_1. The red points are the energy minima evaluated by fixing the toroidal radius R_0 (lengths are expressed in healing length units).

analyzed the parameter space $R_0 \in [20; 70]$ and $R_1 \in [1; R_0 - 1]$ in order to have always well defined tori with $R_1 < R_0$ and knots always smaller than two thirds of the system size to limit boundary effects. We observe that in both cases the energy is a monotonically function of the toroidal radius R_0. We have evaluated the energy minima fixing R_0 and plotted as red points. It is evident that in both cases the minima form a straight line in the parameter plane: our best fits give

$$\begin{aligned} R_1^{(2,3)} &= (0.2816 \pm 0.0010)R_0 + (0.807 \pm 0.047) \\ R_1^{(3,2)} &= (0.5811 \pm 0.0022)R_0 + (0.05 \pm 0.10) \end{aligned} \tag{9}$$

for the $\mathcal{T}_{2,3}$ and $\mathcal{T}_{3,2}$ cases respectively. We plot the energy minimum values as a function of R_0 in Figure 4 in order to compare between the two cases. For fix values of R_0, the energy of the

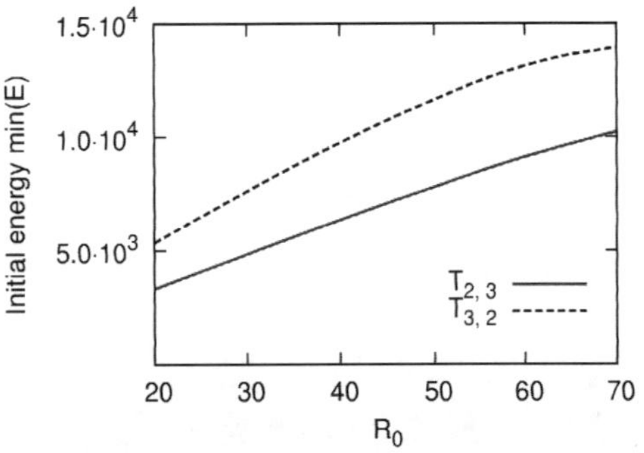

Figure 4. Computed excitation energy minima with respect to R_0 in the range of second parameter $R_1 \in [1; R_0 - 1]$ for the $\mathcal{T}_{2,3}$ and $\mathcal{T}_{3,2}$ knots - R_1 is in this interval in order to have always a properly defined torus (lengths are expressed in healing length units).

$\mathcal{T}_{2,3}$ knot appears to be much less with respect to the $\mathcal{T}_{3,2}$ knot one: this is somehow expected as the former propagates much slower than the latter [Proment *et al.* (2012)]. One could for completeness obtain numerically the momentum-energy relations for different $\mathcal{T}_{p,q}$ knots and compare them to the same relation for the perfect vortex rings [Jones & Roberts (1982)] but is out of our present scope.

4. Knot dynamics

The easiest measurable quantities giving information about the knot dynamics are its centre of mass $\mathbf{r}_{CM}(t) = [x_{CM}(t), y_{CM}(t), z_{CM}(t)]$ and its characteristic size $\mathbf{D} = (D_x, D_y, D_z)$. These can be defined as

$$\mathbf{r}_{CM}(t) = \frac{\int_V \mathbf{r}\, H(\rho_{th} - |\psi(\mathbf{r},t)|^2)\, dV}{\int_V H(\rho_{th} - |\psi(\mathbf{r},t)|^2)\, dV} \tag{10}$$

and

$$\mathbf{D}(t) = \max_{\mathbf{r},\mathbf{r}' \in V} \left[\mathbf{r}\, H(\rho_{th} - |\psi(\mathbf{r},t)|^2) - \mathbf{r}'\, H(\rho_{th} - |\psi(\mathbf{r}',t)|^2)\right] \tag{11}$$

respectively, being $\mathbf{r}' = (x', y', z')$, $H(\cdot)$ the Heaviside step function and $\rho_{th} = 0.2\,\bar{\rho}$.

Torus knots propagate essentially along the torus axis (here the z axis) twisting around it. Thus, the translation can be quantified measuring the evolution of the centre of mass component z_{CM}, which is plotted in Figure 5, for different torus knots. It appears evident that $\mathcal{T}_{p,q}$ knots

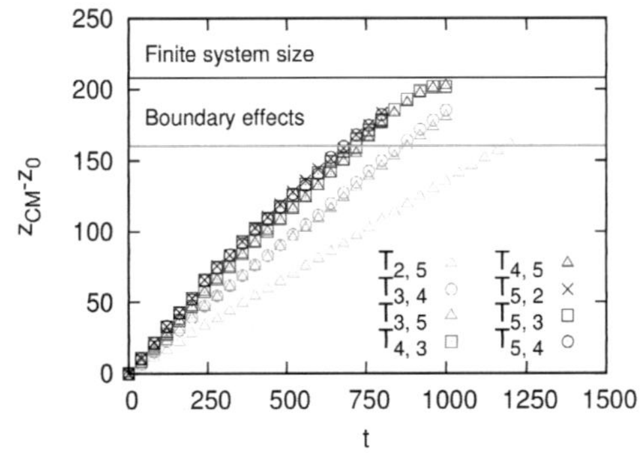

Figure 5. Evolution of the displacement in the axisymmetric torus z direction of different $\mathcal{T}_{p,q}$ knots (lengths are expressed in healing length units). Knots having the same number of p twists are shown using the same colour, while the same symbol is used to underline the same number of q twists. Estimated boundary and finite size effects are also shown.

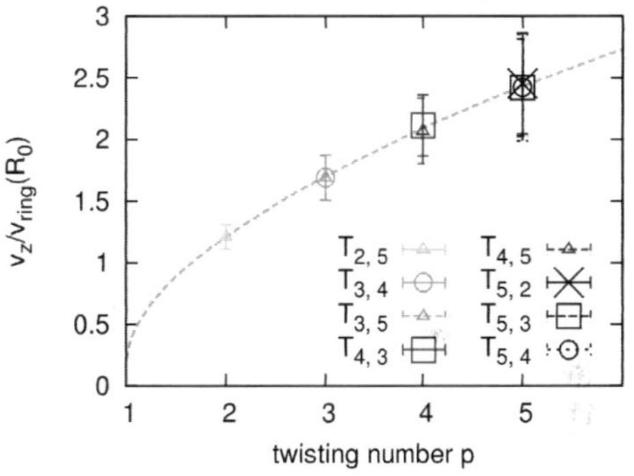

Figure 6. Estimated z velocity component v_z normalised using $v_{\text{ring}}(R_0)$ for different $\mathcal{T}_{p,q}$ knots. Knots having the same number of p twists are shown using the same colour, while the same symbol is used to underline the same number of q twists. The grey dashed line is the best fit with the function $f(p) = (p-1)^a + c$, with $a = 0.5747 \pm 0.0026$ and $c = 0.210 \pm 0.011$.

with equal twisting number p (here plotted using the same colour) have the same motion along the z axis; moreover, to a first approximation, this motion is linear. The estimated velocity component v_z for different $T_{p,q}$ knots is shown in Figure 6, normalised by the velocity of a vortex ring of radius R_0. Following Donnelly (1991), the ring velocity is

$$v_{\text{ring}}(R) = \frac{\kappa}{4\pi R}\left[\ln\left(\frac{8R}{\xi}\right) - 0.615\right], \tag{12}$$

where we have set $\xi = 1$ and $\kappa = 2\pi$ to be coherent with our non-dimensional GPE (2). Confirming previous observations, it results that the torus knot translational velocity depends only on the twisting number p, being for toroidal and poloidal radi $R_0 = 20$ and $R_1 = 4$ always greater than $v_{\text{ring}}(R_0)$. Moreover, we found that the behaviour of $v_z/v_{\text{ring}}(R_0)$ with respect to the twisting number p is very well fitted by the function $f(p) = (p-1)^a + c$, plotted with dashed grey line, with $a = 0.5747 \pm 0.0026$ and $c = 0.210 \pm 0.011$ coefficients.

During the evolution the characteristic knot size visibly varies. In Figure 7 we plot the characteristic size component D_z for the different knots analysed, rescaled with their respective initial size. It appears that for every knots considered the rescaled size component D_z oscillates

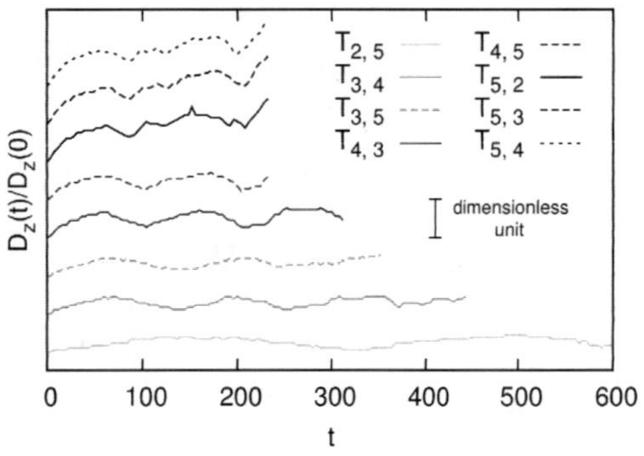

Figure 7. Evolution of the characteristic size D_z rescaled with the corresponding initial size for different quantum vortex knots. A dimensionless unit is also plotted in order to appreciate the fluctuations of around 25% of each rescaled size.

(around 25%) with a characteristic frequency that seems to depend only on the twisting number p. We underline that especially for knots having high twisting number p, $\mathcal{T}_{5,q}$ for instance, the characteristic size D_z also grows in time.

5. Decaying mechanisms

In Proment *et al.* (2012) we have shown that initial $\mathcal{T}_{2,3}$ and $\mathcal{T}_{3,2}$ torus knots having $R_0 = 20$ and $R_1 \geq 8$ do not preserve their shape during the evolution and decay after several reconnection events. The decay product is a set of vortex rings which have different radii and carry Kelvin waves on them. In general we observed that knots having twisting number p produce (at least) p rings, with a decay mechanism that seems to be peculiar to the initial topology.

We report here a mere observation on how the more topologically complicated $\mathcal{T}_{p,q}$ torus knots shown in Figure 1 having $R_0 = 20$ and $R_1 = 12$ break (visit `http://youtu.be/whxqwdp_ogA` for a complete movie). In all knots tested we observe always two phenomena which we believe to be universal. First, the final products of the breaking event(s) are p vortex rings, which may eventually produce even more rings if their Kelvin waves have enough amplitude to trigger single vortex ring self-reconnections. This is somehow expected, as, far enough from the vortex knot, the initial phase field plotted in a generic plane sOz is similar to the one having p vortex rings axisymetric to the z axis and sufficiently close together. Second, all self-reconnections which break up the initial $\mathcal{T}_{p,q}$ knot by producing the final p rings, are always simultaneous (this may be a consequence of the symmetry of the initial condition) and q in number.

6. Open questions and future perspectives

Many open questions remain on BEC quantum knot static and dynamical properties. We will address in the following the more interesting ones and list what are the future perspectives in this field.

Understanding the stability of knotted structures is probably the most difficult mathematical issue. Many studies have been carried out using the local induction approximation of the Biot-Savart model and recently it has been shown that a $\mathcal{T}_{p,q}$ torus knot with $p > q > 1$ are unstable, while in the case $q > p$ it could be neutrally stable and quickly changes its knot type during the evolution [Calini & Ivey (2011); Calini *et al.* (2011)]. However, it was noted numerically that the full Biot-Savart model has a stabilization effect allowing unstable knots to travel for distances longer with respect to their characteristic size [Ricca *et al.* (1999); Maggioni *et al.* (2010)]. As far as we are aware, no mathematical studies on stability have ever been attempted using the GPE. Is the $\mathcal{T}_{p,q}$ torus knot for a particular set of (p, q) a Hamiltonian or shape-preserving structure?

Regarding non shape-preserving knots, it has been observed that breaking mechanisms seem to follow specific universal rules. At present, only torus knots were investigated but many other examples can be studied including much more complicated topologies like other knots and links. What are the breaking mechanisms for a generic initial topology? What are the intermediate breaking products and the type and number of final ones?

Finally, one would like to understand what are the possibilities to find such kind of topologically non-trivial vortex structures in reality. Can they occur naturally during (or after) a turbulent decay? Otherwise, what are the experimental techniques in superfluids to create them in a controlled and reproducible way? What kind of detecting techniques may be used to detect knotted and linked structures?

Hopefully, in the following years some of these questions will find an answer and new advances in this field may help to solve open problems in related disciplines like classical and magneto-hydrodynamics. For instance, during the final drafting of this work, two different manuscripts addressing helicity conservation during reconnection events have appeared online as preprints [Baggaley (2014); Scheeler *et al.* (2014)].

Acknowledgements
The authors acknowledge the Isaac Newton Institute for Mathematical Sciences in Cambridge and the Julian Schwinger Foundation for supporting there the *Topological Dynamics in the Physical and Biological Sciences* program. D. P. is grateful to A. Baggaley, M. Dennis, D. Kleckner, W. Irvine, D. Maestrini, R. Ricca, and H. Salman for fruitful suggestions and discussions. The research presented in this paper was carried out on the High Performance Computing Cluster supported by the Research and Specialist Computing Support service at the University of East Anglia.

References

Abo-Shaeer, J. R., Raman, C., Vogels, J. M. & Ketterle, W. 2001 Observation of Vortex Lattices in Bose-Einstein Condensates. *Science* **292** (5516), 476–479.

Anderson, B. P., Haljan, P. C., Regal, C. A., Feder, D. L., Collins, L. A., Clark, C. W. & Cornell, E. A. 2001 Watching Dark Solitons Decay into Vortex Rings in a Bose-Einstein Condensate. *Phys. Rev. Lett.* **86** (14), 2926–2929.

Anderson, M. H., Ensher, J. R., Matthews, M. R., Wieman, C. E. & Cornell, E. A. 1995 Observation of Bose-Einstein condensation in a dilute atomic vapor. *Science* **269** (5221), 198–201.

Baggaley, A. W. 2014 Helicity transfer during quantised vortex reconnection. *ArXiv e-prints* , 1403.8121.

Berloff, N. 2004 Padé approximations of solitary wave solutions of the Gross-Pitaevskii equation. *Journal of Physics A: Mathematical and General* **37**, 1617–1632(16).

Bewley, G. P., Paoletti, M. S., Sreenivasan, K. R. & Lathrop, D. P. 2008 Characterization of reconnecting vortices in superfluid helium. *Proceedings of the National Academy of Sciences* **105** (37), 13707–13710.

Calini, A. & Ivey, T. 2011 Stability of small-amplitude torus knot solutions of the localized induction approximation. *Journal of Physics A: Mathematical and Theoretical* **44** (33).

Journal of Physics: Conference Series **544** (2014) 012022 doi:10.1088/1742-6596/544/1/012022

Calini, A., Keith, S. F. & Lafortune, S. 2011 Squared eigenfunctions and linear stability properties of closed vortex filaments. *Nonlinearity* **24** (12).

Davis, K., Mewes, M.-O., Andrews, M. v., Van Druten, N., Durfee, D., Kurn, D. & Ketterle, W. 1995 Bose-Einstein condensation in a gas of sodium atoms. *Physical Review Letters* **75** (22), 3969.

Demokritov, S., Demidov, V., Dzyapko, O., Melkov, G., Serga, A., Hillebrands, B. & Slavin, A. 2006 Bose–Einstein condensation of quasi-equilibrium magnons at room temperature under pumping. *Nature* **443** (7110), 430–433.

Donnelly, R. 1991 **Quantized vortices in Helium II.** , vol. 3. Cambridge Univ Pr.

Jones, C. A. & Roberts, P. H. 1982 Motions in a bose condensate. iv. axisymmetric solitary waves. *Journal of Physics A: Mathematical and General* **15** (8), 2599.

Kasprzak, J., Richard, M., Kundermann, S., Baas, A., Jeambrun, P., Keeling, J., Marchetti, F., Szymańska, M. H., André, R., Staehli, J. L., Savona, V., Littlewood, P. B., Deveaud, B. & Dang, L. S. 2006 Bose–Einstein condensation of exciton polaritons. *Nature* **443** (7110), 409–414.

Klaers, J., Schmitt, J., Vewinger, F. & Weitz, M. 2010 Bose-Einstein condensation of photons in an optical microcavity. *Nature* **468** (7323), 545–548.

Kleckner, D. & Irvine, W. T. M. 2013 Creation and dynamics of knotted vortices. *Nature Physics* **9** (4), 253–258.

Madison, K. W., Chevy, F., Wohlleben, W. & Dalibard, J. 2000 Vortex Formation in a Stirred Bose-Einstein Condensate. *Phys. Rev. Lett.* **84** (5), 806–809.

Maggioni, F., Alamri, S., Barenghi, C. F. & Ricca, R. L. 2010 Velocity, energy, and helicity of vortex knots and unknots. *Phys. Rev. E* **82**, 026309.

Matthews, M. R., Anderson, B. P., Haljan, P. C., Hall, D. S., Wieman, C. E. & Cornell, E. A. 1999 Vortices in a Bose-Einstein Condensate. *Phys. Rev. Lett.* **83** (13), 2498–2501.

Neely, T. W., Samson, E. C., Bradley, A. S., Davis, M. J. & Anderson, B. P. 2010 Observation of Vortex Dipoles in an Oblate Bose-Einstein Condensate. *Phys. Rev. Lett.* **104** (16), 160401.

Pitaevskii, L. & Stringari, S. 2003 **Bose-Einstein Condensation.** , vol. 116. Oxford University Press, USA.

Proment, D., Onorato, M. & Barenghi, C. F. 2012 Vortex knots in a Bose-Einstein condensate. *Phys. Rev. E* **85**, 036306.

Ricca, R., Samuels, D. C. & Barenghi, C. F. 1999 Evolution of vortex knots. *J. Fluid Mech.* **391**, 29–44.

Scheeler, M. W., Kleckner, D., Proment, D., Kindlmann, G. L. & Irvine, W. T. M. 2014 Helicity conservation in topology-changing reconnections: the flow of linking and coiling across scales. *ArXiv e-prints* , 1404.6513.

Weiler, C. N., Neely, T. W., Scherer, D. R., Bradley, A. S., Davis, M. J. & Anderson, B. P. 2008 Spontaneous vortices in the formation of Bose-Einstein condensates. *Nature* **455** (7215), 948–951.

Journal of Physics: Conference Series **544** (2014) 012023 doi:10.1088/1742-6596/544/1/012023

Quantum turbulence in atomic Bose-Einstein condensates

A. J. Allen, N. G. Parker, N. P. Proukakis, and C. F. Barenghi.

Joint Quantum Centre (JQC) Durham-Newcastle, School of Mathematics and Statistics, Newcastle University, Newcastle upon Tyne NE1 7RU, England, UK.

E-mail: joy.allen@ncl.ac.uk, nick.parker@ncl.ac.uk, nikolaos.proukakis@ncl.ac.uk, carlo.barenghi@ncl.ac.uk

Abstract.
Weakly interacting, dilute atomic Bose-Einstein condensates (BECs) have proved to be an attractive context for the study of nonlinear dynamics and quantum effects at the macroscopic scale. Recently, weakly interacting, dilute atomic BECs have been used to investigate quantum turbulence both experimentally and theoretically, stimulated largely by the high degree of control which is available within these quantum gases. In this article we motivate the use of weakly interacting, dilute atomic BECs for the study of turbulence, discuss the characteristic regimes of turbulence which are accessible, and briefly review some selected investigations of quantum turbulence and recent results. We focus on three stages of turbulence - the generation of turbulence, its steady state and its decay - and highlight some fundamental questions regarding our understanding in each of these regimes.

1. Introduction

Turbulence in ordinary fluids such as air or water consists of rotational eddies of different sizes which we term vortices. Vortices therefore are a hallmark signature of a turbulent flow (Barenghi *et al.*, 2001). In superfluids, quantum vortices differ from their classical counterparts because of the quantization of circulation. This means that the rotational motion of a superfluid is constrained to discrete vortices which all have the same core structure. Turbulence in superfluid helium has been the subject of many recent experimental and theoretical investigations (Skrbek & Sreenivasan, 2012). Recently, experimentalists have been able to visualise individual vortex lines and reconnection events using tracer particles (Fonda *et al.*, 2014). Weakly interacting, dilute atomic Bose-Einstein condensates (henceforth referred to as BECs) present a distinct platform to view and probe quantum turbulence. A key feature here is the ability to directly resolve the structure of individual vortices and in turn the dynamics of a turbulent vortex tangle (Henn *et al.*, 2009). As a result of the quantized nature of vorticity, quantum turbulence in superfluid helium and in BECs can be viewed as a simpler, idealized analog of turbulence in ordinary fluids, and opens the possibility of studying problems which may be relevant to our general understanding of turbulence.

2. Why atomic Bose-Einstein condensates?

Since their first generation in 1995 (Davis *et al.*, 1995; Anderson *et al.*, 1995), atomic BECs have been used to study a wide variety of nonlinear dynamics, for example, solitons, vortices and four-wave mixing (Kevrekidis & Carretero-Gonzalez, 2008). A merit of exploiting BECs as a testbed of nonlinear physics lies with the immense control and flexibility they offer. For example:

- *Trapping geometry, shape and dimensionality*
  Optical and magnetic fields can be employed to precisely create a potential landscape for the atoms in the BEC, which in turn enables control of the shape and effective dimensionality of the system (Görlitz *et al.*, 2001). A basic requirement of these gases is confinement in space to prevent contact with hot surfaces. This is typically provided by magnetic traps which are harmonic in shape and have the form (Pethick & Smith, 2002)

$$V_{\text{ext}}(\mathbf{r}) = \frac{1}{2} m\omega(x^2 + y^2 + z^2), \tag{1}$$

  where ω is the trapping frequency and m the mass of the atom. This type of trap results in an atomic cloud with a radial density profile which resembles an inverted parabola.
  If a harmonic trap is used which is very strongly confining in one direction, for example

$$V_{\text{ext}}(\mathbf{r}) = \frac{1}{2} m\omega(x^2 + y^2 + \epsilon z^2), \tag{2}$$

  where $\epsilon \gg 1$, the dynamics in that direction are inhibited and the system becomes effectively two-dimensional (2D). By changing ϵ, one can easily change the effective dimensionality, which is particularly important in turbulence (2D turbulence is very different from 3D turbulence). In the same way, if the trap is very tightly confining in two directions, the dynamics is mainly in the third direction, and the system is effectively one-dimensional (1D).
  More complicated trapping geometries can be realised, for example a toroidal ring or a periodic optical lattice. Traps can also be made time-dependent by rotating or shaking the trapping potential. Furthermore, one can create localized potentials using optical fields, which can mimic an obstacle and be moved through the system on demand.
- *Interaction strength*
  Typically, the dominant atomic interaction in a BEC is the short-range and isotropic s-wave interaction. Experimentalists can employ magnetic Feshbach resonances (Inouye *et al.*, 1998) to change the strength of these interactions and even their nature, i.e. whether they are attractive or repulsive (Roberts *et al.*, 1998). Furthermore, by using atoms with relatively large magnetic dipole moments, e.g. ^{52}Cr, it is possible to create a BEC where the atoms also experience significant dipole-dipole interactions, which are long-range and anisotropic, and greatly modify the static and dynamical properties of the system (Lahaye *et al.*, 2009).
- *Vortex core optical imaging*
  The healing length which characterizes the vortex core size is typically around 10^{-7}m in a BEC (c.f. 10^{-10}m in superfluid Helium). By expanding the BEC (following release from the trapping potential), the vortex can be directly imaged and resolved via optical absorption (Madison *et al.*, 2000; Raman *et al.*, 2001). Advanced real time imaging of condensates containing vortices has also recently been developed (Freilich *et al.*, 2010), allowing the precession of vortices to be observed.

In the limit of zero temperature and weak interactions, the evolution equation for the macroscopic condensate wavefunction, $\phi(\mathbf{r}, t)$, is a form of nonlinear Schrödinger equation, commonly known as the Gross-Pitaevskii Equation (GPE) (Pethick & Smith, 2002):

$$i\hbar \frac{\partial \phi(\mathbf{r}, t)}{\partial t} = \left[-\frac{\hbar^2}{2m} \nabla^2 + V_{\text{ext}}(\mathbf{r}, t) + g|\phi(\mathbf{r}, t)|^2 - \mu \right] \phi(\mathbf{r}, t), \tag{3}$$

where $\mathbf{r}$ is the position in space, t is time and $\hbar$ is Planck's constant divided by 2π. The GPE provides a good quantitative description of the dynamics of a BEC over all lengthscales available, from the vortex core to the system size, up to temperatures of approximately $T \simeq 0.5T_c$ (where the critical temperature T_c is of the order of μK in typical experiments). At the right-hand side we recognise the kinetic energy term, $(-\hbar^2/2m)\nabla^2$, the trapping potential $V_{\text{ext}}(\mathbf{r},t)$ (which in general may be time-dependent), the interaction term, $g|\phi(\mathbf{r},t)|^2$ where $g = 4\pi\hbar^2 a_s/m$ and a_s is the 3D $s-$wave scattering length, and the chemical potential μ. The GPE can be almost exactly mapped to the classical Euler equation; the small difference, namely the quantum stress, regularizes the solutions, preventing singularities which may arise in an Euler fluid (Barenghi, 2008). The GPE is a practically exact model in the limit of zero temperature, where essentially all of the atoms exist in the Bose-Einstein condensate phase. In many experiments the condensate exists at well below the BEC transition temperature such that this approximation is justified. Extensions of the GPE to include the effect of thermal atoms provide a more complete (albeit not exact) physical model of a real BEC (Jackson *et al.*, 2009) (see e.g. Proukakis & Jackson (2008) for an in depth review of finite temperature models).

However BECs suffer an important limitation. The systems which can be currently created in the laboratory contain a small number of atoms, typically 10^3 to 10^9, hence do not sustain the number of quantum vortices present in helium experiments. For example, up to a few hundred vortices have been achieved in the largest quasi-2D BECs (Abo-Shaeer *et al.*, 2002). This brings to light the issue of length scales. A defining feature of classical turbulence, besides nonlinearity, is the huge number of length scales which are excited. The range of length scales available in turbulent superfluid helium at very small temperatures is perhaps even larger, since short wavelength helical waves along the vortices can be generated by nonlinear interactions, producing a turbulent cascade called the Kelvin wave cascade (Vinen, 2006). A simple question arises: *can a BEC, containing a limited number of vortices, really become turbulent?* Our tentative answer is yes. Numerical results thus far (Nore *et al.*, 1997; Berloff & Svistunov, 2002; Kobayashi & Tsubota, 2005*b*; Yepez *et al.*, 2009) suggest that kinetic energy is distributed over the length scales in agreement with the $k^{-5/3}$ Kolmogorov scaling which is observed in ordinary turbulence (where k is the wavenumber) even over this small range of length scales. Therefore, the study of turbulence in a BEC represents an exciting opportunity to probe a new regime residing somewhere between chaos and turbulence.

In the remainder of this paper, we aim to identify some important open questions about turbulence in BECs; where appropriate we will review some of the work which has been carried out to date.

3. Quantum Turbulence in atomic BECs, where are we?

The following is an extensive, but by no means exhaustive, list of aspects yet to be understood regarding turbulence in atomic BECs. To structure our discussion, we distinguish the evolution of turbulent flow into three stages, namely;

(i) The generation of the turbulence. *What are the most effective and efficient ways to generate turbulence? Does the way in which the turbulence is generated affect the 'type' of turbulence created?*

(ii) The statistical steady state. *Are there universal features of turbulence, for instance, is the Kolmogorov energy spectrum present? What are the statistics of the turbulence velocity field?*

(iii) The decay of the turbulence. *How does the turbulence decay? What is the best way to measure the decay?*

(i) The generation of the turbulence

To understand the generation of a vortex tangle in a quantum gas, we must first understand how individual vortices are nucleated. The very first creation of such a vortex took place in a two-component condensate and was driven by the rotation of one component around the other. The subsequent removal of the inner component resulted in the formation of a hollow core of a singly quantized vortex (Anderson *et al.*, 2000). Further techniques for generating vortex structures soon followed, including the creation of vortex rings following the "snake instability" decay of a dark soliton (Anderson *et al.*, 2001), phase imprinting (Leanhardt *et al.*, 2002) and by a rapid quench through the transition temperature for the onset of Bose-Einstein condensation, i.e. the Kibble-Zurek mechanism (Weiler *et al.*, 2008; Freilich *et al.*, 2010).

However, to generate a large number of vortices in the system at any one time, two other techniques have proved to be more effective:

(i) *Rotation* of an anisotropic BEC excites surface modes leading to the nucleation of vortices at the edge which then drift into the bulk of the BEC. If the rotation is performed about only one axis, a vortex lattice is created (Hodby *et al.*, 2001; Abo-Shaeer *et al.*, 2002, 2001; Madison *et al.*, 2000, 2001). In 3D, if the rotation is performed about more than one axis, a vortex tangle has been predicted to form (Kobayashi & Tsubota, 2007).

(ii) *A moving (cylindrical) obstacle*, such as that created by the potential from a blue-detuned laser beam moving through a quantum fluid, generates pairs of vortices in its wake when its speed exceeds a critical value (Raman *et al.*, 2001). Recently, this method has been used to generate and study a collection of vortex dipoles in a 2D BEC (Neely *et al.*, 2010).

Both methods generate a large number of vortices, in 2D as well as in 3D. However, one can bypass the initial transient period and begin with a nonequilibrium state of vortices. Experimentally, this can be achieved via imprinting a phase profile onto the condensate via laser beams, as performed by Leanhardt *et al.* (2002) for generating single vortices of arbitrary charge. The use of such imprinting to generate a vortex tangle has been implemented theoretically (White *et al.*, 2010), with the resulting tangle similar to that depicted in Fig 3.

Our preliminary results with method (ii) (laser stirring), suggest that it is possible to generate a large number of vortices; we have found qualitative evidence that, by moving the obstacle along different paths, we can change the isotropy of the resulting tangle of vortices (Allen *et al.*, 2014).

Fig. 1 (top row) shows the density isosurface of a 3D spherical condensate at three instants in time, after the condensate has been stirred for a time $t_{\rm stir}$, along a circular path with a Gaussian-shaped laser stirrer oriented in the $z-$direction. These results are based on simulations of the GPE, Eq. (3). For a simple measure of the isotropy of the tangle, we plot the projected vortex lengths L_x, L_y and L_z in each Cartesian direction (bottom part of Fig. 1). All projected lengths rapidly increase during the stirring period ($t < t_{\rm stir}$); after the laser has been removed ($t > t_{\rm stirr}$), L_x, L_y and L_z all decrease for a short period of time. Later, only the vortex lengths L_x and L_y in the transverse (x and y) directions further decrease, whereas L_z remains approximately constant because the vortex tangle decays into a regular lattice, as it is apparent from the final density isosurface (Top row, c)). This is as expected: in stirring the condensate circularly we impart angular momentum about the z-axis, and it is well known that the ground state of a superfluid with sufficient angular momentum features a lattice of regularly-spaced, vortex lines aligned along the z-axis.

In Fig. 2 the vortex length is shown when the stirring takes place, for the same amount of time, along a Figure-eight path. Again, the vortex length increases over the duration of the laser stirring ($t < t_{\rm stir}$); however, after the laser is removed ($t > t_{\rm stir}$), the tangle decays isotropically, i.e. all projected lengths L_x, L_y and L_z decay together. For the Figure-eight path, the laser also moves through the centre of the condensate, where the density is higher, the vortices which

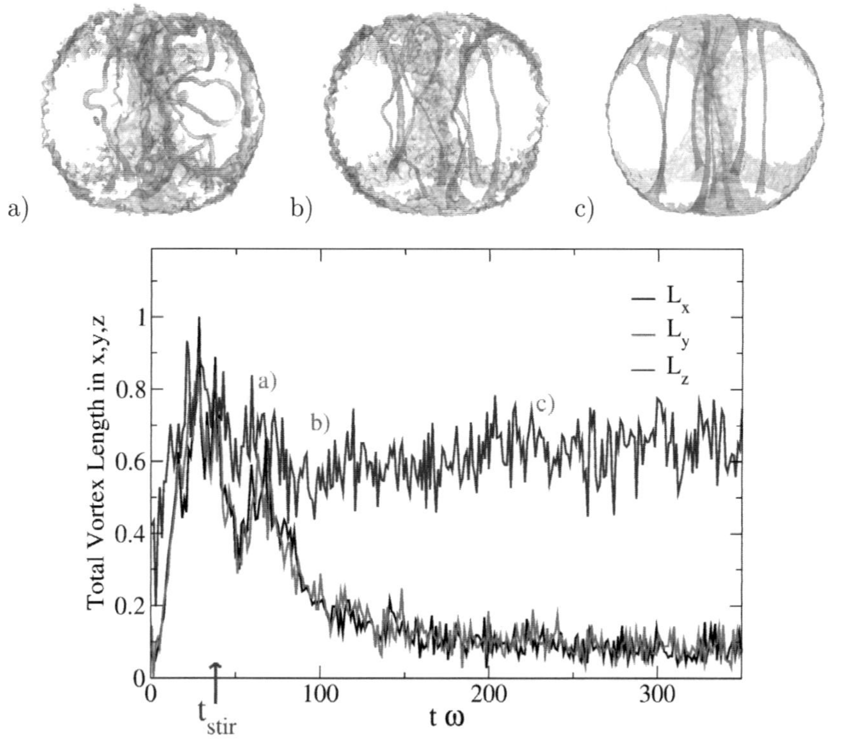

Figure 1. Top row: Density isosurfaces of a 3D spherical BEC at times $t\omega$ = a) 60, b) 100 and c) 240 after stirring the condensate along the $z-$direction in a circular path. We see here that the surface plot picks up the vortex cores as well as some of the condensate edge. The resulting vortex length in each direction is shown over time in the bottom part of this figure where it has been normalised by the peak vortex length.

are generated are longer than those generated at the edge of the condensate (Allen *et al.*, 2014); therefore, this laser stirring path is more efficient at creating a dense, random vortex tangle than simply stirring the condensate in a circular fashion.

(ii) Statistical state

Once the turbulence state has been created (by whichever means), its steady state properties can be investigated. BECs offer the possibility to study 2D and 3D turbulence and the cross-over region between the two (Parker & Adams, 2005). We now review our current understanding of the properties of a turbulent tangle of vortices in a BEC, first in 2D and then in 3D.

Quantum turbulence in 2D

In 2D classical turbulence, the conservation of enstrophy dictates that energy must flow from small scales of energy injection to large scales forming, for example, large clusters of vortices (Kraichnan, 1967). This inverse cascade is thought to underly Jupiter's giant Red Spot and has been experimentally examined in classical fluids (Sommeria *et al.*, 1988; Marcus, 1988) (for a review see Kellay & Goldburg (2002)). Attempts to observe the inverse cascade effect in quantum gases have led to modelling vortex generation (Parker & Adams, 2005; White *et al.*, 2012; Fujimoto & Tsubota, 2011; Reeves *et al.*, 2013), and to experiments on the dynamics of vortex dipoles created by a moving potential (Neely *et al.*, 2010, 2013).

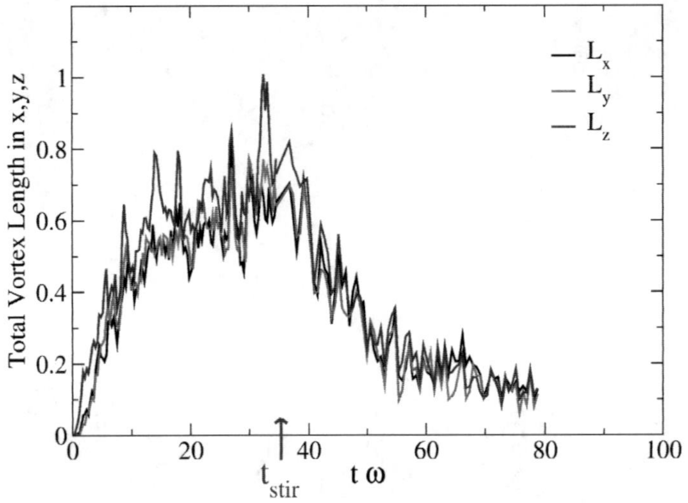

Figure 2. Vortex length, normalised by the peak length, in the x, y and z directions for a stirrer along the $z-$direction, moving in a Figure-eight path.

Numasato *et al.* (2010), evolved the 2D GPE to a turbulent state by initially imposing a random phase on the wavefunction. They did not observe a reverse cascade but rather a direct cascade. They argued that since the total number of vortices, and therefore the enstrophy, is not conserved in simulations of the GPE because of vortex pair annhilation, the inverse energy cascade is irrelevant for 2D quantum turbulence.

Conversely, Reeves *et al.* (2013) reported the numerical observation of the inverse cascade. They solved the 2D damped GPE (DGPE) and generated a turbulent state by imposing the fluid to flow past 4 stationary potential barriers. The speed of the flow was sufficiently high ($v \simeq 0.822c$, where c is the sound speed of the quantum fluid) so as to create many vortices and thereby a turbulent flow. Using a statistical algorithm they measured how prone 'like-winding' vortices, i.e. vortices of the same charge, were to cluster together depending on the amount of dissipation imposed. They found that an intermediate level of damping led to small clusters of like-winding vortices being formed and inferred an inverse energy cascade from analysis of the incompressible energy spectrum (energy associated with vortices). However, this analysis was carried out in the wake of the obstacle where like-winding vortices are naturally in the vicinity of each other. To make this result more convincing, a larger box for analysis would be desirable.

Similar work by White *et al.* (2012) implemented a rotating elliptical paddle to generate large numbers of vortices. Again, clustering of like-winding vortices was observed, but no inverse cascade was reported.

A possible reason for the lack of definitive evidence for or against the inverse cascade in quantum gases is their relatively small size, i.e. the systems which were studied lacked enough length scales. However, this limitation has not prevented the observation of the direct Kolmogorov-3D cascade in systems of similar size (Nore *et al.*, 1997; Kobayashi & Tsubota, 2005*a*,*b*; Yepez *et al.*, 2009; Kobayashi & Tsubota, 2007).

Quantum turbulence in 3D

In their seminal work, Nore *et al.* (1997) used the GPE to investigate 3D quantum turbulence in a homogeneous box by evolving an initial, large scale Taylor Green vortex. By decomposing the velocity field into divergence-free and curl-free parts, they obtained incompressible (associated with vortex motion) and compressible (associated with acoustic excitations) kinetic energy spectra respectively. They showed that the incompressible kinetic energy spectrum is similar to the classical $k^{-5/3}$ Kolmogorov energy spectrum at scales down to the intervortex spacing. Similarly, Berloff & Svistunov (2002) began with a non-equilibrium state, generated by imprinting a random phase on the equilibrium wavefunction of a homogeneous box. They then observed the evolution of the quantum turbulence by solving the GPE and further allowing the system to evolve to phase coherence. Similarly, Kobayashi & Tsubota (2005*a*,*b*) imprinted a random phase on the equilibrium wavefunction of a homogeneous box and evolved according to the GPE. They introduced a dissipative term which only acted on scales smaller than the healing length to represent thermal dissipation in the system. They also obtained a decaying incompressible energy spectrum which has the Kolmogorov power law over the inertial range. In order to clarify the extent of this range, statistical steady turbulence was created by a moving random potential which continuously injected energy into the system at large scales; a damping term removed energy at small length scales. They found that the inertial range was slightly narrower for the continuously forced turbulence because the moving potential sets the energy containing range. Further to these methods, White *et al.* (2010) imprinted a staggered vortex array onto a harmonically trapped BEC and evolved the system according to the GPE (see Fig. 3). In this work, they calculated the probability density function of the velocity components and found that (in both 2D and 3D) it is not a Gaussian like in ordinary turbulence, in agreement with experimental results obtained in superfluid helium (Paoletti *et al.*, 2008). This is an important observation which distinguishes quantum from classical turbulence.

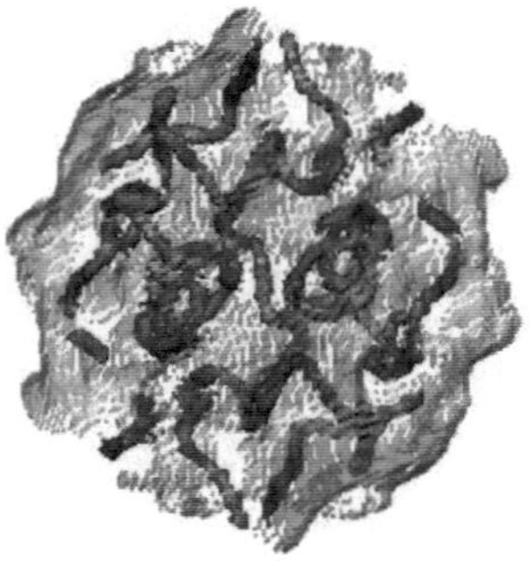

Figure 3. 3D turbulent state of a harmonically trapped BEC. Condensate edge shown by blue shading, turbulent vortex tangle shown in purple (White *et al.*, 2010). Figure courtesy of Angela C. White.

Using a similar method of imprinting, (Yepez *et al.*, 2009) performed impressively large simulations in a homogeneous box using a quantum lattice gas algorithm (up to 5760^3 grid points) and resolved scales smaller than the vortex core radius.

Most of the methods of generating quantum turbulence discussed so far in this section have a common aim: the incompressible energy spectrum of quantum turbulence after an initial, turbulent state has been set up (with the exception of (Kobayashi & Tsubota, 2005*a*,*b*)). However, the group of Tsubota have also carried out simulations where they dynamically create a vortex tangle by solving the DGPE with combined rotation along two axis of a harmonically trapped BEC. They found that by changing the ratio between the rotation frequencies in both directions, they could generate a vortex lattice or a more disordered array of vortices which formed a vortex tangle in which individual vortices appear to be nucleated with no preferred direction. They measured the incompressible energy spectrum and found it to be consistent with Kolmogorov law.

Experimentally, a small vortex tangle has been created in a harmonically trapped BEC through the combination of rotation and an external oscillating perturbation by Bagnato's

Journal of Physics: Conference Series **544** (2014) 012023 doi:10.1088/1742-6596/544/1/012023

group (Henn *et al.*, 2009, 2010; Seman *et al.*, 2011; Shiozaki *et al.*, 2011). They noticed that upon expansion of the condensate the usual inversion of aspect ratio of the gas (Mewes *et al.*, 1996; Castin & Dum, 1996) did not happen. This effect could be a possible signature of the creation of a tangle of vortices. A further observation from this experiment was the identification of different regimes corresponding to the strength and duration of the external oscillatory potential. For small amplitude excitations, the only effect was a bending of the main axis of the cloud, irrespective of the duration of the oscillation. Increasing the amplitude of the oscillation caused regular vortices to be nucleated, with the number of vortices increasing monotonically with the duration of the external excitation. Increasing the excitation time further led to a turbulent vortex regime, and at very long hold times, the condensate fragmented or granulated.

This brings us to the final stage of the evolution of turbulence, the decay.

(iii) The decay of the turbulence

In classical turbulence, the cascade of kinetic energy over the length scales terminates at some very short scale where viscosity dissipates kinetic energy into heat. The absence of viscosity in quantum fluids means there must exist other mechanisms of energy dissipation. The most likely is acoustic emission. When two vortices reconnect, some energy is lost in the form of sound. Reconnections are also thought to create high frequency Kelvin waves on vortices. It is thought that, in superfluid helium, Kelvin waves interact nonlinearly and create shorter and shorter waves, until sound waves are emitted at high frequency (Vinen, 2006). This energy transfer is called the Kelvin wave cascade.

Experiments in superfluid helium show that, depending on the scale at which energy is injected, the decay of the turbulence can be in one of two forms (Baggaley *et al.*, 2012):

(i) *'Semiclassical' or 'Kolmogorov' turbulence:* The vortex tangle seems polarised and structured over many length scales. This type of turbulence is generated when the forcing is at length scales larger than the average intervortex spacing. In this regime, the vortex length L decays as $L \simeq t^{-3/2}$, which is consistent with the decay of a Kolmogorov spectrum.

(ii) *'Ultraquantum' or 'Vinen' turbulence:* The scale of the forcing is less than the intervortex spacing, the vortex tangle seems random and possesses a single length scale, and the vortex length decays as $L \simeq t^{-1}$ (Walmsley & Golov, 2008).

How to measure the decay in a BEC is an open question. White *et al.* (2010) studied the decay of the turbulent tangle by numerically monitoring the vortex length, L, over time. They showed that the line length increases initially as reconnections take place before decaying over time. By further solving the dissipative GPE, they confirmed that thermal dissipation leads to a faster decay of line length but could not clearly distinguish between $L \simeq t^{-1}$ or $L \simeq e^{-\alpha t}$ behaviour (α being some decay parameter). However, is vortex linelength the best measure for this decay? Or can we again look to the incompressible energy spectrum to visualise the decay of vortices and draw some conclusions from both of these quantities? Furthermore, there is the question of how this is best achieved experimentally. The images taken of condensate density are typically column-integrated over the imaging direction which means that depth information becomes lost and an extraction of the true 3D vortex line length is not possible. Just as the attenuation of second sound is used to measure vortex length in helium, what surrogate measures of vortex line length are accessible experimentally in BECs? Such questions, both fundamental and practical in nature, will provide a rich source of research in these systems in the future.

4. Summary

We have discussed weakly interacting, dilute atomic Bose-Einstein condensates as tools for understanding the nature of quantum turbulence, motivated largely by the huge degree of control they offer. Even though the range of lengthscales excited in these systems is much less than in superfluid helium, the direct Kolmogorov energy cascade has been predicted to exist and the regimes of turbulence accessible is vast and interesting in its own right. We have distinguished three distinct phases of quantum turbulence - the generation of turbulence, its steady state and its decay - and briefly reviewed work done in understanding these so far, whilst highlighting fundamental questions about each phase.

Acknowledgments

We acknowledge A. C. White for useful discussions and for providing us with an image for use in this article. This work was supported by the grant EP/I019413/1 from the EPSRC.

References

Abo-Shaeer, J. R., Raman, C. & Ketterle, W. 2002 Formation and decay of vortex lattices in Bose-Einstein condensates at finite temperatures. *Phys. Rev. Lett.* **88** (7), 070409.

Abo-Shaeer, J. R., Raman, C., Vogels, J. M. & Ketterle, W. 2001 Observation of vortex lattices in Bose-Einstein condensates. *Science* **292**, 476.

Allen, A. J., Parker, N. G., Proukakis, N. P. & Barenghi, C. F. 2014 Isotropic vortex tangles in trapped atomic Bose-Einstein condensates via laser stirring. *Phys. Rev. A* **89**, 025602.

Anderson, B. P., Haljan, P. C., Regal, C. A., Feder, D. L., Collins, L. A., Clark, C. W. & Cornell, E. A. 2001 Watching dark solitons decay into vortex rings in a Bose-Einstein condensate. *Phys. Rev. Lett.* **86**, 2926–2929.

Anderson, B. P., Haljan, P. C., Wieman, C. E. & Cornell, E. A. 2000 Vortex precession in Bose-Einstein condensates: observations with filled and empty cores. *Phys. Rev. Lett.* **85**, 2857–2860.

Anderson, M. H., Ensher, J. R., Matthews, M. R. & Wieman, C. E. 1995 Observation of Bose-Einstein condensation in a dilute atomic vapor. *Science* **269**, 198.

Baggaley, A. W., Barenghi, C. F. & Sergeev, Y. A. 2012 Quasiclassical and ultraquantum decay of superfluid turbulence. *Phys. Rev. B* **85**, 060501.

Barenghi, C. F. 2008 Is the Reynolds number infinite in superfluid turbulence? *Physica D: Nonlinear Phenomena* **237**, 2195 – 2202.

Barenghi, C. F., Donnelly, R. J. & Vinen, W. F. (ed.) 2001 **Quantized Vortex Dynamics and Superfluid Turbulence**. Springer, Berlin.

Berloff, N. G. & Svistunov, B. V. 2002 Scenario of strongly nonequilibrated Bose-Einstein condensation. *Phys. Rev. A* **66**, 013603.

Castin, Y. & Dum, R. 1996 Bose-Einstein condensates in time dependent traps. *Phys. Rev. Lett.* **77**, 5315–5319.

Davis, K. B., Mewes, M. O., Andrews, M. R., van Druten, N. J., Durfee, D. S., Kurn, D. M. & Ketterle, W. 1995 Bose-Einstein condensation in a gas of sodium atoms. *Phys. Rev. Lett.* **75**, 3969–3973.

Fonda, E., Meichle, D. P., Ouellette, N. T., Hormoz, S. & Lathrop, D. P. 2014 Direct observation of Kelvin waves excited by quantized vortex reconnection. *Proceedings of the National Academy of Sciences* .

FREILICH, D. V., BIANCHI, D. M., KAUFMAN, A. M., LANGIN, T. K. & HALL, D. S. 2010 Real-time dynamics of single vortex lines and vortex dipoles in a Bose-Einstein condensate. *Science* **329** (5996), 1182–1185.

FUJIMOTO, K. & TSUBOTA, M. 2011 Nonlinear dynamics in a trapped atomic Bose-Einstein condensate induced by an oscillating gaussian potential. *Phys. Rev. A* **83**, 053609.

GÖRLITZ, A., VOGELS, J. M., LEANHARDT, A. E., RAMAN, C., GUSTAVSON, T. L., ABO-SHAEER, J. R., CHIKKATUR, A. P., GUPTA, S., INOUYE, S., ROSENBAND, T. & KETTERLE, W. 2001 Realization of Bose-Einstein condensates in lower dimensions. *Phys. Rev. Lett.* **87**, 130402.

HENN, E. A. L., SEMAN, J. A., ROATI, G., MAGALHÃES, K. M. & BAGNATO, V. S. 2010 Generation of vortices and observation of quantum turbulence in an oscillating Bose-Einstein condensate. *Journal of Low Temperature Physics* **158**, 435–442.

HENN, E. A. L., SEMAN, J. A., ROATI, G., MAGALHÃES, K. M. F. & BAGNATO, V. S. 2009 Emergence of turbulence in an oscillating Bose-Einstein condensate. *Phys. Rev. Lett.* **103**, 045301.

HODBY, E., HECHENBLAIKNER, G., HOPKINS, S. A., MARAGÒ, O. M. & FOOT, C. J. 2001 Vortex nucleation in Bose-Einstein condensates in an oblate, purely magnetic potential. *Phys. Rev. Lett.* **88**, 010405.

INOUYE, S., ANDREWS, M. R., STENGER, J., MIESNER, H.-J., STAMPER-KURN, D. M. & KETTERLE, W. 1998 Observation of feshbach resonances in a Bose-Einstein condensate. *Nature* **392**, 151.

JACKSON, B., PROUKAKIS, N. P., BARENGHI, C. F. & ZAREMBA, E. 2009 Finite-temperature vortex dynamics in Bose-Einstein condensates. *Phys. Rev. A* **79** (5), 053615.

KELLAY, H. & GOLDBURG, W. I. 2002 Two-dimensional turbulence: a review of some recent experiments. *Reports on Progress in Physics* **65** (5), 845.

KEVREKIDIS, P. G. & CARRETERO-GONZALEZ, R. (ed.) 2008 **Emergent Nonlinear Phenomena in Bose-Einstein Condensates**. Springer, Berlin.

KOBAYASHI, M. & TSUBOTA, M. 2005*a* Kolmogorov spectrum of quantum turbulence. *Journal of the Physical Society of Japan* **74** (12), 3248–3258.

KOBAYASHI, M. & TSUBOTA, M. 2005*b* Kolmogorov spectrum of superfluid turbulence: numerical analysis of the gross-pitaevskii equation with a small-scale dissipation. *Phys. Rev. Lett.* **94**, 065302.

KOBAYASHI, M. & TSUBOTA, M. 2007 Quantum turbulence in a trapped Bose-Einstein condensate. *Phys. Rev. A* **76**, 045603.

KRAICHNAN, R. H. 1967 Inertial ranges in two-dimensional turbulence. *Physics of Fluids* **10** (7), 1417–1423.

LAHAYE, T., MENOTTI, C., SANTOS, L., LEWENSTEIN, M. & PFAU, T. 2009 The physics of dipolar bosonic quantum gases. *Reports on Progress in Physics* **72** (12), 126401.

LEANHARDT, A. E., GÖRLITZ, A., CHIKKATUR, A. P., KIELPINSKI, D., SHIN, Y., PRITCHARD, D. E. & KETTERLE, W. 2002 Imprinting vortices in a Bose-Einstein condensate using topological phases. *Phys. Rev. Lett.* **89**, 190403.

MADISON, K. W., CHEVY, F., BRETIN, V. & DALIBARD, J. 2001 Stationary states of a rotating Bose-Einstein condensate: Routes to vortex nucleation. *Phys. Rev. Lett.* **86**, 4443–4446.

MADISON, K. W., CHEVY, F., WOHLLEBEN, W. & DALIBARD, J. 2000 Vortex formation in a stirred Bose-Einstein condensate. *Phys. Rev. Lett.* **84**, 806–809.

MARCUS, P. 1988 Numerical simulation of jupiter's great red spot. *Nature* **331**, 693 – 696.

MEWES, M. O., ANDREWS, M. R., VAN DRUTEN, N. J., KURN, D. M., DURFEE, D. S. & KETTERLE, W. 1996 Bose-Einstein condensation in a tightly confining dc magnetic trap. *Phys. Rev. Lett.* **77**, 416–419.

NEELY, T. W., BRADLEY, A. S., SAMSON, E. C., ROONEY, S. J., WRIGHT, E. M., LAW, K. J. H., CARRETERO-GONZÁLEZ, R., KEVREKIDIS, P. G., DAVIS, M. J. & ANDERSON, B. P. 2013 Characteristics of two-dimensional quantum turbulence in a compressible superfluid. *Phys. Rev. Lett.* **111**, 235301.

NEELY, T. W., SAMSON, E. C., BRADLEY, A. S., DAVIS, M. J. & ANDERSON, B. P. 2010 Observation of vortex dipoles in an oblate Bose-Einstein condensate. *Phys. Rev. Lett.* **104**, 160401.

NORE, C., ABID, M. & BRACHET, M. E. 1997 Decaying Kolmogorov turbulence in a model of superflow. *Physics of Fluids* **9** (9), 2644–2669.

NUMASATO, R., TSUBOTA, M. & L'VOV, V. S. 2010 Direct energy cascade in two-dimensional compressible quantum turbulence. *Phys. Rev. A* **81**, 063630.

PAOLETTI, M. S., FISHER, M. E., SREENIVASAN, K. R. & LATHROP, D. P. 2008 Velocity statistics distinguish quantum turbulence from classical turbulence. *Phys. Rev. Lett.* **101**, 154501.

PARKER, N. G. & ADAMS, C. S. 2005 Emergence and decay of turbulence in stirred atomic Bose-Einstein condensates. *Phys. Rev. Lett.* **95**, 145301.

PETHICK, C. & SMITH, H. 2002 **Bose-Einstein condensation in dilute gases**. Cambridge University Press.

PROUKAKIS, N. P. & JACKSON, B. 2008 Finite-temperature models of Bose-Einstein condensation. *Journal of Physics B: Atomic, Molecular and Optical Physics* **41** (20), 203002.

RAMAN, C., ABO-SHAEER, J. R., VOGELS, J. M., XU, K. & KETTERLE, W. 2001 Vortex nucleation in a stirred Bose-Einstein condensate. *Phys. Rev. Lett.* **87**, 210402.

REEVES, M. T., BILLAM, T. P., ANDERSON, B. P. & BRADLEY, A. S. 2013 Inverse energy cascade in forced 2D quantum turbulence. *Phys. Rev. Lett* **110**, 104501.

ROBERTS, J. L., CLAUSSEN, N. R., BURKE, J. P., GREENE, C. H., CORNELL, E. A. & WIEMAN, C. E. 1998 Resonant magnetic field control of elastic scattering in cold ^{85}rb. *Phys. Rev. Lett.* **81**, 5109–5112.

SEMAN, J., HENN, E., SHIOZAKI, R., ROATI, G., POVEDA-CUEVAS, F., MAGALHÃES, K., YUKALOV, V., TSUBOTA, M., KOBAYASHI, M., KASAMATSU, K. & BAGNATO, V. 2011 Route to turbulence in a trapped Bose-Einstein condensate. *Laser Physics Letters* **8** (9), 691–696.

SHIOZAKI, R., TELLES, G., YUKALOV, V. & BAGNATO, V. 2011 Transition to quantum turbulence in finite-size superfluids. *Laser Physics Letters* **8** (5), 393–397.

SKRBEK, L. & SREENIVASAN, K. R. 2012 Developed quantum turbulence and its decay. *Physics of Fluids* **24** (1), 011301.

SOMMERIA, J., MEYERS, S. D. & SWINNEY, H. L. 1988 Laboratory simulation of jupiter's great red spot. *Nature* **331**, 689 – 693.

VINEN, W. 2006 An introduction to quantum turbulence. *Journal of Low Temperature Physics* **145**, 7–24.

WALMSLEY, P. M. & GOLOV, A. I. 2008 Quantum and quasiclassical types of superfluid turbulence. *Phys. Rev. Lett.* **100**, 245301.

Journal of Physics: Conference Series **544** (2014) 012023 doi:10.1088/1742-6596/544/1/012023

WEILER, C., NEELY, T. W., SCHERER, D. R., BRADLEY, A. S., DAVIS, M. J. & ANDERSON, B. P. 2008 Spontaneous vortices in the formation of Bose-Einstein condensates. *Nature* **455**, 948.

WHITE, A. C., BARENGHI, C. F. & PROUKAKIS, N. P. 2012 Creation and characterization of vortex clusters in atomic Bose-Einstein condensates. *Phys. Rev. A* **86**, 013635.

WHITE, A. C., BARENGHI, C. F., PROUKAKIS, N. P., YOUD, A. J. & WACKS, D. H. 2010 Nonclassical velocity statistics in a turbulent atomic Bose-Einstein condensate. *Phys. Rev. Lett.* **104**, 075301.

YEPEZ, J., VAHALA, G., VAHALA, L. & SOE, M. 2009 Superfluid turbulence from quantum Kelvin wave to classical Kolmogorov cascades. *Phys. Rev. Lett.* **103**, 084501.

Journal of Physics: Conference Series **544** (2014) 012024 doi:10.1088/1742-6596/544/1/012024

Abelian and non-Abelian Hopfions in all odd dimensions

Eugen Radu[1] **and Tigran Tchrakian**[2,3]

[1] Institut für Physik, Universität Oldenburg, Postfach 2503 D-26111 Oldenburg, Germany
[2] School of Theoretical Physics – DIAS, 10 Burlington Road, Dublin 4, Ireland
[3] Department of Computer Science, Maynooth University, Ireland

E-mail: eugen.radu@uni-oldenburg.de, tigran@stp.dias.ie

Abstract. We extend the definition of the topological charge pertaining to the CP^1 (*i.e.* O(3)) Skyrme-Fadde'ev Hopfion on $\mathbb{R}^3$, to candidates for topological charges of $\mathbb{C}\mathrm{P}^n$ sigma models on $\mathbb{R}^{2n+1}$, for all n. For this, the Abelian composite connections of the $\mathbb{C}\mathrm{P}^n$ sigma models are employed. In higher dimensions ($n \geq 1$) it turns out that such charges, described by the non-Abelian composite connections of suitable Grassmannian sigma models, can also be constructed. A concrete discussion of the non-Abelian case for $n = 2$ is presented.

1. Introduction

Hopfions are static "soliton like" solutions to the field equations of nonlinear sigma models. With the usual solitons namely instantons of the Yang-Mills (YM) systems in all even dimensions and of Skyrmions of nonlinear sigma models in all dimensions, monopoles of YM-Higgs (YMH) systems in all dimensions three and higher, and vortices of YMH and gauged sigma models systems in two dimensions Manton & Sutcliffe (2004), have finite energy and are topologically stable. Likewise, Hopfions have finite energy and are topologically stable. To date, the only example are the solutions Faddeev & Niemi (1996) and Battye & Sutcliffe (1998) to the Skyrme-Fadde'ev $O(3)$ sigma model on $\mathbb{R}^3$. For a review see Manton & Sutcliffe (2004), Radu & Volkov (2008).

The salient qualitative difference between the afore mentioned solitons and the Hopfion is this. The topological charge densities of the former are by construction total divergence. For Yang-Mills (YM) and complex sigma models, and YMH models, they are $Chern - Pontryagin$ densities or their decendants, respectively. For the nonlinear sigma models with real sphere-valued fields they are "essentially total divergence". Such densities are not explicitly total diveregence. But when they are subjected to the variational princliple, with the sigma model constraint taken account of by a Lagrange multiplier term, they yield no nontrivial Euler-Lagrange equations. They share this property with total divergence densities. In the case of the Hopfion, the topological charge density is instead the $Chern - Simons$ (CS) density, which is **not** total divergence. It follows that Hopfions can exist only in odd space dimensions, where CS densities are defined.

The Chern-Simons density pertaining to Hopfions, which is defined in terms of the gauge connection and curvature, depends on the scalar sigma model fields. These connections and curvatures are the *composite* quantities constructed from the sigma model fields appropriate to any given example. The most transparent way of doing this is to employ a "complex sigma model" (*e.g.* complex projective, quaternionic, Grassmannian, *etc.*) enabling the definitions of the *composite* connection and curvature necessary for the definition of the CS density.

Given that the topological charge of the Hopfion is the volume integral of the Chern-Simons density, the lower bound on the energy of the Skyrme-Fadde'ev Hopfion on $\mathbb{R}^3$ was established in Vakulenko & Kapitansky (1979), using a method that employs Sobolev space norms. Note that the analysis of Vakulenko & Kapitansky (1979) is not a proof of existence, and that this Hopfion is constructed numerically Faddeev & Niemi (1996); Battye & Sutcliffe (1998). A proof of existence can be found in Lin & Yang (2004).

To qualify as a topological charge density, this Chern-Simons density must become a total divergence which it is not *a priori*. This can be achieved by subjecting it to suitable symmetries which render it total divergence. This is perhaps the most striking difference between the usual solitons and Hopfions. The topological charges of the former exist subject to no symmetries, while that of the Hopfions exist only for systems constrained by the appropriate symmetry. Our aim in the present talk is strictly restricted to this task, namely to find suitable symmetries which render the Chern-Simons density in the given (odd) dimension, a total divergence.

To this end, our prescription here to achieve this aim is the imposition of multi-azimuthal symmetries. Specifically, it is the imposition of azimuthal symmetry in each of the n, 2-planes in $\mathbb{R}^{2n+1}$. (This is the case with the familiar Skyrme-Fadde'ev Hopfion on $\mathbb{R}^3$ with axial symmetry imposed in the $x-y$ plane.) Multi-azimuthal symmetry in $\mathbb{R}^{2n+1}$ eliminates n azimuthal angles, each in one of the n, 2-dimensional subspaces (planes), resulting in $(n+1)$-dimensional residual subsystems.

Our prescription hinges on positing an adequate Ansatz that results in the residual Chern-Simons density in a $(n+1)$-dimensional space being parametrised by $(n+1)$ independant functions of the $(n+1)$ residual 'coordinates'. The equality of the number independant functions in the residial system, and the number of the residual 'coordinates' can result in the residual CS being total divergence.

Finally, it should be noted that this aim of rendering the residual Chern-Simons density a total divergence can be achieved only when the sigma model constraint is satisfied. This is automatic if the Ansatz in question is parametrised in terms of functions that satisfy this constraint. Often however it is convenient to use more relaxed parametrisations, in which case the corresponding criterion is to show that the residual CS density is "essentially total divergence", subjecting this density to the variational principle taking account of the (residual) constraint *via* a Lagrange multiplier term, results in *trivial* Euler-Lagrange equations.

The above account of topological charges of Hopfions relies entirely on the existence of suitable composite connections and curvatures of the nonlinear sigma model in question. In this sense, both Abelian and non-Abelian systems can be considered. It is this context that we have used the nomenclature of Abelian and non-Abelian Hopfions in the title. The talk consists of two main sections, **2** and **3**, describing our prescriptions for Abelian and non-Abelian Hopfions respectively. Each of these sections is subdivided in a subsection defining the appropriate nonlinear sigma model, followed by the imposition of symmetries yielding the desired toplogical charge densities. Section **4** gives a brief summary.

2. Abelian Hopfions on $\mathbb{R}^{2n+1}$

The description of the arbitrary n case in the Abelian case is much more transparent than the non-Abelian case that follows. We describe the three cases $n=1,2,3$, which point to the generic case readily.

Journal of Physics: Conference Series **544** (2014) 012024 doi:10.1088/1742-6596/544/1/012024

2.1. CP^n models on $\mathbb{R}^{2n+1}$

We start with the generic structure of models that can support Abelian Hopfion on $\mathbb{R}^{2n+1}$. These are the $\mathbb{C}$P^n sigma models on $\mathbb{R}^{2n+1}$ described by complex $n-$tuplets

$$Z = \begin{bmatrix} z_1 \\ z_2 \\ .. \\ .. \\ z_{n+1} \end{bmatrix} \equiv z_a \quad ; \quad \bar{Z} = \begin{bmatrix} \bar{z}^1 \\ \bar{z}^2 \\ .. \\ .. \\ \bar{z}^{n+1} \end{bmatrix} \equiv \bar{z}^a, \quad \text{with } a = 1, 2, ..., n+1\,, \tag{1}$$

subject to the constraint

$$Z^\dagger\, Z \equiv \bar{z}^a\, z_a = 1\,, \tag{2}$$

taking their values in $\frac{U(n+1)}{U(n)\times U(1)}$, such they are described by $2n$ real parameters that parametrise Z on $\mathbb{R}^{2n+1}$. In (1), $\bar{z}^a$ is the complex conjugate of z_a, transforming with an index that is *contravariant* to the *covariant* index of z_a, and $Z^\dagger$ in (2) is the transpose of $\bar{Z}$. This leads to the definition of the *projection operator*

$$P = \left(\mathbb{1} - Z\, Z^\dagger\right) \equiv \left(\delta_a{}^b - z_a\, \bar{z}^b\right)\,. \tag{3}$$

The most interesting feature of these models is that when the field Z is subjected to a *local* $U(1)$ gauge transformation $g = \mathrm{e}^{i\Lambda(x)}$, the constraint (2) is invariant under

$$Z \to Z\, g\,. \tag{4}$$

As a consequence the quantity defined as

$$B_i = i\, Z^\dagger \partial_i Z \quad , \quad i = 1, 2, ..., 2n+1 \tag{5}$$

transforms like an *Abelian composite connection* under $g(\Lambda)$,

$$B_i \to B_i \pm \partial_i \Lambda\,,$$

which lead to the definitions of the covariant derivative of Z and and the *Abelian curvature* of this connection,

$$D_i Z = \partial_i\, Z + B_i\, Z\,, \tag{6}$$
$$G_{ij} = \partial_i\, B_j - \partial_j\, B_i\,, \tag{7}$$

with $D_i Z$ transforming covariantly under the action of g, and G_{ij} invariantly.

The Abelian Chern-Simons (CS) density on $\mathbb{R}^{2n+1}$ is then readily defined in terms of the quantities (7) and (6).

$$\Omega_{\rm CS} \simeq \varepsilon_{i_1 i_2 \dots i_{2n+1}}\, B_{i_{2n+1}}\, G_{i_1 i_2}\, G_{i_3 i_4} \dots G_{i_{2n-1} i_{2n}}\,. \tag{8}$$

It is these densities, subject to the appropriate symmetries, that play the role of topological charges stabilising Hopfions. That $\Omega_{\rm CS}$ is not a total divergence is clear.

Subjecting the density

$$\Omega_{\rm CS} + \lambda\,(1 - Z^\dagger\, Z)$$

to arbitrary variations of $Z^\dagger$ (or Z), where the Lagrange multiplier term accounts for the costraint (2), one has

$$\varepsilon_{i_1 i_2 \dots i_{2n+1}}\, D_{i_{2n+1}} Z\, G_{i_1 i_2}\, G_{i_3 i_4} \dots G_{i_{2n-1} i_{2n}} = 0\,. \tag{9}$$

Eqn. (9) trivialises only under the appropriate symmetries, *i.e.* (8) becomes a density that is "essentially total divergence", as required of a topological charge density.

Journal of Physics: Conference Series **544** (2014) 012024 doi:10.1088/1742-6596/544/1/012024

2.2. Residual Abelian CS density subject to n-fold azimuthal symmetry

We treat the three cases $n = 1, 2, 3$ below by positing an Ansatz which imposes the appropriate symmetries that renders (8) a total divergence.

In the $n = 1$ case, the field (1) is subjected to axial symmetry in the (x_1, x_2) plane of $\mathbb{R}^3$,

$$Z = \begin{bmatrix} a+ib \\ c\,e^{in\varphi} \end{bmatrix} \equiv \begin{bmatrix} \sin\frac{f}{2}\,e^{i\alpha} \\ \cos\frac{f}{2}\,e^{in\varphi} \end{bmatrix} \tag{10}$$

where the functions a, b, c, f and α all depend on both $\rho = \sqrt{|x_\alpha|^2}$ and $z \equiv x_3$, $\alpha = 1, 2$.

Substituting the azimuthally symmetric Ansatz (10) in the Abelian Chern–Simons density on $\mathbb{R}^3$

$$\Omega^{(3)}_{\rm CS} \;=\; \varepsilon_{mij}\,B_m\,G_{ij}\,. \tag{11}$$

yields the simple expression

$$\Omega^{(3)}_{\rm CS} = -4\,\frac{n}{\rho}\,c\cdot\det\begin{vmatrix} a & b & c \\ a_{,\rho} & b_{,\rho} & c_{,\rho} \\ a_{,z} & b_{,z} & c_{,z} \end{vmatrix}\,. \tag{12}$$

The usual notation $a_{,\rho} = \partial_\rho\, a$, *etc.* is used here. It is easy to verify that this quantity is "essentially total divergence" when varied subject to the constraint $a^2 + b^2 + c^2 = 1$. Further, using the polar parametrisation in (10), it becomes a total divergence, since the constraint is automatically satified in that case.

Requiring the field configurations in question have the asymptotic values

$$\lim_{r\to\infty} f(r,\theta) = 0\,, \qquad \lim_{r\to\infty} \alpha(r,\theta) = m\,\pi\,, \tag{13}$$

one finds the topological charge

$$Q = \frac{8}{3}\,i\,n\,m\,\pi^2\,. \tag{14}$$

In the $n = 2$ case, the field Z on $\mathbb{R}^5$ is subjected to the bi-azimuthal symmetry

$$Z = \begin{bmatrix} a+ib \\ c_1\,e^{in_1\varphi} \\ c_2\,e^{in_2\chi} \end{bmatrix} \equiv \begin{bmatrix} \sin\frac{1}{2}f\,e^{i\alpha} \\ \cos\frac{1}{2}f\ \sin g\,e^{in_1\varphi} \\ \cos\frac{1}{2}f\ \cos g\,e^{in_2\chi} \end{bmatrix} \tag{15}$$

where $\rho = \sqrt{|x_\alpha|^2}$, $\sigma = \sqrt{|x_A|^2}$ with $\alpha = 1, 2$, $A = 3, 4$ and $z \equiv x_5$. φ and χ are the azimuthal angles in the (x_1, x_2) and (x_3, x_4) planes respectively, (n_1, n_2) being the winding (vortex) numbers of these planes respectively. All functions (a, b, c, d) or $(f, , g, \alpha)$ in (15) depend on (ρ, σ, z).

Substitution of the bi-azimuthally symmetric Ansatz (15) in the Abelian Chern–Simons density on $\mathbb{R}^5$

$$\Omega^{(5)}_{\rm CS} \;=\; \varepsilon_{mijkl}\,B_m\,G_{ij}\,G_{kl} \tag{16}$$

yields the simple expression

$$\Omega^{(5)}_{\rm CS} = 32\,\frac{n_1 n_2}{\rho\sigma}\,c_1\,c_2\cdot\det\begin{vmatrix} a & b & c_1 & c_2 \\ a_{,\rho} & b_{,\rho} & c_{1,\rho} & c_{2,\rho} \\ a_{,\sigma} & b_{,\sigma} & c_{1,\sigma} & c_{2,\sigma} \\ a_{,z} & b_{,z} & c_{1,z} & c_{2,z} \end{vmatrix}\,. \tag{17}$$

which when varied subject to the constraint $a^2+b^2+c_1^2+c_2^2=1$ turns out to be "essentially total divergence". Again, using the polar parametrisation in (15), this becomes a total divergence.

Requiring the field configurations in question have the asymptotic values

$$\lim_{r\to\infty} f = 0 \quad , \quad \lim_{r\to\infty} g = \theta \quad , \quad \lim_{r\to\infty} \alpha = m\,\pi\,, \tag{18}$$

where $0 \le \theta \le \frac{\pi}{2}$ is the polar angle in the (ρ,σ) quarterplane, one finds the topological charge

$$Q_{\rm CS}^{(5)} = -12\,n_1\,n_2\,m\,\pi^3\,. \tag{19}$$

Finally, in the $n=3$ case, one subjects the field Z on $\mathbb{R}^7$ is to the tri-azimuthal symmetry

$$Z = \begin{bmatrix} a+ib \\ c_1\,e^{in_1\varphi} \\ c_2\,e^{in_2\chi} \\ c_3\,e^{in_3\xi} \end{bmatrix} \equiv \begin{bmatrix} \sin\frac{1}{2}f\,e^{i\alpha} \\ \cos\frac{1}{2}f\,\sin g\,\cos h\,e^{in_1\varphi} \\ \cos\frac{1}{2}f\,\sin g\,\sin h\,e^{in_2\chi} \\ \cos\frac{1}{2}f\,\cos g\,e^{in_3 l\xi} \end{bmatrix} \tag{20}$$

in terms of the variables $\rho=\sqrt{|x_\alpha|^2}$, $\sigma=\sqrt{|x_A|^2}$, $\tau=\sqrt{|x_a|^2}$ with $\alpha=1,2$, $A=3,4$, $a=5,6$ and $z\equiv x_7$. φ, χ and ξ are the azimutal angles in the (x_1,x_2), (x_3,x_4) and (x_5,x_6) planes respectively, (n_1,n_2,n_3) being the winding (vortex) numbers of each plane respectively. All functions (a,b,c,d,e) or $(f,,g,h,\alpha)$ in (20) depend on (ρ,σ,τ,z).

Substitution of the tri-azimuthally symmetric Ansatz (20) in the Abelian Chern–Simons density on $\mathbb{R}^7$

$$\Omega_{\rm CS}^{(7)} \;=\; \varepsilon_{pijklmn}\,B_p\,G_{ij}\,G_{kl}\,G_{mn} \tag{21}$$

yields the simple expression

$$\Omega_{\rm CS}^{(7)} = 96\,\frac{n_1\,n_2\,n_3}{\rho\sigma\tau}\,c_1\,c_2\,c_3\cdot\det\begin{vmatrix} a & b & c_1 & c_2 & c_3 \\ a_{,\rho} & b_{,\rho} & c_{1,\rho} & c_{2,\rho} & c_{3,\rho} \\ a_{,\sigma} & b_{,\sigma} & c_{1,\sigma} & c_{2,\sigma} & c_{3,\sigma} \\ a_{,\tau} & b_{,\tau} & c_{1,\tau} & c_{2,\tau} & c_{3,\tau} \\ a_{,z} & b_{,z} & c_{1,z} & c_{2,z} & c_{3,z} \end{vmatrix}\,. \tag{22}$$

which when varied subject to the constraint $a^2+b^2+c_1^2+c_2^2+c_3^2 = 1$ turns out to be "essentially total divergence". Again, using the polar parametrisation in (20), this becomes a total divergence.

Requiring the field configurations in question have the asymptotic values

$$\lim_{r\to\infty} g = \theta_1 \quad , \quad \lim_{r\to\infty} h = \theta_2 \quad , \quad \lim_{r\to\infty} \alpha = m\,\pi\,, \tag{23}$$

where $0 \le \theta_1 \le \frac{\pi}{2}$ and $0 \le \theta_2 \le \frac{\pi}{2}$ are the polar angles of the octant sphere (ρ,σ,τ), one finds the topological charge

$$Q_{\rm CS}^{(7)} = 60\,i\,n_1\,n_2\,n_3\,m\,\pi^4\,. \tag{24}$$

From the results (12), (17) and (22) for $n=1$, $n=2$ and $n=3$, the form of the corresponding result for arbitary n follows by induction. The appropriate imposition of symmetry is the application of azimuthal symmetry in each of the n planes in $\mathbb{R}^{2n+1}$. The resulting reduced subsystem will now be an $n+1$ dimensional system of PDE's, parametrised by $n+2$ functions

$$a\,,\,b\,,\,c_1\,,\,c_2\,,\ldots,c_n$$

only $n+1$ of which are independent, subject to the sigma model constraint

$$a^2+b^2+c_1^2+c_2^2+\cdots+c_{n-1}^2+c_n^2=1\,. \tag{25}$$

The reduced Chern-Simons density will then take the form

$$\Omega_{\rm CS}^{(2n+1)}\simeq\frac{n_1\,n_2\ldots n_n}{\rho_1\rho_2\ldots\rho_n}\,c_1\,c_2\ldots c_n\cdot\det\left|\begin{array}{ccccccc} a & b & c_1 & c_2 & \cdot & \cdot & c_n \\ a_{,\rho_1} & b_{,\rho_1} & c_{1,\rho_1} & c_{2,\rho_1} & \cdot & \cdot & c_{n,\rho_1} \\ a_{,\rho_2} & b_{,\rho_2} & c_{1,\rho_2} & c_{2,\rho_2} & \cdot & \cdot & c_{n,\rho_2} \\ \cdot & \cdot & \cdot & \cdot & \cdot & \cdot & \cdot \\ \cdot & \cdot & \cdot & \cdot & \cdot & \cdot & \cdot \\ a_{,\rho_n} & b_{,\rho_n} & c_{1,\rho_n} & c_{2,\rho_n} & \cdot & \cdot & c_{n,\rho_n} \\ a_{,z} & b_{,z} & c_{1,z} & c_{2,z} & \cdot & \cdot & c_{,z} \end{array}\right|\,, \tag{26}$$

where of course $z=x_{n+1}$.

Again by induction, it follows that the charges (14), (19) and (24) for $n=1$, $n=2$ and $n=3$, that

$$Q_{\rm CS}^{(2n+1)}\simeq i^{(2n+1)}\,n_1\,n_2\ldots n_n\,m\;\pi^{n+1}\,, \tag{27}$$

where $n_1\,,n_2,\ldots,n_n$ are the vorticities in the n, 2-planes in $\mathbb{R}^{2n+1}$.

3. non-Abelian Hopfions on $\mathbb{R}^{2n+1}$

The description of the arbitrary n case in the non-Abelian case is much less transparent than the Abelian case that preceded. We describe the two cases $n=1,2$ in detail, demonstrating that the Chern-Simons densities become "essentially total divergence" when subjected to the appropriate symmetries. But our discussion of the $n=3$ case is less detailed and is aimed at giving a convincing demonstration of this property, only qualitatively.

What is more, is we shall see (in detail) that the non-Abelian $n=1$ case reduces to the corresponding Abelian case.

3.1. $2^{n+1}\times 2^n$ Grassmannian models on $\mathbb{R}^{2n+1}$

Here, we start with the definition of models that may support non-Abelian Hopfions on $\mathbb{R}^{2n+1}$. These are Grassmannian sigma models on $\mathbb{R}^{2n+1}$ described by $2^{n+1}\times 2^n$ scalar fields with complex entries

$$Z=\left[\begin{array}{c} z_1 \\ z_2 \end{array}\right] \tag{28}$$

where z_1 and z_2 are complex $2^n\times 2^n$ matrices subject to the constraint

$$Z^\dagger\,Z=\mathbb{1}_{2^n\times 2^n}\,. \tag{29}$$

and the $2^{n+1}\times 2^{n+1}$ quantity

$$\Pi=\left(\mathbb{1}-Z\,Z^\dagger\right) \tag{30}$$

is a projection operator.

This constraint is invariant under the action of the local *unitary* non-Abelian gauge transformation g acting on Z

$$Z\to Z\,g\quad,\quad Z^\dagger\to g^\dagger\,Z^\dagger\,. \tag{31}$$

Here, the unitary matrix g is chosen to be an element of $SO(2n+2)$ in the $2^n\times 2^n$ chiral Dirac matrix representation.

Journal of Physics: Conference Series **544** (2014) 012024 doi:10.1088/1742-6596/544/1/012024

The invariance (31) of the constraint (29) leads to the definition of the non-Abelian (anti-Hermitian) *composite* connection

$$B_i = Z^\dagger \partial_i Z\,. \tag{32}$$

transforming like

$$B_i \to g^{-1} B_i\, g + g^{-1}\partial_i\, g\,.$$

There follow the definitions of the covariant derivative of Z and the (composite) non-Abelian curvature

$$D_i Z = \partial_i Z - Z\, B_i \tag{33}$$

$$G_{ij} = \partial_{[i} B_{j]} + [B_i, B_j] \tag{34}$$

which under the action of g transform covariantly as

$$D_i Z \to D_i Z\, g\,, \tag{35}$$

$$G_{ij} \to g^{-1} G_{ij}\, g\,. \tag{36}$$

Unlike in the Abelian case where the Chern-Simons density (8) can be defined compactly for all $2n+1$ dimensions, in the non-Abelian case the expression for arbitrary n is rather formal and less transparent. For this reason we list here the non-Abelian Chern-Simons densities on $\mathbb{R}^{2n+1}$, up to $n = 3$.

These are, for $(n = 1\,, D = 3)$ and $(n = 2\,, D = 5)$

$$\Omega_{\rm CS}^{(1)} = \varepsilon_{ijk} \mathrm{Tr}\, B_k \left(G_{ij} - \frac{2}{3} B_i B_j \right)\,, \tag{37}$$

and

$$\Omega_{\rm CS}^{(2)} = \varepsilon_{ijklm}\, \mathrm{Tr}\, B_m \left(G_{ij} G_{kl} - G_{ij} B_k B_l + \frac{2}{5} B_i B_j B_k B_l \right)\,, \tag{38}$$

respectively. For $n \geq 3$ there are multiple distinct definitions for the CS density, each characterised by the number of traces in its definition. For $(n = 3\,, D = 7)$, there are two possibilities; one definition with a single trace and another one with double trace. These are

$$\Omega_{\rm CS}^{(3)} = \varepsilon_{ijklmnp} \mathrm{Tr}\, B_p \Big(G_{ij} G_{kl} G_{mn} - \frac{4}{5} G_{ij} G_{kl} B_m B_n - \frac{2}{5} G_{ij} B_k G_{lm} B_n + \frac{4}{5} G_{ij} B_k B_l B_m B_n - \frac{8}{35} B_i B_j B_k B_l B_m B_n \Big)\,, \tag{39}$$

$$\tilde{\Omega}_{\rm CS}^{(3)} = \varepsilon_{ijklmnp} \mathrm{Tr}\, B_p \left(G_{mn} - \frac{2}{3} B_m B_n \right) \cdot (\mathrm{Tr}\, G_{ij} G_{kl})\,. \tag{40}$$

Subjecting these densities to variations of $Z^\dagger$ (or Z)

$$\Omega_{\rm CS}^{(n)} + \Lambda\, (\mathbb{1} - Z^\dagger\, Z)$$

Λ being the (now matrix valued) Lagrange multiplier, the resulting *nontrivial* gauge covariant equations in the above examples are

$$\varepsilon_{ijk}\, D_k Z\, G_{ij} = 0 \tag{41}$$

$$\varepsilon_{ijklm}\, D_m Z\, G_{ij}\, G_{kl} = 0 \tag{42}$$

$$\varepsilon_{ijklmnp}\, D_p Z\, G_{ij}\, G_{kl}\, G_{mn} = 0 \tag{43}$$

$$\varepsilon_{ijklmnp}\, (\mathrm{Tr}\, G_{ij} G_{kl}) \cdot D_p Z\, G_{mn} = 0 \tag{44}$$

which trivialise only under the appropriate symmetries, *i.e.* $\Omega_{\rm CS}^{(n)}$ become "essentially total divergence", whence they qualify as candidates for topological charge densities.

3.2. Residual non-Abelian CS densities subject to n-fold azimuthal symmetry

As in the Abelian case, we treat the three cases $n = 1, 2, 3$ below by positing suitable Ansätze that impose the appropriate symmetries that qualify the densities (37), (38) and (39)-(40) as topological charge densities. We also use the same notation for the residual coordinates as in the Abelian case.

The $n = 1$ case is special in the sense that the topological charge density in that case trivialises to the corresponding Abelian one.

Before examining each n example in detail, we define the spin matrices employed in stating the respective Ansätze. As in the Abelian case presented above, the symmetries imposed will be the azimuthal symmetries in each of the n planes in $\mathbb{R}^{2n+1}$.

The chiral Dirac representations of the $SO(2n+2)$ algebra are

$$\Sigma_{ij}^{(+)} = -\frac{1}{4}\Sigma_{[i}\tilde{\Sigma}_{j]} \quad \text{or} \quad \Sigma_{ij}^{(-)} = -\frac{1}{4}\tilde{\Sigma}_{[i}\Sigma_{j]} \tag{45}$$

in terms of the spin matrices

$$\Sigma_i = \frac{\mathbb{I} + \Gamma_{2n+3}}{2}\Gamma_i \quad \text{and} \quad \tilde{\Sigma}_i = \frac{\mathbb{I} + \Gamma_{2n+3}}{2}\Gamma_i \,,$$

Γ_{2n+3} being the chiral matrix in $2n+2$ dimensions.

In the $n = 1$ case, the field (28) is subjected to axial symmetry in the (x_1, x_2) plane of $\mathbb{R}^3$,

$$Z = \begin{bmatrix} z_1 \\ z_2 \end{bmatrix} = \begin{bmatrix} a\,\mathbb{I} + 2\,b\,\Sigma_{34} \\ c\,n^\alpha\,\tilde{\Sigma}_\alpha \end{bmatrix}, \tag{46}$$

the functions (a, b, c) depending on (ρ, z) as in the Abelian case and with the unit vector n^α

$$n^\alpha = \begin{pmatrix} \cos\, n\,\varphi \\ \sin\, n\,\varphi \end{pmatrix} \tag{47}$$

n being the winding (vortex) number in the (x_1, x_2) plane.

It turns out that the Ansatz (46) leads to an Abelian composite connection $B_i = (B_\alpha, B_z)$, so our prescription cannot supply a non-Abelian Hopfion in three dimensions.

$$\begin{aligned} B_\alpha &= 2\left[(a\,b_{,\rho} - b\,a_{,\rho})\,\hat{x}_\alpha + \frac{n}{\rho}\,c^2\,(\hat{x}\varepsilon)_\alpha\right]\Sigma_{12}\,, \\ B_z &= 2\,(a\,b_{,z} - b\,a_{,z})\,\Sigma_{12}\,, \end{aligned}$$

whose commutators

$$[B_i, B_j] = [B_i, B_z] = 0.$$

The composite curvature G_{ij} is then Abelian, and coincides with the previously constructed Abelian case.

In the $n = 2$ case, the field (28) is subjected to azimuthal symmetry in the (x_1, x_2) and (x_3, x_4) planes of $\mathbb{R}^5$,

$$Z = \begin{bmatrix} z_1 \\ z_2 \end{bmatrix} = \begin{bmatrix} a\,\mathbb{I} + 2\,b\,\Sigma_{56} \\ c_1\,n_1^\alpha\,\tilde{\Sigma}_\alpha + c_2\,n_2^A\,\tilde{\Sigma}_A \end{bmatrix}, \tag{48}$$

the functions (a, b, c_1, c_2) depending on (ρ, σ, z) as in the corresponding Abelian case, with $\alpha = 1, 2$, $A = 3, 4$ and $z \equiv x_5$. φ and χ are the azimuthal angles in the (x_1, x_2) and (x_3, x_4) planes respectively, such that the unit vectors n^α and n_2^A are parametrised as

$$n_1^\alpha = \begin{pmatrix} \cos\, n_1\,\varphi \\ \sin\, n_1\,\varphi \end{pmatrix} \quad n_2^A = \begin{pmatrix} \cos\, n_2\,\chi \\ \sin\, n_2\,\chi \end{pmatrix}, \tag{49}$$

(n_1, n_2) being the winding (vortex) numbers in each of the two planes respectively.

Substitution of (48) in the CS density (38) yields

$$\Omega_{\rm CS}^{(5)} = 96\, i\, \frac{n_1 n_2}{\rho\sigma}\, c_1\, c_2\, [-4 + 3(c_1^2 + c_2^2)] \cdot \det \begin{vmatrix} a & b & c_1 & c_2 \\ a_{,\rho} & b_{,\rho} & c_{1,\rho} & c_{2,\rho} \\ a_{,\sigma} & b_{,\sigma} & c_{1,\sigma} & c_{2,\sigma} \\ a_{,z} & b_{,z} & c_{1,z} & c_{2,z} \end{vmatrix} . \qquad (50)$$

Note that this non-Abelian density (50) differs qualitatively from the corresponding Abelian one (17) due to the appearance of the prefactor

$$[-4 + 3(c_1^2 + c_2^2)] .$$

When (50) is varied subject to the constraint $a^2 + b^2 + c_1^2 + c_2^2 = 1$, it turns out to be "essentially total divergence". using the polar parametrisation in (15), this becomes a total divergence.

Again, requiring the field configurations in question have the asymptotic values (18) as in the Abelian case, one finds the topological charge

$$Q_{\rm CS}^{(5)} = -12\, n_1\, n_2\, m\, \pi^3 .$$

4. Summary

The scope of this talk is restricted to the consideration of the topological charges of possible Hopfions in $2n+1$ dimensions. Since the topological charges are the volume integrals of Chern-Simons densities in the appropriate dimension. The latter however are not total divergence, and to qualify as topological charge densities they must be subjected to the appropriate symmetries that render them total divergences. This is the main technical task carried out.

The prescription used to achieve this is the imposition of azimuthal (axial) symmetry in each of the n, 2-dimensional planes inside $\mathbb{R}^{2n+1}$.

The Chern-Simons densities in question are defined my means of the composite connections of the complex sigma model appropriate to the dimension. Composite connections can be both Abelian and non-Abelian, depending on the choice of sigma model. Here, the Abelian connections employed in $2n+1$ dimensions are those of the $\mathbb{C}P^n$ models, while the non-Abelian ones pertain to $2^{n+1} \times 2^n$ (complex) Grassmannian sigma models. We have hence used the nomenclature of Abelian and non-Abelian Hopfions in the title.

In the Abelian case, it has been possible to carry out this task for the case of arbitrary n, by induction. In the non-Abelian case, our consideration here have been restricted to $n = 2$, *i.e.* $\mathbb{R}^5$, due to technical complexity. It turns out also, that $n = 2$, *i.e.* 5, is the lowest dimension that can admit a non-Abelian Hopfion topological charge.

Post Script The above invoked criterion, namely that the Chern-Simons density reduces to total divergence, eliminates the possibility of $U(1)$ gauging of the CP^1 Hopfion on $\mathbb{R}^3$.

The Covariant derivative of the $U(1)$ gauged CP^1 field is

$$D_i Z = \partial_i Z + i(B_i \mathbb{I} + A_i \sigma^3) Z$$

where A_i is the Maxwell connection, and B_i the composite connection (5). This extended connection and its curvature are

$$\begin{aligned} & B_i \mathbb{I} + A_i \sigma^3 \\ & G_{ij} \mathbb{I} + F_{ij} \sigma^3 , \quad F_{ij} = \partial_{[i} A_{j]} . \end{aligned}$$

The resulting Chern-Simons density is

$$\tilde{\Omega}_{\rm CS} = \Omega^{(3)}_{\rm CS}[B_i] + \varepsilon_{ijk}\, A_i\, F_{ij}\,, \tag{51}$$

where $\Omega^{(3)}_{\rm CS}$ is the CS density for the composite connection B_i, (11).

While $\Omega^{(3)}_{\rm CS}$ in (51) reduces to a total divergence when subjected to symmetry, the second term in (51) is does not. Consequently, the CS density of the $U(1)$ gauged CP^1 model does not qualify as a topological charge density of that (putative) Hopfion.

Acknowledgments
We would like to thank to Olaf Lechtenfeld, Muneto Nitta, Yakov Shnir and Mikhail Volkov for clarifying discussions. E.R. gratefully acknowledges support by the DFG, in particular, also within the DFG Research Training Group 1620 "Models of Gravity".

References

Manton, N. S., Sutcliffe, P. 2004 *Topological solitons*, Cambridge, UK: Univ. Press.

Faddeev, L. D., Niemi, A. J. 1996 Knots and particles. *Nature* **387**, 58–61.

Battye, R. A., Sutcliffe, P. M. 1998 Knots as stable soliton solutions in a three-dimensional classical field theory *Phys. Rev. Lett.* **81** (22), 4798-4801.

Radu, E., Volkov, M. 2008 Stationary ring solitons in field theory Knots and vortons. *Phys. Rept.* **468** (4), 101–151.

Vakulenko, A. F., Kapitansky, L. V. 1979 Stability of solitons in $S(2)$ in the nonlinear sigma model . *Sov. Phys. Dokl.* **24**, 433–434.

Lin, L., Yang, Y. 2004 Existence of energy minimizers as stable knotted solitons in the Faddeev model. *Commum. Math. Phys.* **42** (2), 273–203.

Journal of Physics: Conference Series **544** (2014) 012025 doi:10.1088/1742-6596/544/1/012025

The spectrum of tightly knotted flux tubes in QCD

Roman V. Buniy
Chapman University, Schmid College of Science, Orange, CA 92866
E-mail: roman.buniy@gmail.com

Jason Cantarella
Department of Mathematics, University of Georgia, Athens, GA 30602
E-mail: jason.cantarella@gmail.com

Thomas W. Kephart
Department of Physics and Astronomy, Vanderbilt University, Nashville, TN 37235
E-mail: tom.kephart@gmail.com

Eric Rawdon
Department of Mathematics, University of St. Thomas, St. Paul, MN 55105
E-mail: ericrawdon@gmail.com

Abstract. We fit the observed J^{++} mesonic mass spectrum with one and two parameter models in terms of energies for tightly knotted and linked chromoelectric QCD flux tubes. We predict a possible new state at approximately 1190 MeV and many new states above 1690 MeV.

1. Introduction

Many experiments can be interpreted as signatures of unusual mesonic states, i.e., bosonic hadrons that are not pure $q\bar{q}$ [Nakamura *et al.*, 2010]. Our interest is in one of the most studied and least understood classes of particles in QCD–glueballs—states with no valence quarks, e.g., string loops à la Nielsen-Olesen [Nielsen & Olesen, 1973] like closed flux tubes [1]. Here we model all J^{++} mesonic states, i.e., all f_J and f'_J states listed by the Particle Data Group (PDG) [Nakamura *et al.*, 2010] (f states, for brevity), as knotted or linked chromoelectric QCD flux tubes [Buniy & Kephart, 2003]. Hence we will use the term "glueball" loosely as a shorthand for any J^{++} state in QCD.

2. Model

In [Buniy & Kephart, 2003] two of us argued how to generalize various classical ideas from plasma physics to a semiclassical model of knotted and linked configurations in QCD. Here we discuss a

[1] We assume confinement is due to monopole condensation in terms of 't Hooft's dual confinement picture, with chromoelectric flux tubes of uniform cross section.

Journal of Physics: Conference Series **544** (2014) 012025 doi:10.1088/1742-6596/544/1/012025

major reanalysis carried out recently by the present authors [Buniy *et al.*, 2014]. Several recent developments are included in the new analysis. First and foremost are the new set of ideal knot and link lengths obtained in [Ashton *et al.*, 2011]. In 2002 when [Buniy & Kephart, 2003] was written, only a handful of knot and link length were known, and some of those only to an accuracy in the 5–10% range. Ashton et al., [Ashton *et al.*, 2011] provides us with hundreds of ideal knot and link lengths to better than 1% accuracy. Since we deal with knotted physical flux tubes, not ideal knots, there are curvature correction due to energy consideration from the bending of the tubes and these corrections have now been included [Buniy *et al.*, 2014]. Angular momentum is treated differently because of the new assumption that the tube diameter is smaller and in the ~ 0.1 fm range. More details can be found in [Buniy *et al.*, 2014].

Let us first recall that Maxwell's equations for an ideal plasma imply that flux lines are locked into the plasma flow. This means that if the flux lines are knotted or linked, then this flux line topology is conserved as the flow evolves. A classical consequence of conservation of topology involves the concept of helicity and its conservation in a plasma [Woltier, 1958; Moffatt, 1969, 1985]. Helicity in our context corresponds to the degree of Gaussian linking of QCD flux tubes. Shortening while keeping flux tube topology fixed leads to tight knots [Katritch *et al.*, 1996; Woltier, 1999; Pieranski & Przybyl, 2001; A. Stasiak & Pieranski, 1998] in QCD.

It is interesting to contrast knots in QCD with knots in plasma physics [Moffatt, 1990]. In the QCD case, which is a quantum (or at least semiclassical) system, the tube diameters are fixed. The length of a tube can shorten by loosing energy via particle emission. E.g., in a typical tightening process a loose knot K^* would undergo a transition like $K^* \to K + N\pi$, i.e., to become a tight knot K plus some number N of pions. In the plasma case the knot can tighten and at the same time conserve energy by changing its diameter. Diameter change is allowed classically.

There is an infinite family of topologically different link types distinguished by topological invariants. Many of the simpler configurations have been tabulated; see [Doll & Hoste, 1991] and [Hoste *et al.*, 1998]. However, regardless of which invariants are used to identify the configurations, the "frozen-in field" hypothesis implies that tube topology is conserved. Hence, we conjecture that, to zeroth order, the ground states of all systems of flux tubes, with any type of nontrivial linking, are the states with the shortest length tubes. We also consider self-linked (knotted) flux tubes, where the associated tight knot ground states are analogously defined. The simplest example of a nontrivially knotted flux tube has the form of the trefoil knot 3_1. This configuration has dimensionless knot energy (i.e., length/diameter) which has been numerically calculated to be $\varepsilon_0(3_1) \approx 16.3715$ [Pieranski & Przybyl, 2012], but note that there are no known analytic forms for the lengths of any tight knots or links with nonplanar elements.

We consider a high energy hadron-hadron collision in the process of rehadronization, where there are baryons, mesons and quantized fluxes confined to tubes. If the tubes are open, with quarks and anti-quarks at their ends, then they are excited baryon or meson states. Our interest is in closed linked and knotted tubes. As a key part of our model, we identify all the f states as knotted or linked QCD chromoelectric flux tubes. The topological quantum numbers are what stabilizes the configurations, so we assume that non-topological (i.e., unknotted/unlinked) closed flux tube are too unstable to have measurable widths. A tightly knot/link K will have a corresponding state $f(K)$. Nontrivial knotting and linking leads to quasi-stable generalized minimum energy states. The generalized minimum energy state is the one that minimizes $\epsilon_0(K)$. We will see below that the approximation of a fixed energy per unit length can be improved by an analysis of the effect of field rearrangement within a bent tube.

To be more specific we now summarize the model assumptions:

(i) There is a one-to-one correspondence between f states and tightly knotted and linked chromoelectric flux tubes.

(ii) The flux is quantized with one flux quantum per tube.

(iii) Knotted and linked flux tubes are stabilized by topological quantum numbers.
(iv) The tube diameter is in the $\sim$ 0.1 fm range. (This corresponds to a string tension of approximately 400 MeV, which agrees with lattice estimates.)
(v) The quantity J in an f_J or f'_J state is the intrinsic angular momentum of the associated knotted solitonic solution of the QCD field equations.
(vi) The relaxation to a tight state configuration (via processes where no topology change is involved) is faster than its decay rate (via processes with topology change) for an f state, i.e., $\tau_{\text{relax}} \ll \tau_{\text{decay}}$.

One modification from [Buniy & Kephart, 2003] is that we now assume J is the intrinsic rather than the rotational angular momentum since the tube diameter is now assumed to be smaller and hence the rotational energy level spacing is larger, $\sim$ 500 MeV, as opposed to a few MeV for the thicker tubes assumed in [Buniy & Kephart, 2003]. The other significant modification is that we correct the energy due to tube curvature and include estimated errors due to other physical corrections, see below.

To proceed with the model we identify ideal knot and link lengths with glueballs and/or predicted glueballs, where we include all f states. The lightest candidate is the $f_0(500)$, which we identify with the shortest knot or link, i.e., the Hopf link 2^2_1; the $f_0(980)$ is identified with the next shortest knot or link, in this case the trefoil knot 3_1, etc.

Our initial one-parameter fit of the data is shown in Figure 1. The slope is $\Lambda_{\text{tube}} = 57$ MeV and $\chi^2 = 84$. (The χ^2 test is a measure of goodness of fit, and small values on the order of one per data point indicate the fit assumptions are compatible with the data. We define $\chi^2 = \Sigma_i \frac{O_i - E_i}{E_i}$ where the O_i are observed values and E_i are expected values from the assumptions of the fit. For more details see [Nakamura *et al.*, 2010].) The fit is poor mainly because of the constraint imposed by the very small error bars on the masses of the $f_2(1270)$ and the $f_1(1285)$. We will now see how to improve the fit when ideal tubes are replaced with physical tubes. Note we are already assuming errors of 3% on the knot lengths in the figure which anticipates this replacement.

3. Curvature corrections

So far we have assumed uniform flux across the cross section of the tubes, but this is not necessarily the case for curved tubes. This leads us to define a new energy functional for tubes which we call the "flux tube energy" as discussed below.

The magnetic field of an ideal toroidal solenoid with fixed flux falls like $1/\rho$ from the symmetry axis. To see this, we hold the flux Φ fixed and vary the field to find the functional form of the energy W for a toroidal solenoid. The general form of the energy in toroidal coordinates is $W = \frac{1}{2}\int_D B^2 \rho\, dz\, d\rho\, d\alpha$. The $dzd\rho$ integration runs over the cross section of the tube D. Since B is independent of α, the α integration gives $W = \pi \int_D B^2 \rho\, dz\, d\rho$. The flux through D is $\Phi = \int_D B\, dz\, d\rho$. Now we vary W with respect to B while holding Φ fixed which is equivalent to considering $\delta(W - \lambda\Phi) = 0$, where λ is a Lagrange multiplier. For unit vector n normal to the cross section, the variation of B gives $\int_D (2\pi B\rho - \lambda n) \cdot \delta B\, dz\, d\rho = 0$, which vanishes for arbitrary δB only if $B(\rho) = \frac{\lambda n}{2\pi\rho}$. We find λ from the requirement $\Phi = \text{const}$, which gives $B = \frac{\Phi}{\rho I}$, $W = \frac{\pi\Phi^2}{I}$ with $I = \int_D \frac{dz\, d\rho}{\rho}$.

To calculate the integral I over the cross section of a torus of major radius R_2 and minor radius R_1, it is convenient to introduce polar coordinates (r, θ) with the origin at the center of disk D, plus a toroidal angle α. The result of integration over D is $I = 2\pi\left[R_2 - \sqrt{R_2^2 - R_1^2}\right]$,

Journal of Physics: Conference Series **544** (2014) 012025 doi:10.1088/1742-6596/544/1/012025

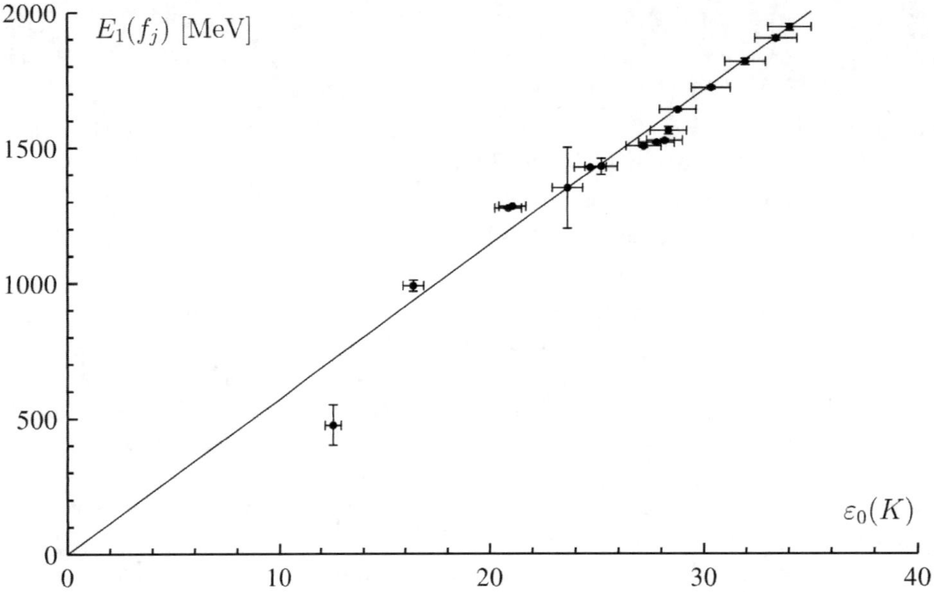

Figure 1. Fit using uncorrected knot/link lengths: Fit of the f_J states data to tight knot and link lengths (ropelengths). Errors are shown for the states, but they are too small to be visible for the lengths of ideal knots and links, however we include a 3% error in the knot energies due to the fact that we are dealing with physical knots and links; see discussion in the text. Non-fitted knots and links are not shown. Here $\chi^2 = 84$ is rather high. This is due to the tension in the data caused by the incompatability of the small experimental errors in the masses of the $f_0(980)$, the $f_2(1270)$ and the $f_2(1285)$ with the fit of the rest of the data.

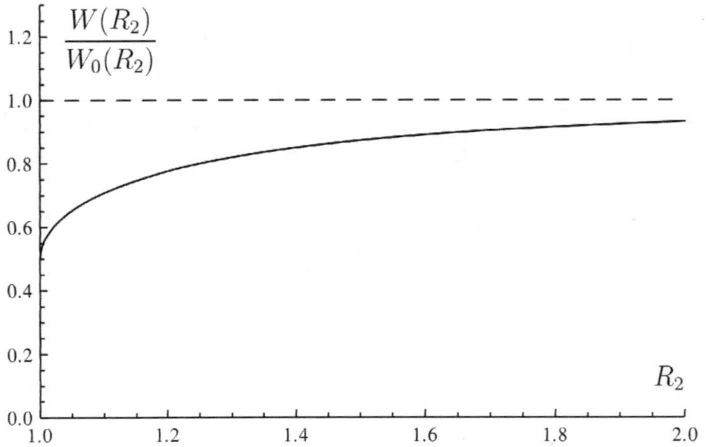

Figure 2. The function $W(R_2)/W_0(R_2)$ (solid curve) and its approximation (dashed line) for $R_2/R_1 \gg 1$ and $R_1 = 1$.

which leads to

$$W(R_2) = \frac{\Phi^2}{2\left[R_2 - \sqrt{R_2^2 - R_1^2}\right]}.$$

Journal of Physics: Conference Series **544** (2014) 012025 doi:10.1088/1742-6596/544/1/012025

The analogous result for the straight cylinder of length $2\pi R_2$ is $W_0(R_2) = \frac{\Phi^2 R_2}{R_1^2}$, and so the ratio

$$\frac{W(R_2)}{W_0(R_2)} = \frac{(R_1/R_2)^2}{2[1-\sqrt{1-(R_1/R_2)^2}]},$$

the graph of which is plotted in Fig. 2, defines our new dimensionless energy $\varepsilon(K)$ for the toroidal solenoid.

More generally, for an embedded tube K of fixed radius R_1 and parametric centerline curve $\gamma(s)$ with curvature $\kappa(s)$, we define the flux tube energy $\varepsilon(K)$ by the integral

$$\varepsilon(K) = \frac{1}{2\pi R_1^2}\left(L + \int_0^L \sqrt{1 - R_1^2\kappa^2(s)}\, ds\right), \tag{1}$$

where $L = \int_\gamma ds$ is the length of the center line. (In numerical studies of tight knots and links, it is observed that the integral in Eq. (1) is typically $\sim \frac{\sqrt{3}L}{2}$ which translates into an $\sim$7% correction of the energy from the ropelength value.)

We minimize the original $\epsilon_0(K)$ energy numerically using `ridgerunner` [Ashton *et al.*, 2011], and then compute the $\varepsilon(K)$ energy for the $\epsilon_0(K)$-minimizing configurations on the grounds that the difference between $\varepsilon(K)$-minimizing and $\epsilon_0(K)$-minimizing configurations are likely to be small. In Figure 3 we have histogrammed the shortest 72 knots and links after curvature corrections have been applied.

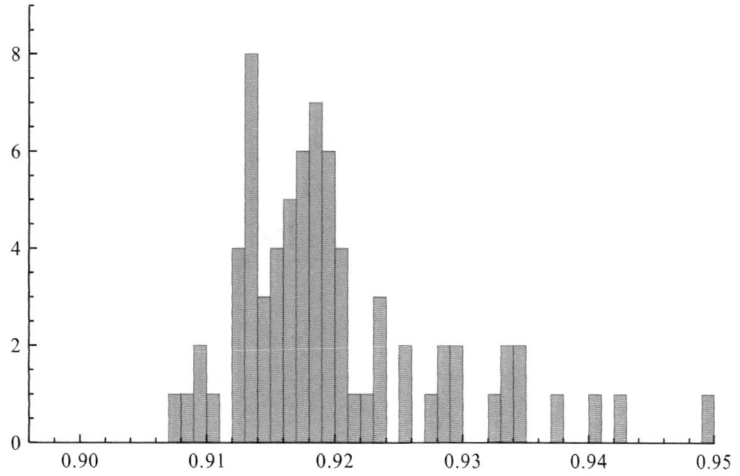

Figure 3. Histogram of the magnitudes of the curvature corrections to the first 72 knot and link lengths.

As an example of a case where corrections can be calculated exactly, consider the chain of three unknots $2_1^2\#2_1^2$ which has length $6\pi + 2 \approx 20.8496$ and has $R_2 = 2R_1$ in curved regions and $R_2 \to \infty$ in straight sections. We find an overall corrected value

$$\varepsilon(2_1^2\#2_1^2) = \frac{1}{4(2-\sqrt{3})}6\pi + 2 \approx 0.933013(6\pi + 2) \approx 19.4529$$

In addition to the curvature corrections, tubes can be constricted due to being wrapped by another section of the tube or distorted by wrapping tightly around another section of tube (like a rope wrapped tightly around a post). We have approximated such corrections and have included them in our error estimates.

Journal of Physics: Conference Series **544** (2014) 012025 doi:10.1088/1742-6596/544/1/012025

4. Results

In our model, the chromoelectric fields are confined to the knotted and linked tubes, each carrying one quantum of conserved flux. The energy is positive and, to first approximation, proportional to the length of the tube l and thus the minimum of the energy is achieved by shortening l (i.e., tightening the knot), subject to the curvature correction discussed above and other corrections discussed below.

We proceed to identify knotted and linked QCD flux tubes, i.e., curvature corrected physical flux tubes, with glueballs and/or predicted glueballs, where we include all f states as described above, but by the fourth knot/link, the ordering begins to be reshuffled due to the curvature corrections, see [Buniy *et al.*, 2014] for details.

All knot and link lengths have been calculated for states corresponding to energies well beyond 2 GeV. Above $\sim$ 2 GeV the number of knots and links grows rapidly, and so the corresponding hadronic states should become dense relative to their typical width. Hence we have confined our investigations to knot lengths corresponding to all known f states below $\sim$ 2 GeV.

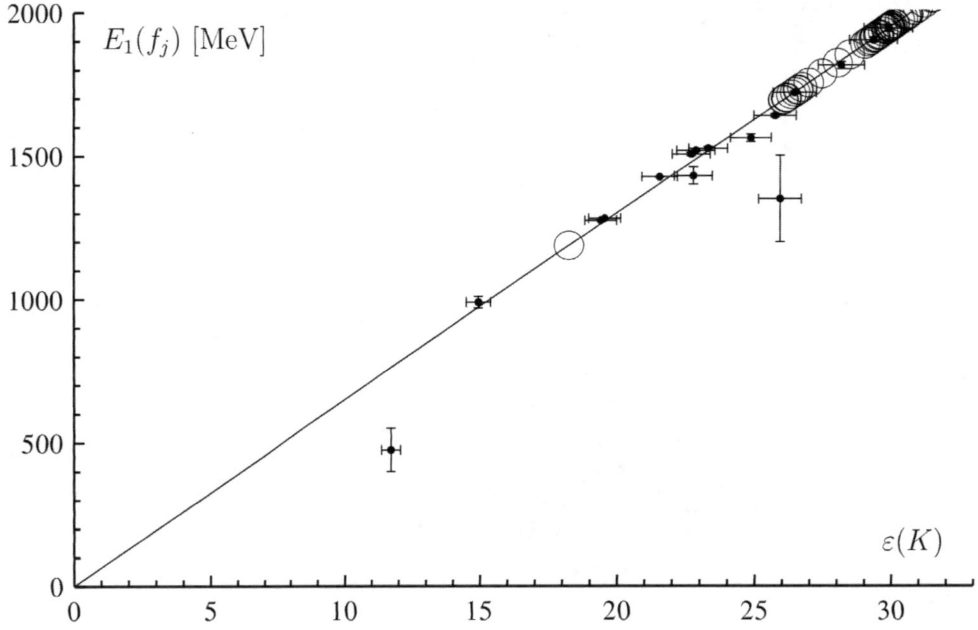

Figure 4. Curvature-corrected lengths: The combined set of measured and fitted states (dots with error bars) and predicted states (circles) for the high $f_0(1370)$ one parameter fit. Note that the fit has been much improved over that in Figure [1]. χ^2 has been reduced to 33. The curvature corrections have brought the $f_0(980)$, $f_2(1270)$ and $f_2(1285)$ in line with the rest of the data.

We will give two interpretations of the data. The first possibility is with the $f_0(1370)$ identified with the 5_1 knot which results in a prediction of a new state near 1190 MeV identified with the 4_1^2 link. Alternatively we can identify the $f(1370)$ with the 4_1^2 which gives our best fit.

We are now ready to discuss the details of the fit possibilities. Since the error on the $f_0(1370)$ mass is rather large ($\pm$150 MeV) it can be identified with several different knots and links, with minimal effect on χ^2. However, the identifications of the other f-states are much more constrained due to the small errors on their masses and available nearby knot lengths. This in turn restricts the identification of the $f_0(1370)$ to two allowed choices with reasonable χ^2s. We call the case discussed above the high-fit, where we identify the $f_0(1370)$ with the 5_1 knot, and

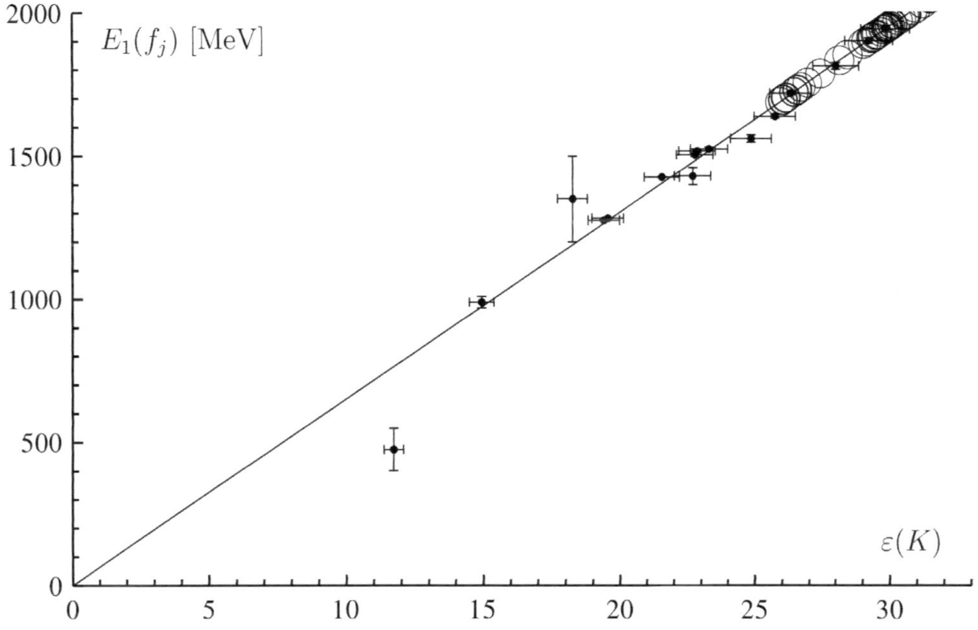

Figure 5. Low $f_0(1370)$ one parameter fit for curvature-corrected lengths: This figure shows the combined set of measured and fitted states (dots with error bars) and predicted states (circles). The reordering of $f_0(1370)$ has provided one further improvement in the fit to $\chi^2 = 28$.

the low-fit where we identify the $f_0(1370)$ with the 4_1^2 link. The high-fit predicts a new state at $\sim$ 1190 MeV, while the low-fit gives a one-to-one match between the first 12 f-states and the first 12 knots and links. The low-fit gives a somewhat better χ^2, but the high-fit gives an acceptable χ^2 and could be required if for instance the 1220 MeV state of [Breakstone *et al.*, 1990] is confirmed making it necessary to free up a low energy knot/link to identify with it.

The low-fit leaves no gaps in the spectrum until we get near 1700 MeV. The χ^2 is somewhat better than that for the high fit, 28 vs 33, and the value of Λ_{tube} is similar to the high-fit case. We find

$$E(K) = \Lambda_{\text{tube}}\, \varepsilon(K), \quad \Lambda_{\text{tube}} = 65.16 \pm 0.61\,\text{MeV},$$

which is our best overall fit to the data. We have tabulated the fitted energies for the first 72 knots and links, identifying all f-states of energy less than 2 GeV, hence giving predictions of many new states at 1690 MeV and above. See [Buniy *et al.*, 2014] for details.

Next let us discuss the statistical significance of our results. First we have done a number of tests to determine if our hypothesis that glueballs are knotted flux tubes is likely to be correct. For example, the well-known Kolmogorov-Smirnov test, which tests distribution functions $\hat{F}(y)$ and $F(y)$ by means of the quantity $\sup_y |\hat{F}(y) - F(y)|$, which in this case are the distribution functions of the glueball masses and calculated knot and link lengths. The resulting p-value for the Kolmogorov-Smirnov test is $p_{\text{KS}} = 0.95$, hence our model is in excellent agreement with the data. The Kolmogorov-Smirnov test is a measure of goodness-of-fit. A number of other goodness-of-fit tests as well as several variance tests all show excellent agreement between model and data [Buniy *et al.*, 2014].

Finally, let us make one final comment about the model. A more conservative approach is to assume only the 0^{++} states, i.e., the f_0 states, correspond to knotted/linked flux tubes. In that case there are only five states to fit and the result is displayed in Figure 6 for comparison, where the identified states are

$$[f_0(500), f_0(980), f_0(1370), f_0(1500), f_0(1710)] \longleftrightarrow [2_1^2, 3_1, 5_1, 5_2, 6_2].$$

The slope is essentially unchanged, the $R^2 = .998$ value is roughly the same. The masses of the predicted states can be easily gotten by rescaling the knot lengths in the table with the new slope parameter $\Lambda_{\text{tube}} = (65.50 \pm 1.81)\,\text{MeV}$. Note that the $f_0(1370)$ has been moved in the ordering to improve the fit. More data is needed to distinguish between the fit of all f_J states and the restricted f_0 fit. The $\chi^2 = 15.62$ for the fit is not particularly good but it only takes

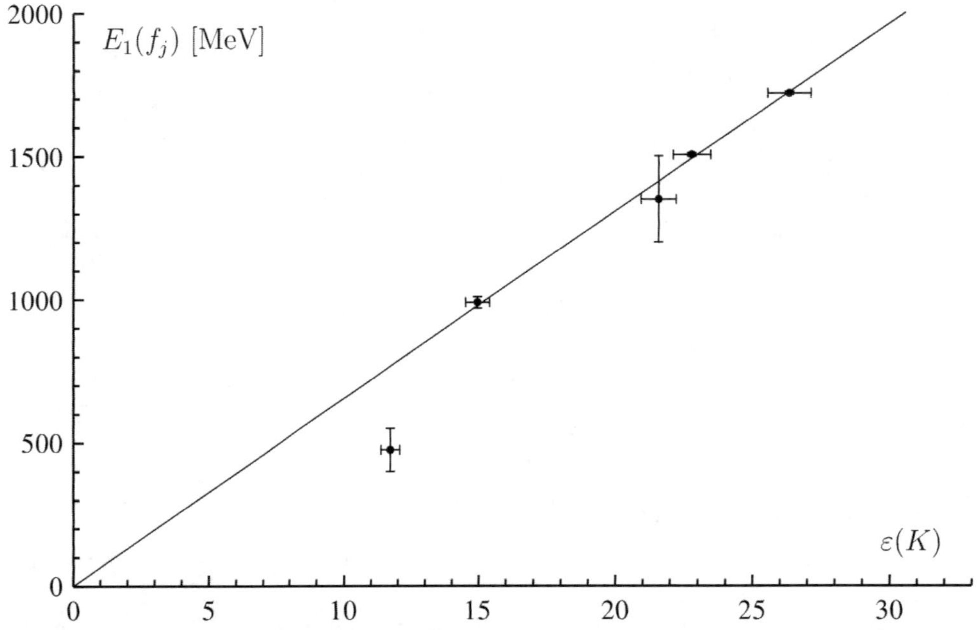

Figure 6. The f_0 states data is fitted to the curvature corrected knot and link data. Errors are shown for the states and estimated to be 3% for the knot/link energy. Non-fitted knots and links are not shown.

replacing the new $f_0(500)$ values with the old $f_0(600)$ mass and error bars to get $\chi^2 = 0.56$. So if the shift to the new PDG $f_0(500)$ mass value is due to mixing and one could extract the unmixed value for the pure gluonic state it is possible that the fit would improve again.

If a sufficient number of f_J states are found, so that they outnumber the total number of knots and links, then this would be evidence to support the restricted fit (since all short knots and links are presumed to be known). However, this is not the case at present.

5. Discussions and conclusions

We have considered hadronic collisions that produce some number of baryons and mesons plus gluonic states in the form of a closed QCD flux tubes. From an initial state, the fields in the flux tubes quickly relax to an equilibrium configuration, which is topologically equivalent to the initial state. (We assume topological quantum numbers are conserved during this rapid process.) The tube radius is set by the confinement scale, so to lowest order the energy of the final state depends only on the topology of the initial state and equals the length l_K of the tube times the average energy per unit length, or the dimensionless knot or link length $\varepsilon_0(K)$ times the

Table 1. Low $f_0(1370)$ curvature corrected fit: Comparison of the glueball mass spectrum and fit energies for $\varepsilon(K)$ less than ~ 29. (See [Buniy *et al.*, 2014] for an extended table.)

State	Mass	K	$\varepsilon(K)$	$E_1(K)$
$f_0(500)$	475 ± 75	2_1^2	11.724	764
$f_0(980)$	990 ± 20	3_1	14.943	974
$f_0(1370)$	1350 ± 150	4_1^2	18.250	1189
$f_2(1270)$	1275.1 ± 1.2	4_1	19.411	1265
$f_1(1285)$	1282.1 ± 0.6	$2_1^2\#2_1^2$	19.556	1274
$f_1(1420)$	1426.4 ± 0.9	5_1	21.559	1405
$f_2(1430)$	≈ 1430	$2_1^2\#3_1$	22.697	1479
$f_0(1500)$	1505 ± 6	5_2	22.779	1484
$f_1(1510)$	1518 ± 5	5_1^2	22.866	1490
$f_2'(1525)$	1525 ± 5	6_3^3	23.309	1519
$f_2(1565)$	1562 ± 13	6_1^2	24.854	1619
$f_2(1640)$	1639 ± 6	7_7^2	25.735	1677
		6_2^2	25.924	1689
		6_1	26.025	1696
		$2_1^2\#4_1^2$	26.046	1697
		$3_1\#3_{1m}$	26.135	1703
		$3_1\#3_1$	26.151	1704
		6_2	26.158	1704
$f_0(1710)$	1720 ± 6	6_1^3	26.327	1715
		$(2_1^2\#2_1^2\#2_1^2)_{\rm kc}$	26.449	1723
		$2_1^2\#4_1$	26.466	1724
		6_3	26.567	1731
		6_3^2	26.590	1733
		7_8^2	26.720	1741
		6_2^3	26.963	1757
		$(2_1^2\#2_1^2\#2_1^2)_{\rm lc}$	27.449	1788
$f_2(1810)$	1815 ± 12	7_1	28.018	1826
		8_7^3	28.152	1834
		8_{19}	28.458	1854

energy scale parameter $\Lambda_{\rm tube}$. While related to $\Lambda_{\rm QCD}$ by constants of order unity, $\Lambda_{\rm tube}$ can be more accurately determined (see above) and hence could be a useful dimensionful parameter in studying other properties of QCD, such as scattering and hadronization processes. The relaxation proceeds through minimization of the field energy. This process occurs via shrinking the tube length and the process halts to form a tight knot or link.

Details of knot excitations would be interesting to investigate, as would other quantum corrections, but at present we do not have a reliable way to estimate these effects, nor do we have a good way to calculate glueball decay rates. However, we do expect high mass glueball production to be suppressed because more complicated non-trivial topological field configurations are statistically disfavored and we also expect higher mass glueballs to be relatively less stable.

In addition to not fitting naturally into the quark model, glueballs have some other characteristic signatures, including enhanced production via gluon rich channels in the central rapidity region, branching fractions incompatible with $q\bar{q}$ decay, very weak coupling to $\gamma\gamma$, and OZI suppression. All the f-states have some or all of these properties. For instance, none have

Table 2. High $f_0(1370)$ curvature corrected fit: Comparison of the glueball mass spectrum and fit energies for $\varepsilon(K)$ less than ~ 30. Except for the $f(4^2_1)$, this table contains only the fitted PDG states. Predictions for other states can be gotten by multiplying the knot energy by the appropriate value of Λ_{tube} as given in [Buniy *et al.*, 2014].

State	Mass	K	$\varepsilon(K)$	$E_1(K)$
$f_0(500)$	475 ± 75	2^2_1	11.724	763
$f_0(980)$	990 ± 20	3_1	14.943	972
$f(4^2_1)$		4^2_1	18.250	1187
$f_2(1270)$	1275.1 ± 1.2	4_1	19.411	1263
$f_1(1285)$	1282.1 ± 0.6	$2^2_1\#2^2_1$	19.556	1272
$f_1(1420)$	1426.4 ± 0.9	5_1	21.559	1403
$f_0(1500)$	1505 ± 6	$2^2_1\#3_1$	22.697	1477
$f_2(1430)$	≈ 1430	5_2	22.779	1482
$f_1(1510)$	1518 ± 5	5^2_1	22.866	1488
$f_2'(1525)$	1525 ± 5	6^3_3	23.309	1516
$f_2(1565)$	1562 ± 13	6^2_1	24.854	1617
$f_2(1640)$	1639 ± 6	7^2_7	25.735	1674
$f_0(1370)$	1350 ± 150	6^2_2	25.924	1687
$f_0(1710)$	1720 ± 6	$(2^2_1\#2^2_1\#2^2_1)_{\text{kc}}$	26.449	1721
$f_2(1810)$	1815 ± 12	7_1	28.018	1823
$f_2(1910)$	1903 ± 9	7_2	29.330	1908
$f_2(1950)$	1944 ± 12	$2^2_1\#5_2$	29.840	1941

substantial branching fractions to $\gamma\gamma$. However, mixing with $q\bar{q}$ isoscalar states can obscure some of these properties. All these observations are in qualitative agreement with the model presented here.

Our high-fit model predicts one new state at 1190 MeV, twelve states concentrated near 1700 MeV and a tower of higher mass states with the next dense concentration starting near 1900 MeV. The low-fit model makes similar prediction except that there is no new state near 1200 MeV. We have argued that there is sufficient tension in the experimental data in these regions to allow the identification of many more states with knots and links. A careful statistical analysis of the data of all f-regions to resolve hidden states is needed.

It would be trivial to extend our predictions to states above 2 GeV. One just takes the knot of appropriate lengths from [Ashton *et al.*, 2011] and scales them by the dimensionful parameter Λ_{tube} from the fit. However, there is insufficient mesonic data above 2 GeV to improve or constrain our fit. More experimental data to test the model in this region would be very welcome.

To conclude, we have given two interpretations of the f-state data with a model of knotted chromoelectric flux tubes in QCD. The first possibility is where the $f(1370)$ is identified with the 5_1 knot. This results in a prediction of a new state around 1190 MeV which is identified with the 4^2_1 link. The other possibility—which also gives our best fit—is to identify the $f(1370)$ with the 4^2_1 link to give a one-to-one matching of the first twelve f_J states with the first twelve knots and links. Experiments could help to resolve which of these two possibilities is the correct choice.

Finally we should point out that there is considerable amount of tension in the f_J data, as indicate by the χ^2s of the individual states quoted by the PDG. We would not expect our fits to be better than the fits of the data on which they are based.

Acknowledgments
We thank the Isaac Newton Institute for hospitality while this work was being completed. RVB thanks Chapman University for support. The work of TWK was supported by U.S. DoE grant number DE-FG05-85ER40226, and that of EJR by NSF DMS #1115722.

References

A. Stasiak, J. Dubochet, V. K. & Pieranski, P. 1998 **Ideal knots and their relation to the physics of real knots**. World Sci. Pub., Singapore.

Ashton, T., Cantarella, J., Piatek, M. & Rawdon, E. J. 2011 Knot Tightening by Constrained Gradient Descent. *Experimental Mathematics* **20**, 57.

Breakstone, A. *et al.* 1990 The Reaction pomeron-pomeron $\rightarrow \pi^+\pi^-$ and an unusual production mechanism for the $f_2(1270)$. *Z.Phys.* **C48**, 569–576.

Buniy, R. V., Cantarella, J., Kephart, T. W. & Rawdon, E. 2014 The tight knot spectrum in QCD. *Phys.Rev.* **D89**, 054513.

Buniy, R. V. & Kephart, T. W. 2003 A Model of glueballs. *Phys.Lett.* **B576**, 127–134.

Doll, H. & Hoste, J. 1991 A tabulation of oriented links. *Math. Computation* **57**, 747.

Hoste, J., Thistlethwaite, M. B. & Weeks, J. R. 1998 The first 1,701,936 knots. *Math. Intelligencer* **20**, 33.

Katritch, V., Bednar, J., Michoud, D., Scharein, R. G., Dubochet, J. & Stasiak, A. 1996 Geometry and physics of knots. *Nature* **384**, 142.

Moffatt, H. K. 1969 The degree of knottedness of tangled vortex lines. *J. Fluid Mech.* **35**, 117.

Moffatt, H. K. 1985 A theorem on force-free magnetic fields. *J. Fluid Mech.* **159**, 359.

Moffatt, H. K. 1990 The energy-spectrum of knots and links. *Nature* **347**, 367.

Nakamura, K. *et al.* 2010 Review of particle physics. *J.Phys.* **G37**, 075021.

Nielsen, H. B. & Olesen, P. 1973 Vortex Line Models for Dual Strings. *Nucl.Phys.* **B61**, 45–61.

Pieranski, P. & Przybyl, S. 2001 *Phys. Rev. E* **64**, 031801.

Pieranski, P. & Przybyl, S. 2012 High resolution portrait of the ideal trefoil knot. Unpublished.

Woltier, L. 1958 A theorem on force-free magnetic fields. *Proc. Nat. Acad. Sci.* **44**, 489.

Woltier, L. 1999 *Proc. Nat. Acad. Sci.* **96**, 4769.

Scaling properties of cosmic (super)string networks

C.J.A.P. Martins

Centro de Astrofísica, Universidade do Porto, Rua das Estrelas, 4150-762 Porto, Portugal

E-mail: Carlos.Martins@astro.up.pt

Abstract. I use a combination of state-of-the-art numerical simulations and analytic modelling to discuss the scaling properties of cosmic defect networks, including superstrings. Particular attention is given to the role of extra degrees of freedom in the evolution of these networks. Compared to the 'plain vanilla' case of Goto-Nambu strings, three such extensions play important but distinct roles in the network dynamics: the presence of charges/currents on the string worldsheet, the existence of junctions, and the possibility of a hierarchy of string tensions. I also comment on insights gained from studying simpler defect networks, including Goto-Nambu strings themselves, domain walls and semilocal strings.

1. Introduction

After a quest of several decades we now know, thanks to the recent LHC results (Aad *et al.*, 2012; Chatrchyan *et al.*, 2012) that fundamental scalar fields are among Natures building blocks. A pressing follow-up question is whether this Higgs-Kibble field has a cosmological role or, more generally, if there are further cosmological scalar field counterparts.

Cosmic defects (Vilenkin & Shellard, 2000; Achucarro & Martins, 2009) are one of the possible signatures of this type of field. Quite apart from this specific motivation, the study of their formation and evolution is interesting since it can span an extremely wide range of energy, length and time scales, but despite this they are always described by the same underlying physical processes. When studying defect evolution in the early universe one is looking at microscopically small defects, which can nevertheless play an important role on cosmological scales. Examples of other contexts where defects have been studied (and seen) include liquid crystals (Chuang *et al.*, 1991) and also tafoni (Garvin & Mouginis-Mark, 1981) which may be seen on the seaside and have also been identified on Mars.

This article provides a brief introduction to (and recent results from) analytic and numerical methods used to study the evolution and cosmological consequences of these networks. (Other contributions in these proceedings describe related aspects.) In particular, in a cosmological context one's main interest is often to look for (late-time, asymptotic) scaling solutions, so this will be our main emphasis in what follows.

Throughout this contribution, when saying that a given defect network is (or is not) scaling, we will be referring to the following 'linear' scaling solutions, for a characteristic lengthscale L and a root-mean squared velocity v

$$L \propto t\,, \tag{1a}$$

$$v \propto \text{const}\,. \tag{1b}$$

The physical relevance of linear scaling (for cosmic strings and superstrings, as well as for semilocal strings and global monopoles) is that it corresponds to the case where the string density remains a constant fraction of the background density, and hence can in principle paly a number of useful cosmological roles. If the characteristic lengthscale grows faster than this the network will eventually disappear, while if it grows more slowly it will eventually dominate the universe.

It should also be emphasised that we will be discussing *cosmic* defect networks: the universe is assumed to be expanding, and often one assumes for simplicity that the scale factor grows as a generic (but constant) power of physical time

$$a(t) \propto t^{\lambda} . \tag{2}$$

This is important especially when one is doing numerical simulations of defect networks. It is well known that some of the solutions that we shall describe only exist in expanding universes. If one simulates networks in Minkowski spacetime, different results will be found. (An example of this will be discussed below.)

Having said that, we should also emphasise that the methods we shall describe are in principle applicable to any defect network. Indeed they can be (and have been) applied, *mutatis mutandis*, to defect networks in other contexts (Martins & Shellard, 1997; Martins *et al.*, 2004).

2. Models for defect networks

Broadly speaking, the *ethos* of analytic modelling of cosmic defect networks is to start from statistical physics but turn it into thermodynamics. This approach was introduced by Kibble (1985), and formulated in a fully quantitative way by Martins & Shellard (1996*b*,*a*, 2002). In other words, one starts out from the knowledge of the microphysics of the defects (given, for the case of strings, by the well-known Goto-Nambu action) an uses this to obtain evolution equations for macroscopic quantities, (possibly) suitably averaged over the network.

The two main macroscopic quantities of interest are a characteristic lengthscale and an averaged (root-mean-square) velocity. The physical interpretation of the latter is fairly straightforward. As for the first, one could in principle interpret it in several different ways:

- The inter-string separation (in other words, a measure of the string density);
- The correlation length (in the usual physics sense) of the strings;
- The curvature radius of the strings.

In the context of an analytic one-scale model, one assumes that these are all equal (or at least of the same order). In the case of cosmic strings, numerical simulations have shown that this is a good assumption for scaling networks. Having said that, one can certainly imagine (and simulate) networks in which the three are significantly different (or evolve differently).

These two macroscopic quantities are the bare minimum required in order to have a quantitative defect evolution model. Naturally, models with additional degrees of freedom (such as charges and/or currents on the string worldsheet) can be constructed. The model complexity will of course increase, but the same formalism will still apply.

Naturally, having an analytic model for the evolution of defect networks has considerable computational advantages, Nevertheless, one must bear in mind that these come at a price. In going from the microphysics to the macrophysics one is necessarily forced to introduce phenomenological parameters, and these cannot be calculated *ab initio*. The only way to determine their values is therefore to measure them directly from numerical simulations, thus calibrating the model. In other words, a detailed (and quantitative) understanding of the evolution of cosmic defects necessarily requires a combination of analytic modelling and high-resolution numerical simulation.

Journal of Physics: Conference Series **544** (2014) 012026 doi:10.1088/1742-6596/544/1/012026

A detailed and rigorous derivation of the evolution equations of the analytic model for cosmic strings can be found in Martins & Shellard (1996*b*,*a*, 2002); a simplified alternative derivation (which has the advantage of being applicable to defect networks of any dimensionality) can be found in Martins & Achucarro (2008). For defects with an n-dimensional worldsheet (monopoles correspond to $n = 1$, strings to $n = 2$, and so forth) the energy density ρ, symmetry breaking scale η and characteristic lengthscale L are related via

$$L^{4-n} = \frac{\eta^n}{\rho}\,. \tag{3}$$

and the evolution equations for L and the velocity v are

$$(4-n)\frac{dL}{dt} = (4-n)HL + v^2\frac{L}{s} + \delta\epsilon\,, \tag{4a}$$

$$\frac{dv}{dt} = (1-v^2)\left(f - \frac{v}{s}\right)\,. \tag{4b}$$

Here $H = \dot{a}/a = \lambda/t$ is the Hubble parameter, f parametrises driving forces (which may be due to curvature, to long-range forces, etc), $\delta\epsilon$ describes energy losses due to interactions, and s is a damping lengthscale (including Hubble damping, friction due to particle scattering, and possibly other mechanisms). Naturally all the art in this modelling depends on how well one can describe these various terms.

It is interesting to point out that cosmic defect network evolution is to a large extent about energy flow from large to small scales (but other mechanisms, such as topological constraints, can also play a role). Defects gradually enter the horizon as the universe expands, and this energy is then lost (through loop production, annihilation or other mechanisms) on substantially small scales. In this sense, defect evolution is similar to a Richardson cascade, which is also discussed (in other contexts) elsewhere in these proceedings.

3. A walk in the zoo

This analytic model can be used to quantitatively confirm more qualitative work that had been done in the past, as well as to study defect evolution in contexts that were beyond the reach of those simpler analyses. In this section we briefly go through one example, by briskly but thoroughly discussing the evolution of networks containing monopoles (on their own or attached to various numbers of strings).

The evolution of networks of local monopoles has been shown to be very sensitive to its available annihilation mechanisms (Preskill, 1979; Martins & Achucarro, 2008). On the other hand, scaling is quite generic for the case of global monopoles, in good agreement with the first generation of numerical simulations (Bennett & Rhie, 1990; Yamaguchi, 2002). It would be highly desirable to have larger (and better resolution) numerical simulations to obtain a better calibration of the model.

Next, consider the case of networks of monopoles attached to one string. These are often referred to as hybrid networks, and they emerge as a result of the symmetry breaking sequence

$$G \to K \times U(1) \to K\,; \tag{5}$$

the monopoles are formed at the first phase transition, while at the second one the strings form connecting monopole-antimonopole pairs. In most circumstances these networks annihilate very quickly (in much less than one Hubble time after string formation) (Martins, 2009). Nevertheless, with considerable fine-tuning one can make them survive much longer than a Hubble time: one requires that both phase transitions occur around the Planck scale, and that the network is not subject to friction due to particle scattering.

Finally we have the case of networks of monopoles attached to two or more strings. There result from the symmetry breaking sequence

$$G \to K \times U(1) \to K \times Z_n ; \tag{6}$$

in the former case ($n = 2$) they are known as necklaces, while for $n \geq 3$ they are known as lattices. In both cases it can be shown that they usually reach linear scaling (Martins, 2010). However, other scaling solutions can also exist, depending on the universe's expansion rate and the network's energy loss mechanism(s).

4. Cosmic strings and walls

Attractor (large-scale) linear scaling solutions for cosmic strings are well established, and in particular they have been confirmed by everyone that simulates them—see Vilenkin & Shellard (2000) and Achucarro & Martins (2009) for overviews. Nevertheless one should bear in mind that large-scale scaling and scaling of the network's small-scale features are two conceptually distinct processes, and one can occur without the other. A further issue (which is not yet fully resolved) is the behaviour of the loop distribution; this is not crucial for our present purposes (although it is discussed elsewhere in these proceedings), but it is highly relevant for gravitational wave bounds.

In the case of cosmic strings the analytic model has been calibrated using both field theory (Moore *et al.*, 2002; Martins *et al.*, 2004) and Goto-Nambu simulations (Martins *et al.*, 2004; Martins & Shellard, 2006). Although this has been highly successful (and led to the adoption of the model for making predictions in a number of contexts, including a part of the Planck data analysis pipeline), it should be kept in mind that the simulations also indicate that extra degrees of freedom (in the form of small-scale 'wiggles') are clearly important if one requires more detailed modelling. It is equally clear that these will be crucial for the case of cosmic superstrings.

We will briefly illustrate the above point with two examples drawn from Martins & Shellard (2006). First, one often uses the intuitive picture that strings are Brownian on scales of the correlation length, but this is quantitatively not correct (for any sensible definition of the string correlation length). Strings are only Brownian outside the horizon; at scale of correlation length they can more accurately be described as a self-avoiding random walk. This is a consequence of the string intercommutings (and the ensuing loop production).

Second, it is clear from Goto-Nambu simulations in expanding universes that string velocities are anticorrelated on the scale of the correlation length. This anticorrelation depends on the expansion rate (it becomes stronger as the expansion rate is increased), and—crucially—it doesn't exist in Minkowski space. Physically this is easy to understand: it is due to string intercommutings, and specifically it is a manifestation of the fact that loop production is radically different with and without expansion. (The latter can also be confirmed by looking at loop family trees from simulations with and without expansion). Anticorrelations are easy to see in Goto-Nambu simulations but can't be seen in field theory simulations, which don't have enough spatial resolution. Additionally, accurate velocity measurements are notoriously difficult in field theory simulations. These anticorrelations are important for a quantitative assessment of the astrophysical signatures of these networks. This is likely to be the main factor explaining the difference (of more than a factor of two) between bounds on the string tension obtained from field theory and Goto-Nambu simulations.

One should also bear in mind that numerical simulations often assume a constant expansion rate (which we have denoted λ), but the real universe goes through several different epochs with different expansion rates. A string network will approach a scaling solution if it evolves sufficiently in the radiation or matter eras (or indeed any epoch with a constant λ). However, the universe goes through a relatively slow radiation-to-matter transition, and for GUT-scale strings the matter-era attractor solution would be approximately reached at a redshift just below $z \sim 1$, which happens to coincide with the onset of dark energy domination (in which context strings no longer scale). The end result is that throughout the entire period of the universe that is relevant for structure formation and CMB anisotropies a cosmic string network is *never* in a linear scaling solution, although it is always relatively close to it. This is important if one wants to make quantitative predictions for their observational consequences—Planck is again a good example.

In passing we can also discuss the case of semilocal strings (which are further discussed in Ana Achúcarro's contribution). These are not purely topological, and can end in a cloud of energy which one can effectively think of as a global monopole. They are stable, both perturbatively and to semi-classical tunnelling into vacuum.

The study of these networks is significantly more complex than that of standard cosmic strings, mostly due to the fact that the dynamics of semilocal segments is highly non-trivial, and further work on this topic is currently in progress. In has already been shown that a scaling attractor exists for some range of model parameters (Nunes *et al.*, 2011). However there can also be regimes where the whole network can disappear, and those where only the monopoles disappear (leaving behind a standard Goto-Nambu string network). Preliminary comparisons with numerical simulations are highly encouraging (a more detailed analysis will be available soon), but the main bottleneck is again measuring the defect velocities.

Also as brief passing remark, we note that there is a possible alternative analytic description of a semilocal network: instead of a network of local strings attached to global monopoles, one can instead consider a network containing two types of strings: the standard ones and also 'gradient strings' accounting for the fact that depending on the field configurations the segments might want to grow and join other segments (instead of shrinking and annihilating). This is, to some extent, analogous to the case of superstring networks.

Finally, for the case of the simplest domain wall networks (described by a single scalar field), an attractor large-scale linear scaling solution is also now well established (Leite & Martins, 2011; Leite *et al.*, 2013), despite earlier claims of deviations from scaling which are now understood to be due to limited dynamical range of earlier simulations. These simulations have also been used to obtain an accurate calibration of the analytic model for domain walls: in this case the two free model parameters have been found to have the values

$$c_w = 0.34 \pm 0.16 \tag{7}$$

for the network energy losses term and

$$k_w = 0.98 \pm 0.07 \tag{8}$$

for the walls curvature parameter. Although the cosmological presence of domain walls is very tightly constrained (since they tend to have pathological consequences), domain walls provide a simple and extremely useful toy model. Being fairly simple to simulate, they allow one to study in detail the various physical processes contributing to defect evolution.

5. How robust is scaling?

So far we have been considering the simplest defect models, and in most cases finding that a scaling solution exists. One may ask whether this result still holds for more complex models. As we will see, the answer seems to be yes: the scaling solution is indeed quite robust and it persists in many cases, although there are certainly mechanisms that can suppress it.

We start by pointing out that explicit energy losses are not absolutely necessary for scaling in a cosmological context: the expansion itself may provide sufficient damping. For strings, scaling will occur for any expansion rate $\lambda > 1/2$, regardless of energy losses (Martins, 2004);

for domain walls, the analogous threshold is $\lambda > 1/4$. Among other things this implies that for strings to scale in the matter era loop production not needed for scaling, as is known since the first generation of numerical simulations in the early 1990s. That said, note that the above paper also shows that in some string-dominated universes the correlation length grows as $L \propto a \propto t$, so this is formally a linear scaling solution, although its physical properties are somewhat different from the standard ones.

The next question has to do with the role of additional degrees of freedom on the string worldsheet. Perhaps the simplest such example is that of a charge, and it has recently been shown by Oliveira *et al.* (2012) that this need not stop scaling. More specifically these solutions fall into two regimes. If the expansion rate is large enough, the charge gets diluted and the standard scaling is recovered. Conversely if the expansion rate is small, the charge stays on the strings and there's no linear scaling solution: instead the correlation length grows more slowly and velocities decay. The transition between these two regimes depends on the balance between network energy losses and the expansion rate: with no energy losses, transition occurs for the matter era ($\lambda = 2/3$), but if energy losses are present this threshold will be lower, and for large enough losses it will be below radiation, meaning that the presence of a charge on the string need not have dramatic consequences.

Similarly, the presence of string junctions need not prevent scaling. This result has in fact been known (even if only in a specific class of models) for a long time (Vachaspati & Vilenkin, 1987). It has also subsequently been obtained for other classes of models (Avelino *et al.*, 2007), and it can occur for any number of coupled scalar fields (under the assumption that all walls have the same tension). A more general exploration of the role of defect junctions is in progress, and will be discussed elsewhere.

Finally, even the assumption of equal tensions can be relaxed: a hierarchy of different tensions need not prevent scaling. Again this result has been known for more than a decade, at least for one specific class of models (McGraw, 1998). The conditions under which it can be generalised are also under current study, and will be discussed elsewhere.

6. Realistic cosmic superstrings

From the discussion above, it should be clear that our approach to modelling realistic cosmic superstring networks is to isolate the various different physical processes that contribute to the dynamics of network, and try to understand the role of each one of them by using suitably simplified (toy) models. One can then put together this information into more accurate models and see what these imply for superstrings.

A number of these physical processes are under control, in the sense that their role is well understood. Specifically, three key examples are

- The dynamical role of monopoles and junctions, as briefly discussed above.
- The role of extra dimensions (if applicable) and topology, as first discussed in Avgoustidis & Shellard (2005).
- The role of the intercommuting probabilities, first considered in a simple way in Martins (2004) and then quantified using numerical simulations in Avgoustidis & Shellard (2006).

On the other hand, another set of relevant physical process have been identified but their understanding still requires more work (which is being done, by many people). Again I will highlight three examples

- The role of additional degrees of freedom (such as charges and currents) needs to be understood in more detail. In these models a second lengthscale must necessarily be introduced, so these will no longer be 'one-scale' models. There are in principle several different ways to do this, which may be relevant for different contexts.

- The behaviour of networks with various tension hierarchies needs to be better understood. A naive approach that has been followed is simply to treat the network as a sum of sub-networks (one for each tension) and to add together 'one-scale' models for each of these, in a minimally coupled way. However this approach (even if it is done in a way that conserves energy, which is not always the case) has its limitations, and may well miss some of the relevant physics.
- More related to the networks' observational consequences, there is the role of the non-trivial velocity correlations. As mentioned above these exist for Goto-Nambu strings, and are expected to persist (with some differences) in superstring networks. Incidentally, there has been a fair amount of recent work on the so-called kinematic constraints on cosmic superstrings (some of which is discussed elsewhere in these proceedings); although this is certainly a possible approach, to my mind it it not ideal: physically the observable effect of the presence of the junctions should be to introduce non-trivial velocity (anti)correlations on the network, and the natural way to model is to use fractal-like tools.

Although a fully self-consistent analysis of the scaling properties of cosmic string networks still remains to be carried out, partial results from various groups (and our own ongoing work) do suggest that full scaling of all the components of the network can certainly occur in significant regions of parameter space. However, there are also regions of parameter space where only the lowest tension string scales (and the heaviest ones decay), and others where no component scales and the network can eventually (at least in some cases) dominate the universe's energy budget. Further work on these issues is ongoing. Which regions of parameter space are natural (or indeed realistic) is a different issue, which is beyond the scope of this analysis.

7. Conclusions

Analytic and numerical work in the past three decades has gradually established the result that scaling attractor solutions are ubiquitous for cosmic defect networks. At least at the qualitative level, the physical reasons behind this are clear: they stem from the usual energy minimisation mechanisms. Having said that, the solution is certainly not universal: in some case thee are additional physical mechanisms that naturally act to suppress it, while in others it can be suppressed at the cost of fine-tuning or otherwise violent assumptions.

An accurate description of the evolution of cosmic defect networks requires quantitative analytic models calibrated by high-resolution and physically realistic numerical simulations. This goal has been achieved for the simplest defect networks (in particular for domain walls and cosmic strings) and will be achieved soon for others (including monopoles and semilocal strings). In the longer term, the goal will be to apply these methods to cosmic superstring networks: currently only toy models of these network have been studied (both analytically and numerically). and more realistic approaches are highly desirable.

Naturally, the ultimate goal of all this work is to improve astrophysical and cosmological constraints. Current astrophysical bounds largely rely on toy-model assumptions, and (as shown by current discrepancy between bounds obtained from Goto-Nambu and field theory simulations) are not more accurate than a factor of two. Improvements are certainly needed, so that more robust analyses can be done as more Planck data becomes available.

Acknowledgments

Many interesting discussions with my collaborators in the work discussed herein (Ana Achúcarro, Tasos Avgoustidis, A. Mafalda Leite, André Nunes, Miguel Oliveira, Paul Shellard and Jon Urrestilla) as well as with the rest of CAUPs Dark Side have shaped my views on this subject, and are gratefully acknowledged.

Our numerical simulations were performed on the COSMOS Consortium supercomputer (within the DiRAC Facility, jointly funded by STFC and the Large Facilities Capital Fund of BIS-UK) and the Advanced Computing Laboratory at University of Coimbra.

This work was done in the context of the project PTDC/FIS/111725/2009 from FCT. CJM is also supported by an FCT Research Professorship, contract reference IF/00064/2012, funded by FCT/MCTES (Portugal) and POPH/FSE (EC).

References

Aad, G. *et al.* 2012 Observation of a new particle in the search for the Standard Model Higgs boson with the ATLAS detector at the LHC. *Phys.Lett.* **B716**, 1–29.

Achucarro, A. & Martins, C. J. A. P. 2009 Cosmic Strings. In *Encyclopedia of Complexity and Systems Science*. Springer New York.

Avelino, P. P., Martins, C. J. A. P., Menezes, J., Menezes, R. & Oliveira, J. C. R. E. 2007 Scaling of cosmological domain wall networks with junctions. *Phys.Lett.* **B647**, 63–66.

Avgoustidis, A. & Shellard, E. P. S. 2005 Cosmic string evolution in higher dimensions. *Phys.Rev.* **D71**, 123513.

Avgoustidis, A. & Shellard, E. P. S. 2006 Effect of reconnection probability on cosmic (super)string network density. *Phys.Rev.* **D73**, 041301.

Bennett, D. & Rhie, S. 1990 Cosmological evolution of global monopoles and the origin of large scale structure. *Phys.Rev.Lett.* **65**, 1709–1712.

Chatrchyan, S. *et al.* 2012 Observation of a new boson at a mass of 125 GeV with the CMS experiment at the LHC. *Phys.Lett.* **B716**, 30–61.

Chuang, I., Yurke, B., Durrer, R. & Turok, N. 1991 Cosmology in the laboratory - Defect dynamics in liquid crystals. *Science* **251**, 1336–1342.

Garvin, J. B. & Mouginis-Mark, P. J. 1981 Tafoni in Basalt Boulders at South Point, Hawaii: Possible Analogs to Pitted Viking Lander Blocks on Mars. In *Lunar and Planetary Institute Science Conference Abstracts*, *Lunar and Planetary Institute Science Conference Abstracts*, vol. 12, pp. 330–332.

Kibble, T. W. B. 1985 Evolution of a system of cosmic strings. *Nucl.Phys.* **B252**, 227.

Leite, A. M. M. & Martins, C. J. A. P. 2011 Scaling Properties of Domain Wall Networks. *Phys.Rev.* **D84**, 103523.

Leite, A. M. M., Martins, C. J. A. P. & Shellard, E. P. S. 2013 Accurate Calibration of the Velocity-dependent One-scale Model for Domain Walls. *Phys.Lett.* **B718**, 740–744.

Martins, C. J. A. P. 2004 Scaling laws for non-intercommuting cosmic string networks. *Phys.Rev.* **D70**, 107302.

Martins, C. J. A. P. 2009 Evolution of Hybrid Defect Networks. *Phys.Rev.* **D80**, 083527.

Martins, C. J. A. P. 2010 Evolution of Cosmic Necklaces and Lattices. *Phys.Rev.* **D82**, 067301.

Martins, C. J. A. P. & Achucarro, A. 2008 Evolution of local and global monopole networks. *Phys.Rev.* **D78**, 083541.

Martins, C. J. A. P., Moore, J. N. & Shellard, E. P. S. 2004 A Unified model for vortex string network evolution. *Phys.Rev.Lett.* **92**, 251601.

Martins, C. J. A. P. & Shellard, E. P. S. 1996*a* Quantitative string evolution. *Phys.Rev.* **D54**, 2535–2556.

Martins, C. J. A. P. & Shellard, E. P. S. 1996*b* String evolution with friction. *Phys.Rev.* **D53**, 575–579.

Martins, C. J. A. P. & Shellard, E. P. S. 1997 Averaged methods for vortex string evolution. *Phys.Rev.* **B56**, 10892.

Martins, C. J. A. P. & Shellard, E. P. S. 2002 Extending the velocity dependent one scale string evolution model. *Phys.Rev.* **D65**, 043514.

Martins, C. J. A. P. & Shellard, E. P. S. 2006 Fractal properties and small-scale structure of cosmic string networks. *Phys.Rev.* **D73**, 043515.

McGraw, P. 1998 Evolution of a nonAbelian cosmic string network. *Phys.Rev.* **D57**, 3317–3339.

Moore, J. N., Shellard, E. P. S. & Martins, C. J. A. P. 2002 On the evolution of Abelian-Higgs string networks. *Phys.Rev.* **D65**, 023503.

Nunes, A. S., Avgoustidis, A., Martins, C. J. A. P. & Urrestilla, J. 2011 Analytic Models for the Evolution of Semilocal String Networks. *Phys.Rev.* **D84**, 063504.

Oliveira, M. F., Avgoustidis, A. & Martins, C. J. A. P. 2012 Cosmic string evolution with a conserved charge. *Phys.Rev.* **D85**, 083515.

Preskill, J. 1979 Cosmological Production of Superheavy Magnetic Monopoles. *Phys.Rev.Lett.* **43**, 1365.

Vachaspati, T. & Vilenkin, A. 1987 Evolution of cosmic networks. *Phys.Rev.* **D35**, 1131.

Vilenkin, A. & Shellard, E. P. S. 2000 **Cosmic Strings and Other Topological Defects**. Cambridge, UK: Cambridge University Press.

Yamaguchi, M. 2002 Scaling property and peculiar velocity of global monopoles. *Phys.Rev.* **D65**, 063518.

Inflation and cosmic (super)strings: implications of their intimate relation revisited

M. Sakellariadou[1]

[1] Department of Physics, King's College London, University of London, Strand, London WC2R 2LS, United Kingdom

E-mail: mairi.sakellariadou@kcl.ac.uk

Abstract. We briefly discuss constraints on supersymmetric hybrid inflation models and examine the consistency of brane inflation models. We then address the implications for inflationary scenarios resulting from the strong constraints on the cosmic (super)string tension imposed from the most recent cosmic microwave background temperature anisotropies data.

1. Introduction

Cosmological inflation Starobinsky (1980); Guth (1980); Linde (1981) offers, by construction, a simple way to evade some of the problems related to the initial conditions that plague the hot big bang model. In addition, inflation can provide a simple mechanism to get adiabatic perturbations that seem to lead to the observed large scale structure and the measured Cosmic Microwave Background (CMB) temperature anisotropies. However, despite efforts over more than three decades, inflation remains a paradigm in search of a model. Apart from the issues related to the onset of inflation Calzetta & Sakellariadou (1992, 1993); Germani *et al.* (2007), particle physics models indicate difficulties in order to satisfy the required properties of the scalar field that would play the rôle of the inflaton.

The main bulk of efforts to realise an inflationary scenario is concentrated on a scalar (single or multiple) field type of inflation. The simple minimal case of a Higgs field driven inflation Bezrukov & Shaposhnikov (2008) faces important difficulties Buck *et al.* (2010) and inconsistencies Barbon & Espinosa (2009); Atkins & Calmet (2011). In the context of supersymmetric Grand Unified Theories (GUTs), hybrid inflationary models end with the formation of cosmic strings Jeannerot *et al.* (2003), which to be compatible with the CMB measurements require fine tuning of the free parameters (couplings and mass scales) of the inflationary potential Rocher & Sakellariadou (2005a,b, 2006); Battye *et al.* (2010). Moreover, the singlets of the Standard Model (SM) seem to encounter difficulties in order to play the rôle of the inflaton, as it has been recently shown Cacciapaglia & Sakellariadou (2013) in the case of the minimal SO(10) model.

One can argue that inflation must be studied in the context of an Ultra-Violet (UV) complete theory and as such, string theory is a well-studied candidate. Within this framework, brane inflation models, which are the ones that have been more extensively studied, lead to the formation of cosmic superstrings whose observational consequences may offer a way of constraining the underlying string theory. However, Supergravity (SUGRA) constraints Komargodski & Seiberg (2009); Dienes & Thomas (2010); Komargodski & Seiberg (2010) imply the existence of severe difficulties for the realisation of at least one class (of the two studied classes) of brane inflationary models Gwyn *et al.* (2011) proposed in the context of IIB string theory. Moreover, constraints are also imposed on the allowed radiation emitted from the produced cosmic superstrings Gwyn *et al.* (2011).

The difficulties in realising a scalar field based inflationary model, which on the one hand is consistent with particle physics and CMB measurements, while on the other hand it satisfies minimal naturalness criteria, may be overthrown if one changes the gravitational sector of the theory, adding for instance an R^2-term Starobinsky (1980); Gottlober *et al.* (1991), without changing its particle content. This approach avoids on the one hand the constraints imposed due to the generically formed cosmic (super)strings, which remain undetectable Ade *et al.* (2013), and on the other hand the (by hand) addition of an extra singlet to play the rôle of the inflaton. Note that it is certainly reasonable to expect corrections to the low-energy effective gravitational action, as for instance within noncommutative spectral geometry Kurkov & Sakellariadou (2014), moreover models with high curvature terms seem to be favoured by the recent Planck CMB measurements Ade *et al.* (2013).

In what follows, we will briefly review the constraints on supersymmetric hybrid inflation, the consistency of brane inflation models, as well as the constraints on the allowed radiation of cosmic superstrings. We will then highlight some open questions which deserve to be addressed given the precision of current astrophysical and particle physics measurements and the plethora of cosmological scenarios.

2. Supersymmetric hybrid inflation

Models based upon $N = 1$ SUSY contain complex scalar fields which often have flat directions in their potential, offering candidates for inflationary models. Within this framework, hybrid inflation driven by F- or D-terms represents the standard inflationary model, which leads Jeannerot *et al.* (2003) generically to cosmic string formation at the end of the inflationary era.

F-term inflation can be naturally accommodated in the framework of GUTs when a GUT gauge group, $\mathrm{G_{GUT}}$, is broken down to the SM gauge group, $\mathrm{G_{SM}}$, at an energy scale $M_{\rm GUT}$ according to the scheme

$$\mathrm{G_{GUT}} \xrightarrow{M_{\rm GUT}} \mathrm{H}_1 \xrightarrow[\Phi_+\Phi_-]{M_{\rm infl}} \mathrm{H}_2 \longrightarrow \mathrm{G_{SM}} \ , \tag{1}$$

where Φ_+, Φ_- is a pair of GUT Higgs superfields in non-trivial complex conjugate representations, which lower the rank of the group by one unit when acquiring non-zero vacuum expectation value. The inflationary phase takes place at the beginning of the symmetry breaking $\mathrm{H}_1 \xrightarrow{M_{\rm infl}} \mathrm{H}_2$. The gauge symmetry is spontaneously broken by adding F-terms to the superpotential. The Higgs mechanism leads generically Jeannerot *et al.* (2003) to Abrikosov-Nielsen-Olesen strings, called F-term strings.

F-term inflation is based on the globally supersymmetric renormalisable superpotential

$$W^{\rm F}_{\rm infl} = \kappa S(\Phi_+\Phi_- - M^2) \ , \tag{2}$$

where S is a GUT gauge singlet left handed superfield and κ, M are two constants (M has dimensions of mass) which can be taken positive with field redefinition. The scalar potential as a function of the scalar complex component of the respective chiral superfields $\Phi_\pm, S$ reads

$$V(\phi_+, \phi_-, S) = |F_{\Phi_+}|^2 + |F_{\Phi_-}|^2 + |F_S|^2 + \frac{1}{2}\sum_a g_a^2 D_a^2 \ . \tag{3}$$

The F-term is such that $F_{\Phi_i} \equiv |\partial W/\partial \Phi_i|_{\theta=0}$, where we take the scalar component of the superfields once we differentiate with respect to $\Phi_i = \Phi_\pm, S$. The D-terms are $D_a =$

$\bar{\phi}_i \, (T_a)^i{}_j \, \phi^j + \xi_a$, with a the label of the gauge group generators T_a, g_a the gauge coupling, and ξ_a the Fayet-Iliopoulos (FI) term. By definition, in the F-term inflation the real constant ξ_a is zero; it can only be non-zero if T_a generates an extra U(1) group. In the context of F-term hybrid inflation the F-terms give rise to the inflationary potential energy density while the D-terms are flat along the inflationary trajectory, thus one may neglect them during inflation.

D-term inflation has been studied a lot in the literature, in particular since it can be naturally implemented within SUSY GUTs, SUGRA, or string theories. The gauge symmetry is spontaneously broken by introducing FI D-terms. In standard D-term inflation, the constant FI term gets compensated by a single complex scalar field at the end of the inflationary era, leading to the formation of cosmic strings, called D-term strings.

Standard D-term inflation is realised within a scheme, such as

$$\mathrm{G_{GUT}} \times \mathrm{U}(1) \xrightarrow{M_{\mathrm{GUT}}} \mathrm{H} \times \mathrm{U}(1) \xrightarrow[\Phi_+\Phi_-]{M_{\mathrm{nfl}}} \mathrm{H} \rightarrow \mathrm{G_{SM}} \ . \tag{4}$$

It is based on the superpotential

$$W = \lambda S \Phi_+ \Phi_- \ , \tag{5}$$

where S, Φ_+, Φ_- are three chiral superfields and λ is the superpotential coupling. It assumes an invariance under an Abelian gauge group $U(1)_\xi$, under which the superfields S, Φ_+, Φ_- have charges 0, +1 and −1, respectively. It also assumes the existence of a constant FI term ξ.

Note that a supersymmetric description of D-term inflation is insufficient since the inflaton field reaches values of the order of at least the Planck mass, even if one concentrates around only the last 60 e-folds of inflation. Hence, the correct D-term inflation analysis should be within the SUGRA framework Rocher & Sakellariadou (2006). Moreover, D-term inflation must be described with a non-singular formulation of supergravity when the superpotential vanishes. To construct a formulation of SUGRA with constant FI terms from superconformal theory, one finds Binetruy *et al.* (2004) that under U(1) gauge transformations in the directions in which there are constant FI terms ξ_α, the superpotential W must transform as $\delta_\alpha W = \eta_{\alpha i} \partial^i W = -i(g\xi_\alpha / M_{\mathrm{Pl}}^2) W$; one cannot keep any longer the same charge assignments as in standard supergravity. D-term inflationary models can be built with different choices of the Kähler geometry Rocher & Sakellariadou (2006) .

Since cosmic strings Sakellariadou (2007) formed within a large class of high energy physics models (SUSY, SUGRA and string theories), one should consider mixed cosmological perturbation models where the dominant rôle is played by the inflaton field and cosmic strings have a small, but not negligible, contribution Bouchet *et al.* (2002). Restricting ourselves to the angular power spectrum we remain in the linear regime, where Bouchet *et al.* (2002),

$$C_\ell = \alpha C_\ell^{\mathrm{I}} + (1-\alpha) C_\ell^{\mathrm{S}} \ ; \tag{6}$$

C_ℓ^{I} and C_ℓ^{S} denote the (COBE normalised) Legendre coefficients due to adiabatic inflaton fluctuations and those stemming from the cosmic strings network, respectively and the coefficient α is a free parameter giving the relative amplitude for the two contributions. Comparing the C_ℓ, given by Eq. (6), with data obtained from the CMB measurements, one imposes constraints on the string tension Bouchet *et al.* (2002); Bevis *et al.* (2008); Urrestilla *et al.* (2011). The most recent Planck collaboration data impose severe constraints on the string contribution. More precisely, denoting by f_{10} the fraction of the spectrum contributed by cosmic strings at a multipole of $\ell = 10$, the Planck data imply Ade *et al.* (2013), for the Nambu-Goto string model that

$$f_{10} < 0.015 \ \text{ leading to } \ G\mu/c^2 < 1.5 \times 10^{-7} \ , \tag{7}$$

Journal of Physics: Conference Series **544** (2014) 012027 doi:10.1088/1742-6596/544/1/012027

with G the Newton's constant and μ the string tension.
The corresponding constraints for the Abelian-Higgs field theory model read Ade *et al.* (2013),

$$f_{10} < 0.028 \;\text{ leading to }\; G\mu/c^2 < 3.2 \times 10^{-7}\ . \tag{8}$$

3. Brane inflation

Within IIB string theory there are two classes of brane inflation models: D3/D7 inflation Dasgupta *et al.* (2002, 2004) and brane-antibrane inflation with D3/$\overline{\text{D3}}$ Kachru *et al.* (2003) being the most studied example. The former is a string realisation of D-term inflation, where the rôle of a FI term is played by a non-self-dual flux on the D7-brane. The two (D3 and D7) branes are attracted to each other due to the breaking of SUSY by the FI term; inflation takes place in an unwarped background. The latter is a string realisation of F-term inflation, which is in general plagued by the η-problem. The brane is attracted to the antibrane due to warping. D3/$\overline{\text{D3}}$ inflation takes place in a warped throat, as for instance the Klebanov-Strassler (KS) geometry; the warping resolves the η-problem of the standard F-term inflation.

To accept the two classes of brane inflation models studied within IIB string theory, one must firstly examine their theoretical consistency. More precisely, SUGRA is a consistent theory, in the sense that there is a globally well-defined Ferrara-Zumino multiplet, provided (i) the FI terms are field-dependent and (ii) the moduli space is non-compact. Assuming the SUGRA constraints remain valid in the full string theory, we will examine the consistency of the two classes of brane inflation and deduce Gwyn *et al.* (2010) severe consequences for the dynamics of the inflationary mechanism in the D3/D7 case. Note that often in brane inflation, it is desirable to leave the brane positions unfixed while the volume of compactification is stabilised. To examine whether the available brane inflation models are consistent with the SUGRA constraints, one has to consider the details of compactification and moduli stabilisation. Only once we are convinced that the brane inflation models satisfy the SUGRA constraints, it makes sense, as a second step, to investigate the observational consequences of the models and test their observational consequences against current measurements.

In the D3/D7 brane inflation model, a D3 brane is parallel to a D7 brane in the four noncompact directions, with the other legs of the D7 wrapping K3. The presence of a non-self-dual flux on the D7 brane leads to SUSY breaking and the appearance of an attractive potential between the branes. The inflaton field is given by $\phi = x^4 + ix^5$ where x^4 and x^5 are the directions perpendicular to both branes, along which they feel the attractive potential. Inflation ends in a waterfall regime, when the open string modes stretched between the branes become tachyonic. Even though the FI term, which is due to a non-self-dual flux on the D7 brane, is field-dependent Gwyn *et al.* (2010), the present version of D3/D7 brane inflation fails to satisfy the SUGRA constraints Komargodski & Seiberg (2010). A successful D3/D7 inflationary scenario requires stabilisation of the Kähler modulus at a large finite value above the SUSY breaking scale, thus rendering the FI term constant Gwyn *et al.* (2011). A possible way out is to assume that quantum corrections will render the FI term to be dependent on the brane separation, however this would render the inflationary mechanism problematic since both the bifurcation point and the Hubble constant during inflation depend on the FI term.

In the D3/$\overline{\text{D3}}$ brane inflation model, the brane-antibrane system lives at a specific point of a Calabi-Yau (CY) manifold, and the net attractive force between the brane and antibrane results from supersymmetry breaking. Here the brane-antibrane separation plays the rôle of the inflaton. To accommodate a sufficient number of e-foldings, the first and second slow roll parameters ϵ and η must be very small and since η cannot be much smaller than unity in a flat space, D3/$\overline{\text{D3}}$ brane inflation is realised in a warped geometry; the corresponding model is referred to as KLMT Kachru *et al.* (2003). Since this inflationary model is a stringy realisation of F-term inflation, there is no FI term, hence the first SUGRA constraint is not applicable.

Journal of Physics: Conference Series **544** (2014) 012027 doi:10.1088/1742-6596/544/1/012027

The $\mathbb{KLMT}$ D3/$\overline{\text{D3}}$ brane inflation model can be made consistent with the second SUGRA constraint, in the sense that the moduli space of the theory can be made non-compact, through a SUSY breaking mechanism which in this model can be set at a scale independent of the volume modulus. Hence, the volume modulus can be fixed below the SUSY breaking scale, something which cannot be achieved for D3/D7 brane inflation.

4. Radiation of cosmic (super)strings

Cosmic superstrings Sakellariadou (2010) produced at the end of brane inflation, can emit gravitational, axionic (Ramond Ramond (RR) or Neveu-Schwarz Neveu-Schwarz (NS NS) particles, since cosmic superstrings are charged under the two-forms B_2, C_2 which are Hodge dual to axionic scalars in four dimensions), or dilatonic radiation. As we have already discussed, to render cosmic strings compatible with the CMB data, the string tension gets severely constrained. In contrast to their gauge theory analogues, cosmic superstrings have in general Planck-scale energies and hence their high tension leads to $G\mu/c^2 \sim 1$ Witten (1985) rendering them inconsistent with the CMB data . One should thus conclude that string inflation models leading to the production of cosmic superstrings with such high tensions must be ruled out. Hence, the IIB string theory motivated inflationary models that could be further examined are those realised within a warped geometry, as for instance brane-antibrane inflation in a throat, or models with large extra dimensions, since only in these cases the cosmic superstring tension gets suppressed.

In what follows we will briefly summarise Gwyn *et al.* (2010) the allowed channels of cosmic superstring radiation in warped backgrounds given our previous discussion (i) on the consistency of brane inflationary models with respect to the SUGRA constraints and (ii) on the constraints on the cosmic superstring tension. Taking into account a consistent compactification, we will see that severe constraints are imposed on the allowed radiation emitted from cosmic superstrings. Let us note that for gauge cosmic strings, gravitational radiation Vachaspati & Vilenkin (1985); Sakellariadou (1990) is the main channel of energy loss, allowing a cosmic string network to reach a scaling solution.

Cosmic superstrings produced at the end of brane inflation can be of three types: F-strings (the fundamental strings of the theory), D-strings (one-dimensional Dirichlet branes), or (p,q) string bound states of p F-strings and q D-strings. Considering a (p,q) string in a KS throat, the gravitational power radiated per unit solid angle is proportional to the square of the tension $T_{(p,q)} = \sqrt{T_{\rm D}^2 + T_{\rm F}^2}$, with $T_{\rm D}$ and $T_{\rm F}$ denoting the tensions of the D-string and F-string, respectively Gwyn *et al.* (2010). Further, since (p,q) strings are charged under the $B_2^{\rm NS}$ and $C_2^{\rm RR}$ 2-forms, they may emit massless RR or NS-NS particles; the former are referred to as axions since the RR 2-form C_2 is Hodge dual in four dimensions to the pseudo-scalar axion.

Considering axion emission from D-strings, it was shown Firouzjahi (2008) that the axionic radiation is

$$P_{\rm RR} = \frac{\Gamma_{\rm RR}\mu_1^2}{\pi^2 g_{\rm s}\beta M_{\rm P}^2} , \tag{9}$$

while the gravitational one reads

$$P_{\rm g} = \Gamma_{\rm g} G \left(\frac{h^2\mu_1}{g_{\rm s}}\right)^2 , \tag{10}$$

leading to the ratio

$$\frac{P_{\rm RR}}{P_{\rm g}} = \left(\frac{8\Gamma_{\rm RR}}{\pi\Gamma_{\rm g}}\right)\frac{g_{\rm s}}{\beta h^4} , \tag{11}$$

where $\Gamma_{\rm RR}, \Gamma_{\rm g}$ are numerical factors of the order $\mathcal{O}(50)$, $g_{\rm s}$ stands for the string coupling, h denotes the warp factor and β parametrises the difference in normalisations between the Chern-Simons and the Einstein-Hilbert term in the presence of warping. Thus, in the limit where $g_{\rm s} \ll 1$ and warping is negligible, power loss by gravitational radiation dominates over axionic radiation. However, in a warped geometry, as in the case of a KS background, since h can be much smaller than unity, one may naively conclude that axionic radiation becomes the dominant one. We will return to this point later in our discussion.

Considering F-strings, the NS-NS particle emission reads

$$P_{\rm NS-NS} = \frac{\Gamma_{\rm NS}\mu_1^2 g_{\rm s}}{\pi^2 \beta M_{\rm P}^2} , \tag{12}$$

which implies that the ratio of the axionic to the gravitational radiation is

$$\frac{P_{\rm NS-NS}}{P_{\rm g}} = \left(\frac{8\Gamma_{\rm NS-NS}}{\pi\Gamma_{\rm g}}\right)\frac{g_{\rm s}^3}{\beta h^4} . \tag{13}$$

Hence, the NS-NS particle emission is suppressed as compared to the RR particle radiation.

The next interesting question is whether particle radiation can be dominant over the gravitational one in the case of (p,q) strings. This is an important issue since it may highlight an observational difference between comic superstrings and their gauge analogues, at least in the case that the former are produced in a warped throat. Let us first remark that, as we have pointed out in Ref. Gwyn *et al.* (2010), the enhancement of RR particle emission claimed for D-strings in Ref. Firouzjahi (2008) is due to the effect of warping. However, as discussed in Ref. Gwyn *et al.* (2010), the allowed radiation channels are constrained from the orientifold projection imposed by a consistent flux compactification. The only known consistent compactification of the Klebanov-Strassler geometry is the flux compactification proposed by Giddings, Kachru and Polchinski (GKP) in Giddings *et al.* (2002), which involves an orientifold projection $\mathcal{O}$ whose action on the NS-NS and RR two-forms reads

$$\mathcal{O}B_2 = -\sigma^\star B_2 \quad \text{and} \quad \mathcal{O}C_2 = -\sigma^\star C_2 , \tag{14}$$

respectively, where $\sigma^\star$ is the pull-back of an isometric and holomorphic involution σ. Noting that the internal symmetry σ acts on the internal manifold while it leaves the four-dimensional non-compact space invariant, one concludes that the NS-NS and RR two-forms with legs in the non-compact directions are projected out Grana *et al.* (2004). This observation has important consequences for the allowed radiation channels. More precisely, since the zero modes of $B_{\mu\nu}$ and $C_{\mu\nu}$ do not appear in the spectrum Copeland *et al.* (2004), there can be no massless RR or NS-NS axions Gwyn *et al.* (2010). As it has been argued in Ref. Gwyn *et al.* (2010), neither D- nor F-strings can lead to significant axionic emission, since by construction of the brane inflationary model within which they are formed, D- and F-strings will not have any massless axionic radiation. However, for a (p,q) string arising from a wrapped D3-brane, the situation may be different. In this case, while NS-NS particle radiation is completely ruled out, RR radiation may be possible since (p,q) strings are also charged under the RR four-form C_4.

However, to determine the allowed modes of radiation by cosmic superstrings within a warped geometry, one should consider not only which modes will survive the orientifold projection, but one should also keep in mind the correct dimensional reduction of those modes. As it has been pointed out in Ref. Underwood (2011), the wave-function of the axionic zero mode is non-trivially modified by warping, however there is only limited progress in understanding this modification. Since the resulting wave-function would affect the magnitude of any radiation in this mode, lack of knowledge of the modified wave-function implies that we are not yet in a position to quantify the amplitude of the radiation.

Finally, cosmic superstrings may also emit dilatonic radiation. As is well-known, massive dilatons are compatible with astrophysical observations only if they acquire a Vacuum Expectation Value (VEV) before nucleosynthesis. Indeed, the dilaton gets a non-trivial potential in the GKP compactification, however the mass of the dilaton located in the throat will be suppressed by the warp factor.

In a warped geometry, the constraints Babichev & Kachelriess (2005) on cosmic superstring tension as a function of the dilaton mass are weakened Sabancilar (2010) since the mass gets suppressed by the warp factor and the dilaton wave-function is localised in the throat with an exponential fall off in the bulk, resulting to an enhancement of the dilaton to matter coupling.

5. Concluding remarks

Cosmic (super)strings seem to be generically formed at the end of an inflationary era in the context of inflationary models built within supersymmetric grand unified theories, supergravity, or even string theory models. These objects decay emitting gravitational waves and particles, the gravitational radiation being, most probably, the dominant decay channel.

As a contradiction to our theoretical expectations and despite various efforts we have not yet identified any such objects in the sky, while temperature anisotropies data constrain severely their potential cosmological rôle in structure formation. As a result, the allowed string tension decreases constantly rendering cosmic (super)strings weaker and therefore diminishing any gravitational implications they may have. However, since the inflationary scale is related to the string tension, one cannot freely constrain the tension of the strings without questioning the implications for the inflationary model within which strings were formed.

Conventional and well-studied inflationary models face some questions, with the onset of inflation and the origin of the inflaton field being the most important ones. To this list we have added one more, namely the compatibility between the constraints on string tension and the validity of the currently well-studied inflationary models. Absence of topological defects may imply lack of knowledge of the thermal history of the universe, which will certainly affect our understanding of inflation.

Let us end by stressing that our conclusion is not to rule out inflation but to emphasise that a number of the currently available models may turn out to be unphysical or inconsistent with the theoretical framework upon which they were based.

References

Starobinsky, A. A. 1980 A new type of isotropic cosmological models without singularity *Physics Letters B* **91**, 99.

Guth, A. H. 1981 The inflationary universe: A possible solution to the horizon and flatness problems *Physical Review D* **23**, 347.

Linde, A. D. 1982 A new inflationary universe scenario: A possible solution of the horizon, flatness, homogeneity, isotropy and primordial monopole problems *Physics Letters B* **108**, 389.

Calzetta, E. & Sakellariadou, M. 1992 Inflation in inhomogeneous cosmology *Physical Review D* **45**, 2802.

Calzetta, E. & Sakellariadou, M. 1993 Semiclassical effects and the onset of inflation *Physical Review D* **47**, 3184.

Germani, C., Nelson. W & Sakellariadou, M. 2007 On the onset of inflation in loop quantum cosmology *Physical Review D* **76**, 043529.

Bezrukov, F. L. & Shaposhnikov, M. 2008 The Standard Model Higgs boson as the inflaton *Physics Letters B* **659**, 703.

Buck, M., Fairbairn, M. & Sakellariadou, M. 2010 Inflation in models with conformally coupled scalar fields: An application to the noncommutative spectral action *Physical Review D* **82**, 043509.

Barbon, J. L. F. & Espinosa, J. R. 2009 On the naturalness of Higgs inflation *Physical Review D* **79**, 081302.

Atkins, M. & Calmet, X. 2011 Remarks on Higgs inflation *Physics Letters B* **697**, 37.

Jeannerot, R., Rocher, J. & Sakellariadou, M. 2003 How generic is cosmic string formation in SUSY GUTs *Physical Review D* **68**, 103514.

Rocher, J. & Sakellariadou, M. 2005 Supersymmetric grand unified theories and cosmology *Journal of Cosmology & Astropartarticle Physics* **0503**, 004.

Rocher, J. & Sakellariadou, M. 2005 D-term inflation, cosmic strings, and consistency with cosmic microwave background measurements *Physical Review Letters* **94**, 011303.

Rocher, J. & Sakellariadou, M. 2006 D-term inflation in non-minimal supergravity *Journal of Cosmology & Astropartarticle Physics* **0611**, 001.

Battye, R. Garbrecht, B. & Moss, A. 2010 Tight constraints on F- and D-term hybrid inflation scenarios *Physical Review D* **81**, 123512.

Cacciapaglia, G. & Sakellariadou, M. 2014 Is F-term hybrid inflation natural within minimal supersymmetric SO(10)? *European Physical Journal C* **74**, 2779.

Komargodski, Z. & Seiberg, N. 2009 Comments on the Fayet-Iliopoulos Term in Field Theory and Supergravity *Journal of High Energy Physics* **0906**, 007.

Dienes, K. R. & Thomas, B. 2010 On the Inconsistency of Fayet-Iliopoulos Terms in Supergravity Theories *Physical Review D* **81**, 065023.

Komargodski, Z. & Seiberg, N. 2010 Comments on Supercurrent Multiplets, Supersymmetric Field Theories and Supergravity *Journal of High Energy Physics* **1007**, 017.

Gwyn, R., Sakellariadou, M. & Sypsas, S. 2011 Theoretical constraints on brane inflation and cosmic superstring radiation *Journal of High Energy Physics* **1109**, 075.

Gottlober, S., Muller, V. & Starobinsky, A. A. 1991 Analysis of inflation driven by a scalar field and a curvature squared term *Physical Review D* **43**, 2510.

Ade, P. A. R., *et al.* Planck Collaboration 2013 Planck 2013 results. XXV. Searches for cosmic strings and other topological defects *arXiv:1303.5085 [astro-ph.CO]*.

Dasgupta, K., Herdeiro, C., Hirano, S. & Kallosh, R. 2002 D3 / D7 inflationary model and M theory *Physical Review D* **65**, 126002.

Dasgupta, K., Hsu, J. P., Kallosh, R., Linde, A. D. & Zagermann, M. 2004 D3/D7 brane inflation and semilocal strings *Journal of High Energy Physics* **0408**, 030.

Kachru, S., Kallosh, R., Linde, A. D., Maldacena, J. M., McAllister, L. P. & Trivedi, S. P. 2003 Towards Inflation in String Theory *Journal of Cosmology & Astropartarticle Physics* **0310**, 013.

Kurkov, M. A. & Sakellariadou, M. 2014 Spectral Regularisation: Induced Gravity and the Onset of Inflation *Journal of Cosmology & Astropartarticle Physics* **01**, 035.

Ade, P. A. R., *et al.* Planck Collaboration 2013 Planck 2013 results. XXII. Constraints on inflation *arXiv:1303.5082 [astro-ph.CO]*.

Binetruy, P., Dvali, G., Kallosh, R. & Van Proeyen A. 2004 Fayet-Iliopoulos Terms in Supergravity and Cosmology *Classical & Quantum Gravity* **21**, 3137.

Sakellariadou, M. 2007 Cosmic strings *Lecture Notes in Physics* **718**, 247.

Bouchet, F. R., Peter, P., Riazuelo, A. & Sakellariadou, M. 2002 Evidence against or for topological defects in the BOOMERanG data ? *Physical Review D* **65**, 021301.

BEVIS, N., HINDMARSH, M., KUNZ, M. & URRESTILLA, J. 2008 Fitting CMB data with cosmic strings and inflation *Physical Review Letters* **100**, 021301.

URRESTILLA, J., BEVIS, N., HINDMARSH, M. & KUNZ, M. 2011 Cosmic string parameter constraints and model analysis using small scale Cosmic Microwave Background data *Journal of Cosmology & Astropartarticle Physics* **12**, 021.

GWYN, R., SAKELLARIADOU, M. & SYPSAS, S. 2010 Cosmic strings from pseudo-anomalous Fayet-Iliopoulos $U(1)_{FI}$ in D3/D7 brane inflation *Journal of High Energy Physics* **1010**, 075.

SAKELLARIADOU, M. 2008 Cosmic Superstrings *Philosophical Transactions of the Royal Society of London A* **366**, 2881.

WITTEN, E. 1985 Cosmic Superstrings *Physics Letters B* **153**, 243.

VACHASPATI, T. & VILENKIN, A. 1985 Gravitational Radiation from Cosmic Strings *Physical Review D* **31**, 3052.

SAKELLARIADOU, M. 1990 Gravitational waves emitted from infinite strings *Physical Review D* **42**, 354; [Erratum-ibid. D **43** (1991) 4150].

FIROUZJAHI, H. 2008 Energy Radiation by Cosmic Superstrings in Brane Inflation *Physical Review D* **77**, 023532.

GIDDINGS, S. B., KACHRU, S. & POLCHINSKI, J. 2002 Hierarchies from Fluxes in String Compactifications *Physical Review D* **66**, 106006.

GRANA, M., GRIMM, T. W., JOCKERS, H. & LOUIS, J. 2004 Soft Supersymmetry Breaking in Calabi-Yau Orientifolds with D-branes and Fluxes *Nuclear Physics B* **690**, 21.

COPELAND E. J., MYERS, R. C. & POLCHINSKI, J. 2004 Cosmic F- and D-strings *Journal of High Energy Physics* **0406**, 013.

UNDERWOOD, B. 2011 A Breathing Mode for Warped Compactifications *Classical & Quantum Gravity* **28**, 195013.

BABICHEV, E. & KACHELRIESS, M. 2005 Constraining cosmic superstrings with dilaton emission *Physics Letters B* **614**, 1.

SABANCILAR, E. 2010 Cosmological Constraints on Strongly Coupled Moduli from Cosmic Strings *Physical Review D* **81**, 123502.

Journal of Physics: Conference Series **544** (2014) 012028 doi:10.1088/1742-6596/544/1/012028

Cosmic superstring networks with Y-junctions: Evolution, B-modes and gravitational waves

D.A. Steer[1]

[1] APC (UMR 7164 - APC, Univ Paris Diderot, CNRS/IN2P3, CEA/lrfu, Obs de Paris, Sorbonne Paris Cité, France), 10 rue Alice Domon et Léonie Duquet, 75205 Paris Cedex 13, France.

E-mail: steer@apc.univ-paris7.fr

Abstract.
We review the properties and evolution of strings networks containing both a spectrum of string tensions as well as Y-junctions, namely a point at which three different strings meet. Such a situation is expected in cosmic superstring networks, where the tension of the different strings is determined by the fundamental string tension μ_F as well as the string coupling g_s. We discuss evolution of such a network, and the proliferation of kinks as they travel through junctions. These effects can leave observational signals on B-modes and gravitational waves, which can then in turn constrain the underlying parameters of the theory.

1. Introduction

The last decade has seen intense effort devoted to trying to embed the inflationary paradigm into a more fundamental theory — and particularly superstring theory. In some of the resulting models (and most successfully in warped 'brane inflation' [Kachru *et al.* (2003)]), when the universe reheats at the end of inflation — thus starting up the big-bang — a network of 'cosmic superstrings' is expected to form. These cosmic superstrings are stable line-like objects of cosmic size whose properties and interactions depend on the two unknown parameters of the underlying superstring theory: the fundamental string coupling g_s, and the fundamental string tension μ_F. In the same way as their field theory counterparts (namely cosmic strings, which have been studied since the pioneering paper [Kibble (1976)]), cosmic superstrings interact with cosmic fluids and leave observational signatures, and hence through experiment one hopes to have a direct window onto physics at the very highest of energy scales. Since different observables depend in different ways on the parameters of the theory g_s and μ_F, a combination of measurements could then determine these parameters. This is illustrated in figure 1 (taken from [Pourtsidou *et al.* (2011)]), which combines forecasted Cosmic Microwave Background (CMB) constraints with ones from the expected stochastic background of Gravitational waves (GW) emitted by a cosmic superstring network.[1]

In this paper we review some of the properties of cosmic superstrings, focusing in particular on the rôle of Y-junctions in the evolution of a tangle of such strings. By definition, a Y-junction

[1] Note that by construction in these models, μ_F is much smaller than the naive expectation of m_s^2, where m_s is the mass scale of string theory. Consequently, $G\mu_F$, the dimensionless number measuring the tension of cosmic superstrings in units of the Newton constant, G, can be lowered below the current observational bounds.

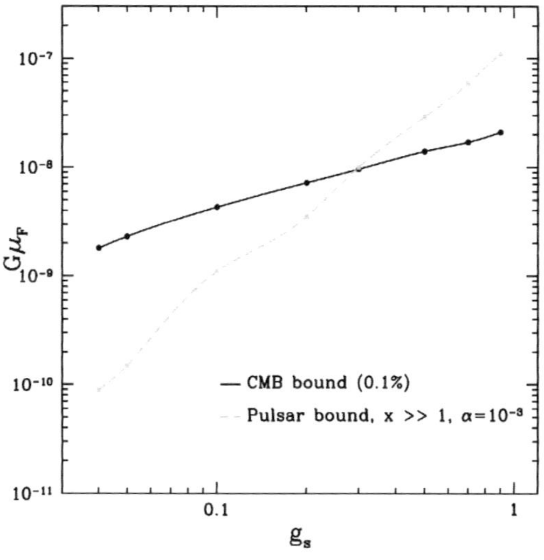

Figure 1. The forecasted bound [Pourtsidou *et al.* (2011)] on μ_F and g_s from CMB measurements based on the hypothesis that cosmic superstrings contribute 0.1% to the CMB temperature anisotropies (solid-black line). The bounds from pulsars (short-dash-green line) on the stochastic background of GWs emitted from a network of cosmic superstrings (assuming 'large loops').

is a point at which two strings meet and bind together for form a third string. These junctions are of course not unique to cosmic superstring networks but can also form in field-theory strings, and this analogy will be exploited in the following. We also focus on two potentially important observational signatures of cosmic superstrings mentioned above — B-modes and Gravitational waves.

2. Local properties of cosmic superstrings

Cosmic superstrings come in different varieties — fundamental (F-) strings and D-strings (D1-branes), as well as (p, q)-strings (where p and q are integer coprime numbers) which are bound states of p F-strings and q D-strings (see e.g. [Copeland *et al.* (2004); Jackson *et al.* (2005)]). In the simplest case, namely when the extra dimensions present in superstring theory are assumed flat and compactified, the tension of a (p_i, q_i) string is given by

$$\mu_i \equiv \mu_{(p_i,q_i)} = \frac{\mu_F}{g_s}\sqrt{p_i^2 g_s^2 + q_i^2} \qquad (0 \le g_s \le 1)\,. \tag{1}$$

The lightest string is therefore the (1,0) string with tension μ_F, and the next lightest is the (0,1) D-string with tension μ_F/g_s. Note that one always has

$$\mu_i + \mu_j \ge \mu_{i+j}\,, \tag{2}$$

so that in particular the $(1, 1)$ FD-string is a bound state of the F- and D-strings.

What is the outcome of a collision between two cosmic superstrings of different types? To answer that question it is simplest to take an alternative view of a (p, q)-string with $q \neq 0$: indeed this can be seen as p units of electric flux dissolved on the world volume of q coincident D-strings. Since collisions must respect charge conservation, it thus follows that if a (p_i, q_i)-string and a (p_j, q_j)-string with $i \neq j$ collide, they can not intercommute. Instead they either go through each other, or exchange partners and form a Y-junction (see figure 2). Notice from (2) that at the Y-junction the 3rd string is a bound state of the 2 colliding strings, and since charges must be conserved at the junction $\sum_{i=1} q_i = 0 = \sum_{i=1} p_i$ [Copeland *et al.* (2008)].

Y-junctions of course can also appear in networks of solitonic strings. The simplest example is that of Abelian-Higgs cosmic strings: in the Type-I regime cosmic strings of all windings

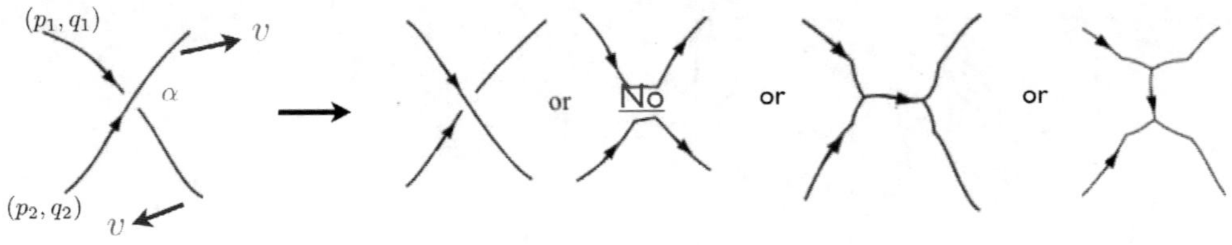

Figure 2. Possible outcomes of the collision between two cosmic strings of different types. Intercommutation is forbidden by charge conservation.

$n \in Z$ are stable, and when two, say $n = 1$ strings collide, they can form a junction (between the two $n = 1$ strings and an $n = 2$ string). Junctions also exist in Ônon-Abelian stringsÕ, for which the fundamental group $\pi_1(\mathcal{M})$ of the manifold of degenerate vacua, which classifies the strings, is non-Abelian, see for e.g. [Hindmarsh & Kibble (1995)]. In both these cases, the strings are uncharged so there is no analogue of charge conservation, and such solitonic strings *may* intercommute.

The dynamics of 3 strings meeting at a junction can be determined from a generalised Nambu-Goto action. Let μ_i $(i = 1, 2, 3)$ denote the tensions of the strings, whose coordinates are $x_i^\mu(\sigma^a)$, and which evolve in a space-time with metric $g_{\mu\nu}$. (Here the two worldsheet coordinates are $\sigma^a = (\tau, \sigma)$, and $\mu = 0, 1, 2, 3$ is a space-time index). Then the action for the system is given by [Copeland *et al.* (2006)]

$$S = -\sum_i \mu_i \int \mathrm{d}\tau \mathrm{d}\sigma \theta\left(s_i(\tau) - \sigma\right) \sqrt{-\det\left(g_{\mu\nu} x^\mu_{i,a} x^\nu_{i,b}\right)} + \sum_i \int \mathrm{d}\tau f_{i\mu} \left[x_i^\mu(s_i(\tau), \tau) - \bar{x}^\mu(\tau)\right] , \quad (3)$$

where the first term is Nambu-Goto action for each of the 3 strings, whilst the last term imposes — through the Lagrange multiplier $f_i^\mu(\tau)$ — that the 3 strings meet at the same point $\bar{x}^\mu(\tau)$. This is the position of the 3 strings when $\sigma = s_i(\tau)$. In flat space-time $g_{\mu\nu} = \eta_{\mu\nu}$, and on imposing the conformal temporal gauge, the resulting equations of motion have been derived in [Copeland *et al.* (2006)] and they can be solved given suitable initial conditions (namely the positions and velocities of the strings at the initial time).

In particular, these equations of motion can be used to determine the region of phase space in which the different collision outcomes in figure 2 can occur. Indeed, as in that figure, consider a straight string of tension μ_1 making an angle α with the positive x-axis, and another of tension μ_2 at angle $-\alpha$. Both strings are in the xy-plane and have a velocity v along the z-axis with opposite directions. Once they intersect, they can potentially form a third string, an Òx-linkÓ which has tension μ_3 (we follow here the notation introduced in [Copeland *et al.* (2007)]; if $\mu_1 = \mu_2$ then the x-link is indeed positioned along the x-axis) or a "y-link". On using the equations of motion derived from (3) for this configuration, kinematic constraints on the parameters (α, v) follow from the requirement that the total length of the progeny string (linking the 2 junctions) must *increase*. When $\mu_2 = \mu_1$, this requirement takes the simple form

$$\alpha < \arccos\left(\frac{\gamma\mu_3}{2\mu_1}\right) \quad (4)$$

where $\gamma = 1/\sqrt{1 - v^2}$, and is shown in figure 3. Notice that no "x-link" is allowed if the velocity of the strings is too large. Numerical experiments with type I abelian-Higgs vortices have shown that when junction formation is kinematically allowed, it dynamically takes place, see figure 4.

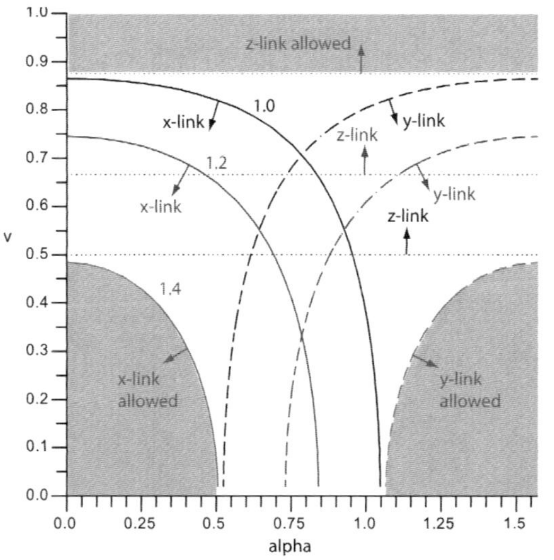

Figure 3. Kinematic constraints for $\mu_1 = \mu_2$. Allowed regions are to the left of the full curves for x-links; and for y-links to the right of the dashed curves. For z-links can form when non-abelian strings collide. The values of μ_3 are 1.4 (red), 1.2 (blue), 1.0 (black). Allowed regions are shaded for the $\mu_3 = 1.4$ case.

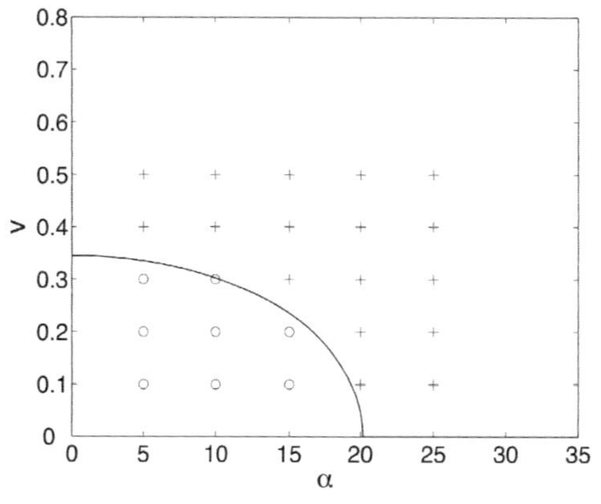

Figure 4. Results from a simulation [Salmi *et al.* (2008)] of the collision of two winding 1 strings in the type I Abelian-Higgs model (with ratio of square of the Higgs to vector masses given by $\beta = 0.36$). The solid line shows the theoretical curve of equality in (4). The + mark the events of single intercommutation, whereas ○ indicates strings merging together to form an x-link, a string with winding number $n = 2$ (as observed in the simulation).

3. Network evolution

The equations of motion derived from (3), together with the kinematic constraints, can be used to study network dynamics.

As a first step, let us ignore Hubble expansion and any energy loss mechanisms. Then the energy in the string network is fixed, but some strings will shorten and others will grow: how fast on average, is this growth or shortening; and how fast on average do the junctions move along a string of tension μ_i? These questions, and others, were addressed in [Copeland *et al.* (2008, 2007)], on assuming that at any Y-junction, the strings are straight with their incoming unit vectors randomly distributed on the unit sphere. Figure 5 shows the average rate of increase of cosmic superstrings of type i, as a function of g_s, where we have solely focused on F-, D- and FD-strings. From (1) for $g_s \ll 1$ it follows $\mu_F \ll \mu_D \sim \mu_{FD}$. Furthermore from figure 5, we see that the rate of increase in length of the F-strings is the largest. We can therefore expect these to be the most populous strings in the network. Furthermore, one can ask if they are sufficiently populous to also dominate the energy density, despite their small tension — we will return to this point below (see also [Rivers & Steer (2008)] who have studied the statistical mechanics of string networks with junctions). When $g_s \simeq 1$, then $\mu_F \sim \mu_D \sim \mu_{FD}/\sqrt{2}$ and one expects similar numbers of F and D strings.

To understand cosmological network dynamics fully, however, it is crucial to include Hubble expansion and loop formation. Furthermore, cosmic superstrings are intrinsically quantum objects which are not strictly localised in 4D — they can fluctuate and explore the extra dimensions, where there is sufficient 'space' for them to miss each other when they go past. To take this into account, the classical equations of motion coming from action (3) (as well as the kinematic constraints discussed above) must be augmented to account for the probability

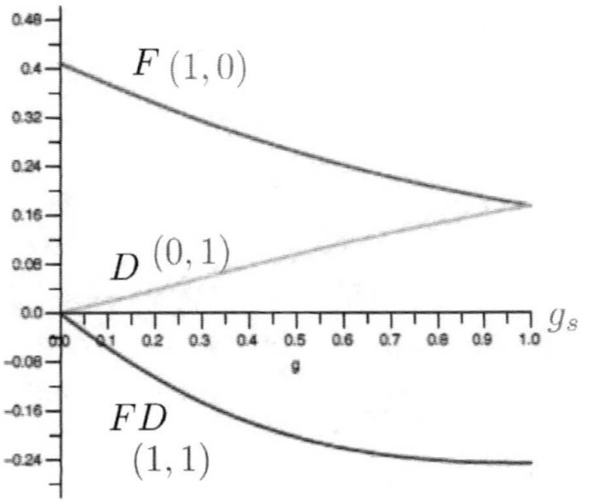

Figure 5. Average rate of increase of length of F-, D- and FD-strings, as a function of g_s, predicted from action (3) in *flat-space*.

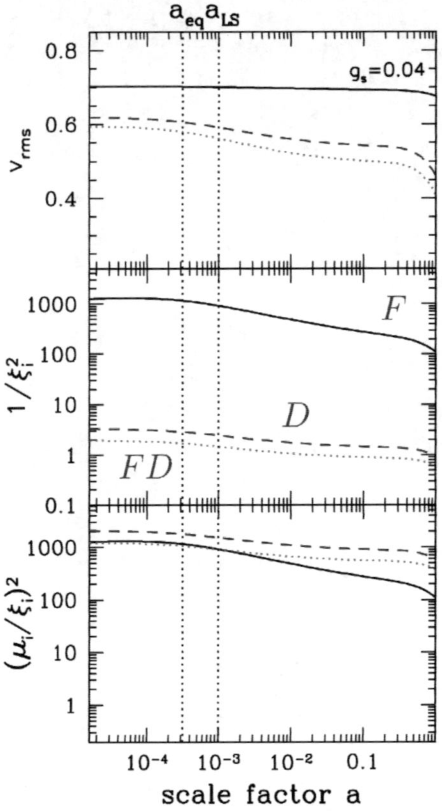

Figure 6. Taking into account energy loss, Hubble expansion, and network dynamics: evolution of the rms velocity v_i, number density $\xi_i^{-2} \equiv (t/L_i)^2$, and the power spectrum density $(\mu_i/\xi_i)^2$.

that string i interacts with string j in the first place. This so-called 'intercommuting probability' P_{ij} is (v, α, g_s)-dependent, and also depends on a further parameter w, which is related to the volume of the extra dimensions. The explicit expressions can be found in for e.g. [Pourtsidou *et al.* (2011)] for different string types. Note that for field theory strings this probability is (essentially always) equal to 1, whereas for cosmic superstrings it can be much smaller than 1.

Given the form of P_{ij}, the kinematic constraints, and the classical equations of motion, one can then attempt to model the network evolution — in [Pourtsidou *et al.* (2011)] we have use a generalised 'velocity dependent 1-scale model' (this was first developed to describe a tangle of type II Abelian-Higgs cosmic strings with no junctions, and only strings of a single tension [Martins & Shellard (1996)]). The variables describing the bulk properties of the system are taken to be (i) the energy density ρ_i in infinite strings of type i, from which one defines a correlation length $\rho_i \equiv \mu_i/L_i^2$, and (ii) the rms velocity v_i of the string of type i. These satisfy the following coupled set of equations (for simplicity we assume that all the energy liberated by the formation of junctions is radiated away, and the dot denotes $\mathrm{d}/\mathrm{d}t$):

$$\dot{\rho}_i = -2\frac{\dot{a}}{a}(1+v_i^2)\rho_i - \frac{c_i v_i \rho_i}{L_i} - \sum_{a,k} \frac{d^k_{ia}\bar{v}_{ia}\mu_i \ell^k_{ia}(t)}{L_a^2 L_i^2} + \sum_{b,\, a\leq b} \frac{d^i_{ab}\bar{v}_{ab}\mu_i \ell^i_{ab}(t)}{L_a^2 L_b^2}, \quad (5)$$

$$\dot{v}_i = (1 - v_i^2)\left[\frac{1}{L_i}\left(\frac{2\sqrt{2}(1-8v_i^6)}{\pi(1+8v_i^6)}\right) - 2\frac{\dot{a}}{a}v_i\right], \qquad (6)$$

where the presence of Y-junctions is reflected in the last two terms in (5); for instance, the penultimate one models the energy loss from the strings of type i due to their collision with strings of type a (with average relative velocity $\bar{v}_{ia}$), resulting in the formation of type k links of average length $\ell_{ia}^k(t)$. This occurs with g_s-dependent probability $d_{ia}^k = d_{ai}^k$ which depends on the $\mathcal{P}_{ia}$ and kinematic constraints. Self-interactions of strings of type i, leading to the formation of loops and thus the removal of energy from the long string network of type i, are quantified by the coefficients c_i in (5) which now depend on $\mathcal{P}_{ii}$ and are thus also g_s-dependent. (Again see [Pourtsidou *et al.* (2011)] for explicit expressions for the different coefficients.)

We have solved Eq. (5-6) numerically for different values of our external parameters g_s and w (and for the 7 lightest strings in the hierarchy given in (1)). The system is found to generically reach an attractor *scaling* solution in which $\xi_i \equiv L_i/t$ and v_i asymptotically approach constant values during the radiation and matter eras. For all studied values of the string coupling g_s, the three lightest strings (F, D, and FD), dominate the two relevant physical quantities — *string number density* $N_i = \xi_i^{-2}$, and the *power spectrum density* $M_i \equiv (\mu_i/\xi_i)^2$ of the network. These are shown in figure 6 together with v_i for $g_s = 0.04$. Just as expected from the discussion above, the lightest F-string is by far the most populous. However, regarding the energy density in the 3 different types of strings, the large number of F-strings does not outweigh their light tension: in fact, the heavier and less numerous D strings are seen to dominate the power spectrum density. Finally figure 7 shows the dependence of v_i, ξ_i and M_i on g_s *at the time of last scattering.* These quantities are crucial to determining the shape of the B-mode power-spectrum.

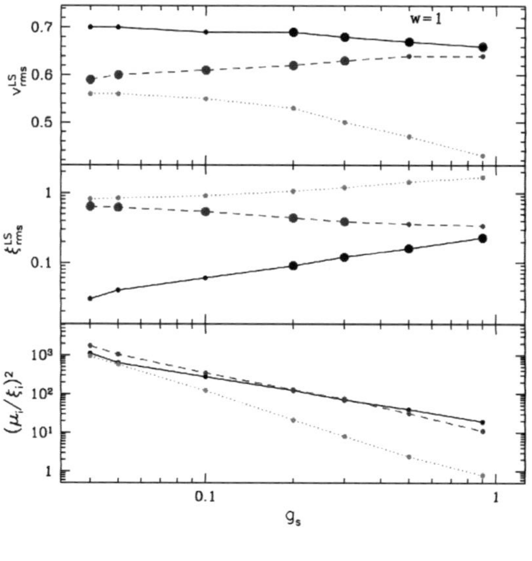

Figure 7. Dependence of the v_i, ξ_i, and M_i at the time of last scattering (LS) as a function of g_s. Labeling corresponds that of figure 6. The string type(s) dominating the power spectrum are shown with oversized dots.

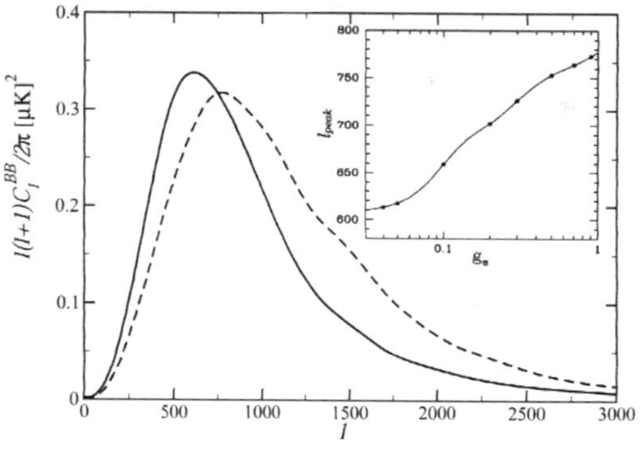

Figure 8. The B-mode power spectra for $g_s = 0.04$ (solid lines) and $g_s = 0.9$ (dashed lines) normalised to $f_s = 0.1$. The insert plot shows the position of the peak as a function of g_s.

4. B-modes

We have examined the imprint of cosmic superstrings on the CMB temperature and polarisation spectra, as a function of g_s and μ_F, by modifying the publicly available code CMBACT so as to allow for strings with multiple tensions whose scaling is modelled by Eqs. (5) and (6), see [Pourtsidou *et al.* (2011); Avgoustidis *et al.* (2011)].

The crucial point to note is that even with a marginal contribution to the TT spectrum of 1%, strings can be a *prominent source* of B-mode polarisation on subdegree scales since strings, unlike inflation, are actively sourcing *vector* mode perturbations of magnitude comparable to the scalar perturbations (see for instance [Pogosian & Wyman (2008)] and references within). The shapes of the spectra depend on the large-scale properties of the network, namely ξ_i and v_i. In particular, the position of the main B-mode peak moves towards lower ℓ for larger correlation length, while the dependence on the rms velocity is nonlinear.

The amplitude of the CMB angular spectra C_ℓ for multi-tension networks is proportional to the sum of the power spectrum densities M_i of each string type: $C_\ell^{strings} \propto M_{\rm total} = \sum_{i=1}^{N} M_i$. This, together with observational data, ultimately constrains the fundamental string tension μ_F, and we have adjusted it such that $f_s = C^{TT}_{strings}/C^{TT}_{total} \leq 0.1$, where $C^{TT}_{strings}$ and C^{TT}_{total} are defined by summing the corresponding C_ℓ's over multipoles $2 \leq \ell \leq 2000$. Figure 1 shows the constraint on μ_F as a function of g_s assuming $f_s = 0.1$ (post-Planck, this number should be reduced).

Figure 8, on the other hand, shows the B-mode spectra for the limiting values $g_s = 0.04$ and $g_s = 0.9$. We see that the peak shifts significantly towards low multipole number as g_s decreases. The reason for this is the dependence of the BB peak position on ξ_i and v_i for the string type i dominating the power spectrum density M_i. Figure 7 shows that the velocity of the dominant string type only varies between 0.6 and 0.7, so the peak position should be determined mostly by the dominant correlation length ξ_i. As we have seen in figure 6, M_i is dominated by F strings for large g_s, but at lower g_s there is a transition and the rare (larger ξ_i), heavy D strings become more important. As g_s decreases, the D-string correlation length increases (Fig. 7, middle panel) and the peak moves further down in ℓ (see Fig. 8).

Our result implies that an observation of the B-mode peak could help to break the degeneracy between μ_F and g_s in the constraint imposed by the normalisation condition $f_s < 0.1$. This degeneracy may also be broken, figure 1, by combining CMB with pulsar timing constraints (see below). Finally, the reader is referred to [Avgoustidis *et al.* (2011)] for a discussion of whether the next generation of B-mode experiments will have the sensitivity to detect the peak, and hence the means to pin down the value of the string coupling constant, a direct observational test of string theory: we feel optimistically positive!

5. Kink proliferation and Gravitational Waves

It is well known that networks of cosmic strings generate a stochastic background of Gravitational Waves (GWs). The reason for this is that a closed loop of length ℓ (with no junctions) oscillates periodically with fundamental period $T = \ell/2 \equiv 2\pi/\omega$, and emits GWs with frequency $\omega_n = n\omega$ ($n \in Z$). As a result the loop decays in a lifetime $\tau = \ell/(\Gamma G\mu)$ where numerical simulations have shown that $\Gamma \sim 50$. Emission is generally dominated by the lowest mode numbers n, and summing over all GWs emitted by the numerous loops formed by the network during the course of its evolution (and taking into account the redshift effects) gives rise to a stationary and nearly Gaussian stochastic GW background. (The reader is referred to [Binetruy *et al.* (2012)] for a detailed review). The stochastic GW background is characterised by

$$h^2\Omega_{GW}(f) = \left(\frac{h^2}{\rho_c}\frac{d\rho_{GW}}{d\ln f}\right)_{\rm today} \tag{7}$$

where h is the uncertainty in the Hubble parameter, and ρ_c the critical density today. It is constrained by pulsar timing experiments: $h^2\Omega_{GW} < 2 \times 10^8$ at $f_{\text{pulsar}} = 1/8\text{yr}^{-1}$ [Jenet *et al.* (2006)]. Figure 1 shows this pulsar bound for cosmic superstrings, where we have only considered loops of F-, D- and FD-strings, and all loops are assumed to contain *no* Y-junctions. The μ_F and g_s-dependence of the constraint shown in figure 1 arises from the dependence of the string tensions on these parameters (Eq. (1)), as well as that of c_j (the loop chopping efficiency in Eq. (5)): as explained above, this latter parameter determines how many loops of type j are formed at time t.[2]

In [Damour & Vilenkin (2000)] it was pointed out that previous estimates of the stochastic GW background may have been too large, since kinks (discontinuities on the tangent vector of a string) and cusps (points where the string instantaneously goes at the speed of light) lead to strong GW bursts — namely short bursts of radiation which, in the case of kinks, are emitted in a fan-shaped set of directions. These infrequent bursts should not be included in the calculation of the stochastic GW background. However, an exciting idea is that their characteristic features may provide a new signature for detecting cosmic strings in the sky.

Regarding this point, we finish by highlighting a potentially new source of GW bursts unique to string networks with junctions. Though not included explicitly in the equations (5)-(6), these networks can also chop off closed loops of strings containing (at least 2) Y-junctions, an example of which is shown in figure 9. One can study the dynamics of such loops using the equations of motion coming from (3), and the important observation is that when a kink arrives at a junction, not only is it reflected, but two transmitted kinks are also formed [Copeland *et al.* (2006)]. From 1 initial kink, we therefore go to 3 kinks, and hence the number of kinks on a closed evolving loops with junctions increases exponentially with time [Binetruy *et al.* (2010*b*)]. The amplitude of the GW burst emitted by a kink is proportional to the sharpness of the kink, and one can show that the number of kinks with amplitude larger than any fixed value also increases exponentially with time. Indeed, following simulations of realistic loops, it is not unusual to find $k \gtrsim 10^4$ kinks of sharpness $\mathcal{O}(1)$ [Binetruy *et al.* (2010*b*)].

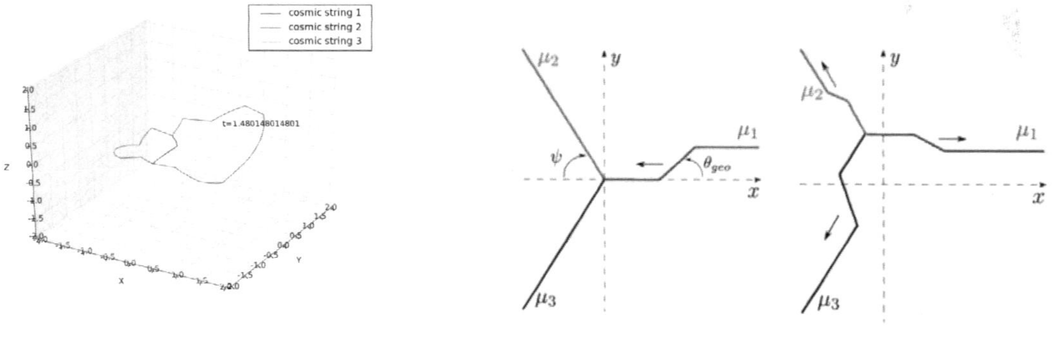

Figure 9. LH panel: a loop with junctions. RH panel: detail of the propagation of a kink through a junction. Here two kinks initially propagate towards the junction, where they are reflected and transmitted to produce 6 final kinks.

What is the effect of these kinks on the GW signal emitted by a network? Let q denote the fraction of loops with junctions relative to loops without junctions. Consistent with a scaling solution, we assume q is constant, and we take $qk \gg 1$. (Furthermore, for simplicity, we consider

[2] Note that all predictions of GW from cosmic strings suffer from important uncertainties, namely the typical size of a loop when it is chopped off the network, and the loop distribution. These uncertainties does not affect the CMB predictions which are determined by the large scale evolution and on which there is much better agreement.

a network in which all strings have the *same* tension, but are allowed to meet a junctions.) One might expect that this large number of kinks might make it easier to detect bursts from string networks with junctions — however, this is not always the case [Binetruy *et al.* (2010*a*)]. Clearly, in order for individual bursts to be resolved by a detector (eg LISA), they must arrive separately in that detector. The large number of kinks on a loop necessarily means that the burst event rate is very large, and if it is too high, the bursts start to superimpose instead of arriving one by one. Their random superposition contributes to the stochastic background of gravitational waves, and when $qk^2 \gg 1$ this is what happens [Binetruy *et al.* (2010*a*)] — kink proliferation on loops with junctions generally makes individual bursts unobservable.[3]

6. Conclusions

In this contribution we have highlighted some of the facets of strings with junctions, starting from their local properties (dynamics and the result of collisions), then discussing the evolution of a network of such strings, and finally considering their observational signatures on the CMB and in Gravitational Waves. If cosmic superstrings really exist, and with a tension that is not too small, then through a combination of observations we conclude that it may be possible to open a window onto physics at the very highest of energy scales.

Acknowledgments

I am grateful to the organisers for the invitation to speak at this very interesting workshop. I would also like to thank all my different collaborators: it has been a great pleasure to work with them on the subjects presented here.

References

Avgoustidis, A., Copeland, E., Moss, A., Pogosian, L., Pourtsidou, A. *et al.* 2011 Constraints on the fundamental string coupling from B-mode experiments. *Phys.Rev.Lett.* **107**, 121301.

Binetruy, P., Bohe, A., Caprini, C. & Dufaux, J.-F. 2012 Cosmological Backgrounds of Gravitational Waves and eLISA/NGO: Phase Transitions, Cosmic Strings and Other Sources. *JCAP* **1206**, 027.

Binetruy, P., Bohe, A., Hertog, T. & Steer, D. 2010*a* Gravitational wave signatures from kink proliferation on cosmic (super-) strings. *Phys.Rev.* **D82**, 126007.

Binetruy, P., Bohe, A., Hertog, T. & Steer, D. A. 2010*b* Proliferation of sharp kinks on cosmic (super-)string loops with junctions. *Phys.Rev.* **D82**, 083524.

Bohe, A. 2011 Gravitational power from cosmic string loops with many kinks. *Phys.Rev.* **D84**, 065016.

Copeland, E., Firouzjahi, H., Kibble, T. & Steer, D. A. 2008 On the Collision of Cosmic Superstrings. *Phys.Rev.* **D77**, 063521.

Copeland, E., Kibble, T. & Steer, D. A. 2006 Collisions of strings with Y junctions. *Phys.Rev.Lett.* **97**, 021602.

Copeland, E., Kibble, T. & Steer, D. A. 2007 Constraints on string networks with junctions. *Phys.Rev.* **D75**, 065024.

Copeland, E. J., Myers, R. C. & Polchinski, J. 2004 Cosmic F and D strings. *JHEP* **0406**, 013.

[3] Note that we have assumed that Γ, the parameter determining the lifetime of a loop is of order 50. However, when there are many radiating kinks on a loop, one might expect Γ to increase: see [Bohe (2011)] for a detailed discussion of this point and its effect on GW emission.

Journal of Physics: Conference Series **544** (2014) 012028 doi:10.1088/1742-6596/544/1/012028

DAMOUR, T. & VILENKIN, A. 2000 Gravitational wave bursts from cosmic strings. *Phys.Rev.Lett.* **85**, 3761–3764.

HINDMARSH, M. & KIBBLE, T. 1995 Cosmic strings. *Rept.Prog.Phys.* **58**, 477–562.

JACKSON, M. G., JONES, N. T. & POLCHINSKI, J. 2005 Collisions of cosmic F and D-strings. *JHEP* **0510**, 013.

JENET, F. A., HOBBS, G., VAN STRATEN, W., MANCHESTER, R., BAILES, M. *et al.* 2006 Upper bounds on the low-frequency stochastic gravitational wave background from pulsar timing observations: Current limits and future prospects. *Astrophys.J.* **653**, 1571–1576.

KACHRU, S., KALLOSH, R., LINDE, A. D., MALDACENA, J. M., MCALLISTER, L. P. *et al.* 2003 Towards inflation in string theory. *JCAP* **0310**, 013.

KIBBLE, T. 1976 Topology of Cosmic Domains and Strings. *J.Phys.* **A9**, 1387–1398.

MARTINS, C. & SHELLARD, E. 1996 Quantitative string evolution. *Phys.Rev.* **D54**, 2535–2556.

POGOSIAN, L. & WYMAN, M. 2008 B-modes from cosmic strings. *Phys.Rev.* **D77**, 083509.

POURTSIDOU, A., AVGOUSTIDIS, A., COPELAND, E., POGOSIAN, L. & STEER, D. 2011 Scaling configurations of cosmic superstring networks and their cosmological implications. *Phys.Rev.* **D83**, 063525.

RIVERS, R. & STEER, D. A. 2008 Statistical mechanics of strings with Y-junctions. *Phys.Rev.* **D78**, 023521.

SALMI, P., ACHUCARRO, A., COPELAND, E., KIBBLE, T., DE PUTTER, R. *et al.* 2008 Kinematic constraints on formation of bound states of cosmic strings: Field theoretical approach. *Phys.Rev.* **D77**, 041701.

Institute of Physics
Dirac House, Temple Back
Bristol BS1 6BE UK

CURRAN ASSOCIATES INC
proceedings.com